Semiconductor Laser Fundamentals

OPTICAL ENGINEERING

Founding Editor

Brian J. Thompson

University of Rochester
Rochester, New York

1. Electron and Ion Microscopy and Microanalysis: Principles and Applications, *Lawrence E. Murr*
2. Acousto-Optic Signal Processing: Theory and Implementation, *edited by Norman J. Berg and John N. Lee*
3. Electro-Optic and Acousto-Optic Scanning and Deflection, *Milton Gottlieb, Clive L. M. Ireland, and John Martin Ley*
4. Single-Mode Fiber Optics: Principles and Applications, *Luc B. Jeunhomme*
5. Pulse Code Formats for Fiber Optical Data Communication: Basic Principles and Applications, *David J. Morris*
6. Optical Materials: An Introduction to Selection and Application, *Solomon Musikant*
7. Infrared Methods for Gaseous Measurements: Theory and Practice, *edited by Joda Wormhoudt*
8. Laser Beam Scanning: Opto-Mechanical Devices, Systems, and Data Storage Optics, *edited by Gerald F. Marshall*
9. Opto-Mechanical Systems Design, *Paul R. Yoder, Jr.*
10. Optical Fiber Splices and Connectors: Theory and Methods, *Calvin M. Miller with Stephen C. Mettler and Ian A. White*
11. Laser Spectroscopy and Its Applications, *edited by Leon J. Radziemski, Richard W. Solarz, and Jeffrey A. Paisner*
12. Infrared Optoelectronics: Devices and Applications, *William Nunley and J. Scott Bechtel*
13. Integrated Optical Circuits and Components: Design and Applications, *edited by Lynn D. Hutcheson*
14. Handbook of Molecular Lasers, *edited by Peter K. Cheo*
15. Handbook of Optical Fibers and Cables, *Hiroshi Murata*
16. Acousto-Optics, *Adrian Korpel*
17. Procedures in Applied Optics, *John Strong*
18. Handbook of Solid-State Lasers, *edited by Peter K. Cheo*
19. Optical Computing: Digital and Symbolic, *edited by Raymond Arrathoon*
20. Laser Applications in Physical Chemistry, *edited by D. K. Evans*
21. Laser-Induced Plasmas and Applications, *edited by Leon J. Radziemski and David A. Cremers*

22. Infrared Technology Fundamentals, *Irving J. Spiro and Monroe Schlessinger*
23. Single-Mode Fiber Optics: Principles and Applications, Second Edition, Revised and Expanded, *Luc B. Jeunhomme*
24. Image Analysis Applications, *edited by Rangachar Kasturi and Mohan M. Trivedi*
25. Photoconductivity: Art, Science, and Technology, *N. V. Joshi*
26. Principles of Optical Circuit Engineering, *Mark A. Mentzer*
27. Lens Design, *Milton Laikin*
28. Optical Components, Systems, and Measurement Techniques, *Rajpal S. Sirohi and M. P. Kothiyal*
29. Electron and Ion Microscopy and Microanalysis: Principles and Applications, Second Edition, Revised and Expanded, *Lawrence E. Murr*
30. Handbook of Infrared Optical Materials, *edited by Paul Klocek*
31. Optical Scanning, *edited by Gerald F. Marshall*
32. Polymers for Lightwave and Integrated Optics: Technology and Applications, *edited by Lawrence A. Hornak*
33. Electro-Optical Displays, *edited by Mohammad A. Karim*
34. Mathematical Morphology in Image Processing, *edited by Edward R. Dougherty*
35. Opto-Mechanical Systems Design: Second Edition, Revised and Expanded, *Paul R. Yoder, Jr.*
36. Polarized Light: Fundamentals and Applications, *Edward Collett*
37. Rare Earth Doped Fiber Lasers and Amplifiers, *edited by Michel J. F. Digonnet*
38. Speckle Metrology, *edited by Rajpal S. Sirohi*
39. Organic Photoreceptors for Imaging Systems, *Paul M. Borsenberger and David S. Weiss*
40. Photonic Switching and Interconnects, *edited by Abdellatif Marrakchi*
41. Design and Fabrication of Acousto-Optic Devices, *edited by Akis P. Goutzoulis and Dennis R. Pape*
42. Digital Image Processing Methods, *edited by Edward R. Dougherty*
43. Visual Science and Engineering: Models and Applications, *edited by D. H. Kelly*
44. Handbook of Lens Design, *Daniel Malacara and Zacarias Malacara*
45. Photonic Devices and Systems, *edited by Robert G. Hunsperger*
46. Infrared Technology Fundamentals: Second Edition, Revised and Expanded, *edited by Monroe Schlessinger*
47. Spatial Light Modulator Technology: Materials, Devices, and Applications, *edited by Uzi Efron*
48. Lens Design: Second Edition, Revised and Expanded, *Milton Laikin*
49. Thin Films for Optical Systems, *edited by François R. Flory*
50. Tunable Laser Applications, *edited by F. J. Duarte*
51. Acousto-Optic Signal Processing: Theory and Implementation, Second Edition, *edited by Norman J. Berg and John M. Pellegrino*
52. Handbook of Nonlinear Optics, *Richard L. Sutherland*
53. Handbook of Optical Fibers and Cables: Second Edition, *Hiroshi Murata*

54. Optical Storage and Retrieval: Memory, Neural Networks, and Fractals, *edited by Francis T. S. Yu and Suganda Jutamulia*
55. Devices for Optoelectronics, *Wallace B. Leigh*
56. Practical Design and Production of Optical Thin Films, *Ronald R. Willey*
57. Acousto-Optics: Second Edition, *Adrian Korpel*
58. Diffraction Gratings and Applications, *Erwin G. Loewen and Evgeny Popov*
59. Organic Photoreceptors for Xerography, *Paul M. Borsenberger and David S. Weiss*
60. Characterization Techniques and Tabulations for Organic Nonlinear Optical Materials, *edited by Mark Kuzyk and Carl Dirk*
61. Interferogram Analysis for Optical Testing, *Daniel Malacara, Manuel Servín, and Zacarias Malacara*
62. Computational Modeling of Vision: The Role of Combination, *William R. Uttal, Ramakrishna Kakarala, Sriram Dayanand, Thomas Shepherd, Jagadeesh Kalki, Charles F. Lunskis, Jr., and Ning Liu*
63. Microoptics Technology: Fabrication and Applications of Lens Arrays and Devices, *Nicholas F. Borrelli*
64. Visual Information Representation, Communication, and Image Processing, *Chang Wen Chen and Ya-Qin Zhang*
65. Optical Methods of Measurement: Wholefield Techniques, *Rajpal S. Sirohi and Fook Siong Chau*
66. Integrated Optical Circuits and Components: Design and Applications, *edited by Edmond J. Murphy*
67. Adaptive Optics Engineering Handbook, *edited by Robert K. Tyson*
68. Entropy and Information Optics, *Francis T. S. Yu*
69. Computational Methods for Electromagnetic and Optical Systems, *John M. Jarem and Partha P. Banerjee*
70. Laser Beam Shaping: Theory and Techniques, *edited by Fred M. Dickey and Scott C. Holswade*
71. Rare-Earth-Doped Fiber Lasers and Amplifiers: Second Edition, Revised and Expanded, *edited by Michel J. F. Digonnet*
72. Lens Design: Third Edition, Revised and Expanded, *Milton Laikin*
73. Handbook of Optical Engineering, *edited by Daniel Malacara and Brian J. Thompson*
74. Handbook of Imaging Materials, *edited by Arthur S. Diamond and David S. Weiss*
75. Handbook of Image Quality: Characterization and Prediction, *Brian W. Keelan*
76. Fiber Optic Sensors, *edited by Francis T. S. Yu and Shizhuo Yin*
77. Optical Switching/Networking and Computing for Multimedia Systems, *edited by Mohsen Guizani and Abdella Battou*
78. Image Recognition and Classification: Algorithms, Systems, and Applications, *edited by Bahram Javidi*
79. Practical Design and Production of Optical Thin Films: Second Edition, Revised and Expanded, *Ronald R. Willey*

80. Ultrafast Lasers: Technology and Applications, *edited by Martin E. Fermann, Almantas Galvanauskas, and Gregg Sucha*
81. Light Propagation in Periodic Media: Differential Theory and Design, *Michel Nevière and Evgeny Popov*
82. Handbook of Nonlinear Optics: Second Edition, Revised and Expanded, *Richard L. Sutherland*
83. Polarized Light: Second Edition, Revised and Expanded, *Dennis Goldstein*
84. Optical Remote Sensing: Science and Technology, *Walter G. Egan*
85. Handbook of Optical Design: Second Edition, *Daniel Malacara and Zacarias Malacara*
86. Nonlinear Optics: Theory, Numerical Modeling, and Applications, *Partha P. Banerjee*
87. Semiconductor and Metal Nanocrystals: Synthesis and Electronic and Optical Properties, *edited by Victor I. Klimov*
88. High-Performance Backbone Network Technology, *edited by Naoaki Yamanaka*
89. Semiconductor Laser Fundamentals, *Toshiaki Suhara*
90. Handbook of Optical and Laser Scanning, *edited by Gerald F. Marshall*

Additional Volumes in Preparation

Semiconductor Laser Fundamentals

Toshiaki Suhara
Osaka University
Osaka, Japan

MARCEL DEKKER

MARCEL DEKKER, INC.　　NEW YORK • BASEL

The Japanese edition of this book was published in the Kyoritsu Advanced Optoelectronics Series (Kyoritsu Shuppan, Tokyo, 1998).

Library of Congress Cataloging-in-Publication Data
A catalog record for this book is available from the Library of Congress.

ISBN: 0-8247-5373-9

This book is printed on acid-free paper.

Headquarters
Marcel Dekker, Inc., 270 Madison Avenue, New York, NY 10016, U.S.A.
tel: 212-696-9000; fax: 212-685-4540

Distribution and Customer Service
Marcel Dekker, Inc., Cimarron Road, Monticello, New York 12701, U.S.A.
tel: 800-228-1160; fax: 845-796-1772

Eastern Hemisphere Distribution
Marcel Dekker AG, Hutgasse 4, Postfach 812, CH-4001 Basel, Switzerland
tel: 41-61-260-6300; fax: 41-61-260-6333

World Wide Web
http://www.dekker.com

The publisher offers discounts on this book when ordered in bulk quantities. For more information, write to Special Sales/Professional Marketing at the headquarters address above.

Current printing (last digit):

10 9 8 7 6 5 4 3 2 1

PRINTED IN THE UNITED STATES OF AMERICA

Preface

Semiconductor lasers are among the most important optoelectronics devices. Remarkable development has been accomplished in the three decades since the first achievement in room-temperature continuous oscillation, which opened the possibility of practical applications of semiconductor lasers. Today, various types of semiconductor lasers are mass-produced and widely used as coherent light sources for a variety of applications, including optical fiber communication systems and optical disk memory systems. Advanced functions and high performance have been realized through distributed feedback lasers and quantum well lasers following the development of Fabry–Perot-type lasers. Accordingly, new applications previously unfeasible (or difficult with other conventional lasers) have been found, and the replacement of gas and solid-state lasers by compact and economical semiconductor lasers is in progress. Thus, semiconductor lasers are indispensable devices of increasing importance. Extensive research and development is being conducted toward specific applications. Remarkable progress is also being made in optoelectronic integrated circuits and integrated photonic devices using semiconductor lasers as the main component.

Implementation and advanced applications of semiconductor lasers require a deep understanding, and high technological expertise, in subareas including materials, crystal growth, device design, microfabrication, and device characterization (all of which comprise the field called semiconductor laser engineering). There already exist a number of authoritative books on semiconductor lasers, as given in the references in Chapter 1. In Chapter 2, the fundamental quantum theory on the interaction of electrons and photons is outlined and summarized in a form that is convenient for the understanding and analysis of semiconductor lasers. Chapter 3 deals with stimulated emission in semiconductors as one of the most important principles for implementation of semiconductor lasers, and explains the basic theory and characteristics of light amplification. Chapter 4 covers theoretical discussions on electron–photon interactions and stimulated emission, and considers characteristics of optical waveguide resonators for laser oscillator implementation. In Chapter 6, rate equation analysis of semiconductor lasers is presented to clarify and explain the static and dynamic characteristics of semiconductor lasers using Fabry–Perot-type semiconductor lasers as a prototype device. Chapter 7 is devoted to distributed feedback lasers and distributed Bragg reflector lasers, which are dynamic single-mode lasers and allow advanced performance. In Chapter 8, semiconductor laser amplifiers are discussed. The Appendixes provide important theoretical topics and experimental techniques.

The chapters were carefully checked for mutual consistency and clarity of context. Efforts were made to give a comprehensive explanation of mathematical formulae including the procedure of the deduction and physical meanings, rather than simple descriptions of the results, in order to ensure full understanding without skipping basic principles or referring to other materials. Almost all the formulae are in such a form that they can actually be used by the readers for analysis and design.

It will give me great satisfaction if this book is helpful to researchers, engineers, and students interested in semiconductor lasers. Finally, I would like to thank the staffs of Kyoritsu Pub., Ltd., and Marcel Dekker, Inc., for their cooperation.

Toshiaki Suhara

Contents

1. Introduction **1**
1.1 Principles and Device Structures of Semiconductor Lasers 1
1.2 Materials for Semiconductor Lasers 9
1.3 Features of Semiconductor Injection Lasers 12
1.4 Applications of Semiconductor Lasers 14
References 16

2. Interaction of Electrons and Photons **17**
2.1 Quantization of Optical Waves and Photons 17
2.2 Interactions of Electrons and Photons 24
2.3 Absorption and Emission of Photons 30
2.4 Population Inversion and Light Amplification 35
References 36

3. Stimulated Emission and Optical Gain in Semiconductors **37**
3.1 Band Structure of Semiconductors and Stimulated Emission 37
3.2 Direct-Transition Model 43
3.3 Gaussian Halperin–Lax Band-Tail Model with the Stern Energy-Dependent Matrix Element 46
3.4 Gain Spectrum and Gain Factor 54
3.5 Spontaneous Emission and Injection Current Density 60
3.6 Density Matrix Analysis 65
References 79

4. Stimulated Emission in Quantum Well Structures **81**
4.1 Electron State in Quantum Well Structures 81
4.2 Direct-Transition Model 92
4.3 Gain Spectrum and Gain Factor 95

4.4 Spontaneous Emission and Injection Current Density 104
4.5 Strained Quantum Wells 107
References 108

5. Semiconductor Heterostructure Optical Waveguides 111
5.1 Outline of Optical Waveguides for Semiconductor Lasers 111
5.2 Fundamental Equations for the Optical Wave 113
5.3 Optical Wave in a Waveguide 116
5.4 Planar Waveguide 119
5.5 Perturbation Theory and the Optical Confinement Factor 134
5.6 Channel Waveguides 139
5.7 Reflection at Waveguide Facets 147
5.8 Waveguide Fabry–Perot Resonator 152
5.9 Far-Field Patterns 159
References 162

6. Characteristics of Semiconductor Lasers 163
6.1 Semiconductor Laser Structure and Outline of Oscillation 163
6.2 Rate Equations 168
6.3 Steady-State Oscillation Characteristics 176
6.4 Modulation Characteristics 187
6.5 Noise Characteristics 200
6.6 Single-Mode Spectrum and Spectrum Linewidth 221
6.7 Ultrashort Optical Pulse Generation 225
References 230

7. Distributed Feedback Lasers 233
7.1 Dynamic Single-Mode Lasers 233
7.2 Coupled-Mode Equations 235
7.3 Distributed Feedback Lasers 243
7.4 Distributed Bragg Reflector Lasers 258
References 266

8. Semiconductor Laser Amplifiers 269
8.1 Gain Spectrum and Gain Saturation 269
8.2 Resonant Laser Amplifiers 271
8.3 Traveling-Wave Laser Amplifiers 273
8.4 Tapered Laser Amplifiers 279
8.5 Master Oscillator Power Amplifier 282
References 283

Appendix 1. *Outline of Density Matrix Analysis* 285
A1.1 Definition of Density Matrix and Expectation Values 285
A1.2 Equation of Motion for the Density Operator 286
Appendix 2. *Density of States for Electrons* 289
A2.1 Three-Dimensional State 290
A2.2 Two-Dimensional State 290
A2.3 One-Dimensional State 290
Appendix 3. *The Kramers–Kronig Relation* 293
Appendix 4. *Experimental Determination of Gain and Internal Loss* 297
Appendix 5. *Spontaneous Emission Term and Factors* 299

Index 303

Semiconductor Laser Fundamentals

1 Introduction

This chapter outlines the working principles and device structures of semiconductor lasers as an introduction to the contents of this book.

1.1 PRINCIPLES AND DEVICE STRUCTURES OF SEMICONDUCTOR LASERS

Semiconductor lasers are devices for oscillation or amplification of an optical wave based on the stimulated emission of photons through optical transition of electrons in a semiconductor. The idea was proposed early in 1957 [1]. Soon after the construction of the fundamental theory of lasers by Schawlow and Townes [2] in 1958, followed by the experimental verifications of laser oscillation in a ruby laser and a He–Ne laser in 1960, the pioneering work on semiconductor lasers was performed [3–5]. In 1962, pulse oscillation at a low temperature in the first semiconductor laser, a GaAs laser, was observed [6–8]. In 1970, continuous oscillation at room temperature was accomplished [9–11]. Since then, remarkable development has been made by the great efforts in different areas of science and technology. Nowadays, semiconductor lasers [12–20] have been employed practically as one of the most important optoelectronic devices and are widely used in a variety of applications in many areas.

The energy of an electron in an atom or a molecule takes discrete values, corresponding to energy levels. Consider two energy levels of energy difference E, and assume that the upper level is occupied by an electron and the lower level is not occupied, as shown in Fig. 1.1(a). If an optical wave of an angular frequency ω that satisfies

$$E = \hbar\omega \tag{1.1}$$

is incident, the electron transition to the lower level takes place with a transition probability, per unit time, proportional to the light intensity.

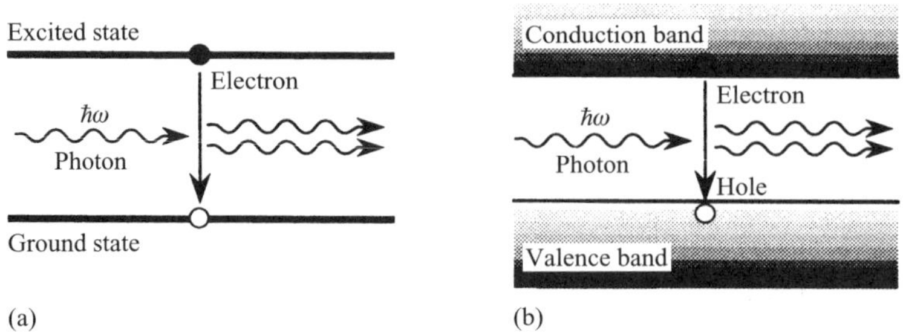

Figure 1.1 Schematic illustrations of stimulated emission of a photon by an optical transition of an electron: (a) stimulated emission by an electron in an atom; (b) stimulated emission by carriers in a semiconductor.

Then a photon of the same mode as the incident wave, i.e., of the same frequency and same propagation direction, is emitted. In Eq. (1.1), h is the Planck constant and $\hbar = h/2\pi$. In a semiconductor, the electron energy levels are not discrete but form a band structure. Assume that there are many electrons in the conduction band and many holes in the valence band, as shown in Fig. 1.1(b). If an optical wave satisfies Eq. (1.1) for E slightly larger than the bandgap energy E_g, electron transition and photon emission take place. These phenomena are called stimulated emission. Photon emission takes place even if there is no incident light. This emission is called spontaneous emission.

On the other hand, if the lower level is occupied by an electron and the upper level is unoccupied, the incident optical wave gives rise to an electron transition in the inverse direction and absorption of an incident photon. Quantum theory shows that the probability of the stimulated emission is the same as that of the absorption. In a system consisting of many electrons in thermal equilibrium, the electron energy distribution obeys Fermi–Dirac statistics; the population of the electrons of higher energy is smaller than that of electrons of lower energy. Therefore, as an overall effect, the optical wave is substantially absorbed. However, if inversion of the population is realized by excitation of the system with continuous provision of energy, stimulated emission of photons takes place substantially, and accordingly optical amplification is obtained. Lasers are based on this substantial stimulated emission.

Population inversion in semiconductors can be realized by producing a large number of electron–hole pairs by excitation of electrons in the valence band up to the conduction band. The excitation can be accomplished by light irradiation or electron-beam irradiation. The method of excitation most effective for implementation of practical laser devices, however, is to form a p–n junction in the semiconductor and to provide forward current through it to inject minority carriers of high energy in the depletion layer near the junction. When the minority carriers, i.e., electrons, are injected into the p-type region from the n-type region, the number of the majority carriers, i.e., holes, increases so as to satisfy electrical neutrality, and the excitation state is obtained. Semiconductor lasers for excitation by current injection in this manner are called injection lasers, diode lasers, or laser diodes (LD). Simple consideration concerning carrier statistics shows that an important requirement to obtain population inversion is that the forward bias voltage V must satisfy

$$eV > \hbar\omega \tag{1.2}$$

For a laser to amplify an optical wave of 830 nm wavelength, for example, the forward-bias voltage should be $V > 1.5\,\mathrm{V}$.

In order to outline the working principle and the device structure of semiconductor lasers, as an example let us take GaAs lasers, which are representative of semiconductor lasers and have the longest history. The first oscillation of injection lasers was obtained in a p–n junction structure consisting of single-crystal material GaAs, i.e., a diode with a homostructure. The laser, however, required current injection of a very large density (greater than 50 kA/cm^2) for lasing, and therefore lasing was limited to pulse oscillation at low temperatures. Continuous oscillation at room temperature and practical performances were accomplished in lasers of double heterostructure (DH), which were developed later. Nowadays, semiconductor lasers usually mean DH lasers. The structure is schematically illustrated in Fig. 1.2. The laser structure consists of a laser-active layer of GaAs with a thickness around 0.1 μm sandwiched between two layers of $Al_xGa_{1-x}As$ with larger bandgap energy and has double heterojunctions. The structure is fabricated by multilayer epitaxy on a GaAs substrate. The GaAs layer and the $Al_xGa_{1-x}As$ layers are called the active layer and cladding layers, respectively. The cladding layers are p doped and n doped, respectively.

The DH structure offers two functions that are very effective for reduction in the current required for laser operation. The first is carrier confinement. As shown in Fig. 1.3(a), the difference in the bandgap energy gives rise to formation of barriers in electron potential, and therefore the injected carriers are confined within the active layer at high densities without diffusion from the junctions. Thus the population inversion required for light amplification can be accomplished with current injection of relatively small density (about 1 kA/cm^2). The second function is optical waveguiding. The cladding layers with larger bandgap energy are almost transparent for optical

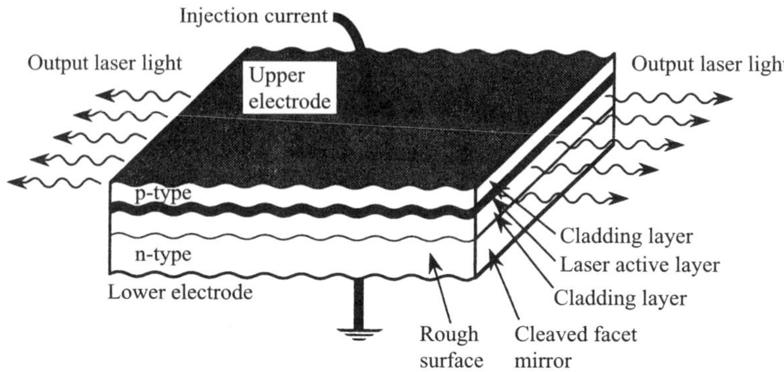

Figure 1.2 Structure of a DH semiconductor injection laser.

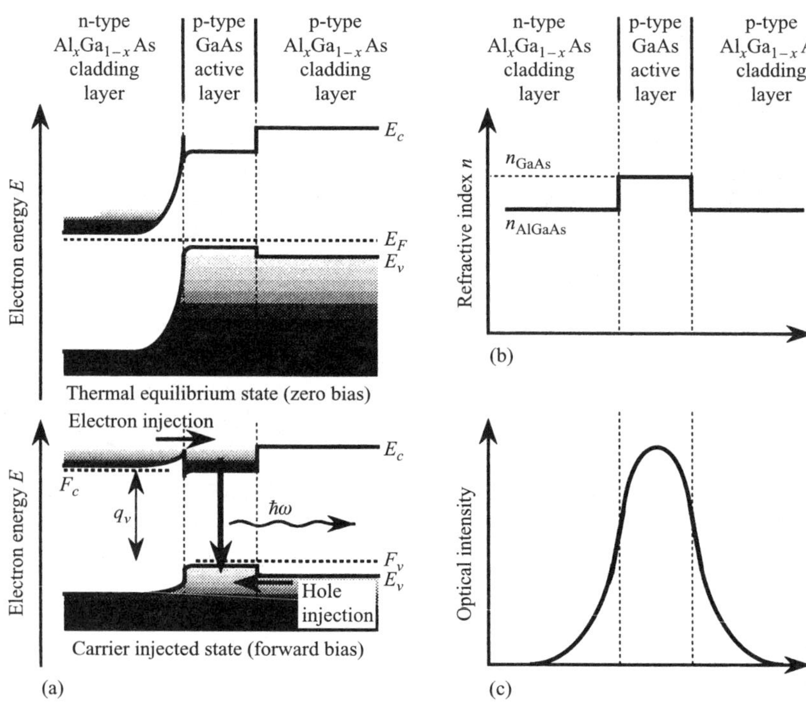

Figure 1.3 Confinement of carriers and optical wave in a DH semiconductor laser: (a) energy band diagram; (b) refractive index distribution; (c) optical intensity distribution.

wavelengths where amplification is obtained in the active layer, and the refractive index of the cladding layers is lower than that of the active layer, as shown in Fig. 1.3(b). Accordingly, the optical wave is confined in the high-index active layer through successive total internal reflection at the interfaces with the cladding layers and propagates along the plane of the active layer. This form of optical wave propagation is called a guided mode. In contrast with a bulk semiconductor of uniform refractive index, where the optical wave diverges owing to diffraction, the DH structure can guide the optical wave as a guided mode, in a region of the thin (less than 1 μm) active layer with optical gain and in its vicinity, over a long (more than several hundred micrometres). Thus the optical wave is amplified very effectively. Let g be the amplification gain factor under the assumption that the optical wave is completely confined and propagates in the active layer, let Γ be the coefficient of reduction due to the penetration of the guided mode into the cladding layers, and let α_{int} be the factor representing the optical losses due to absorption and scattering caused by imperfection of the structure. Then the effective gain in the actual DH structure is given by $\Gamma g - \alpha_{\text{int}}$.

To implement a laser oscillator that generates a coherent optical wave, it is required to provide the optical amplifier with optical feedback. In semiconductor lasers, this can readily be accomplished by cleaving the semiconductor crystal DH structure with the substrate to form a pair of facets perpendicular to the active layer. The interface between the semiconductor and the air serves as a reflection mirror for the optical wave of a guided mode and gives the required feedback. Since the semiconductor has a high index of refraction, a reflectivity as large as approximately 35% is obtained with a simple facet. The structure corresponds to a Fabry–Perot optical resonator (usually constructed with a pair of parallel mirrors) implemented with a waveguide, and therefore the laser of this type is called a Fabry–Perot (FP)-type laser. The optical wave undergoes amplification during circulation in the structure, as shown in Fig. 1.4. Let R be the reflectivity of the facet mirrors, and L be the mirror separation; then the condition for the guided mode to recover its original intensity after a round trip is given by

$$R^2 \exp[2(\Gamma g - \alpha_{\text{int}})L] = 1 \tag{1.3}$$

The condition for the wave to be superimposed with the same phase after the round trip is given by

$$2L = m\lambda \qquad \left(\lambda = \frac{2c\pi}{N_{\text{e}}\omega}\right) \tag{1.4}$$

where λ is the optical wavelength, c the light velocity in vacuum, ω the optical angular frequency, N_{e} the effective index of refraction, and m an

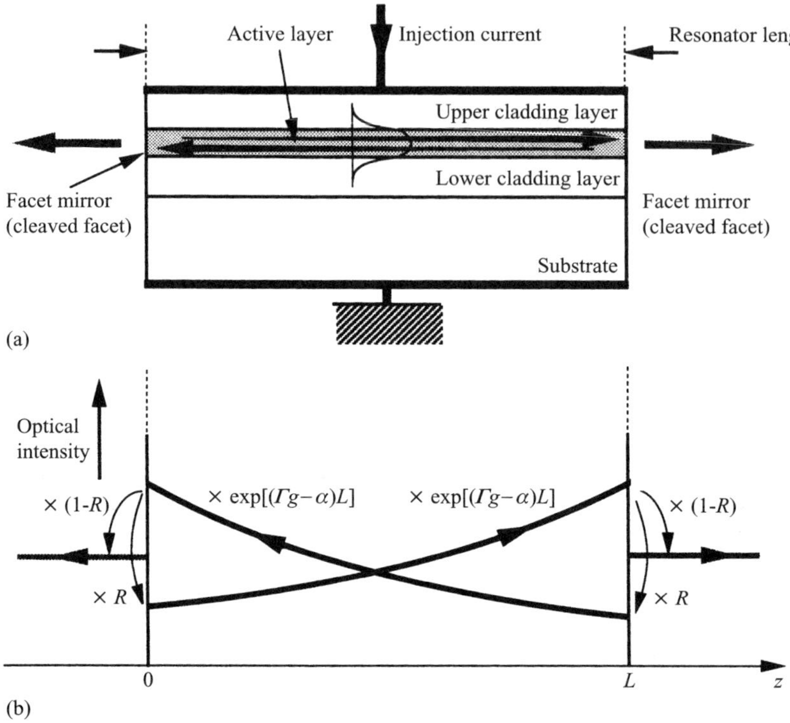

Figure 1.4 Circulation of optical wave in a FP semiconductor laser: (a) cross section including the optical axis; (b) distribution of guided wave power.

arbitrary integer. This is the condition for positive feedback. An optical wave of wavelength satisfying Eq. (1.4) can resonate, since the wave is superimposed with the same phase after an arbitrary number of round trips. When the injection current is increased and the effective gain for one of the resonant wavelengths reaches the value satisfying Eq. (1.3), optical power is accumulated and maintained in the resonator, and the power is emitted through the facet mirrors. This is the laser oscillation, i.e., lasing. The output is a coherent optical wave with only a resonant wavelength component or components. The reflectivity of the facet mirrors is slightly larger for the transverse electric (TE) wave (electric field vector parallel to the active layer) than for the transverse magnetic wave (perpendicular), and therefore FP lasers oscillate with TE polarization. The above two equations describe the oscillation conditions, and the injection current required for obtaining the gain satisfying these conditions is called the threshold current. The resonator length L of typical semiconductor lasers is 300–1000 μm, and the threshold current density for room temperature oscillation is typically of the order of 1 kA/cm^2. When the injection current is increased further, the carriers injected by the increase above the threshold are consumed by recombination associated with stimulated emission of photons. Therefore the optical output power is obtained in proportion to the increase above the threshold.

The mode of the optical wave generated by a laser is generally classified by lateral modes and longitudinal modes. For semiconductor lasers, the lateral mode, i.e., the intensity distribution in the cross section normal to the optical axis, is defined by the waveguide structure. The complexity or instability of the lateral mode gives rise to deterioration of the spatial coherence of the output wave. The longitudinal mode (axial mode), on the other hand, is defined by the distribution, along the direction of propagation (the optical axis), of the standing wave in the resonator. Each longitudinal mode corresponds to each integer m in Eq. (1.4) and constructs components with slightly different wavelengths. Temporal coherence is degraded if several longitudinal modes oscillate simultaneously (multimode lasing) and/or there is fluctuation of modes.

The simplest FP laser of DH structure as shown in Fig. 1.2, where current is injected over whole area of the crystal, is called a broad-area laser. The laser can easily be fabricated and a large output power can be obtained. The broad-area laser, however, suffers from the drawbacks that both the spatial coherence and the temporal coherence are low, since it is very difficult to obtain an oscillation that is uniform over a large width in the active layer along the lateral direction perpendicular to the optical axis, and the laser oscillates in many longitudinal modes. The drawbacks can be removed by restricting the current injection into a narrow stripe region and confining the optical wave also with respect to the lateral direction. A simple method is to fabricate a laser structure where the upper electrode has electrical contacts with the semiconductor only within a stripe region and to inject current only in the region with a width of a few micrometres. Then the optical wave is guided in the region near the axis where the gain is large, and therefore this laser is called a laser of gain guiding type. A better method is not only to restrict the injection region but also to form a channel waveguide where the refractive index is higher in the narrow channel region than in the surrounding areas by microprocessing of the active or cladding layer. The laser is called a laser of index guiding type. Although the gain guiding type is easy to fabricate, it is difficult to stabilize lateral mode(s) over a large injection current range, and the laser involves the drawback of multiple-longitudinal-mode lasing. For the index guiding type, on the other hand, a stable single lateral mode oscillation can be obtained by appropriate design, and an oscillation that can be considered substantially as a single longitudinal mode can be obtained. Because of these advantages, the main semiconductor laser is now of the index guiding type. A typical threshold current of the index guiding laser is 10–30 mA, and continuous oscillation with output power of a few to several tens of milliwatts is obtained with an injection current of a few tens to hundreds of milliamperes.

The FP laser is one of the most important semiconductor lasers that is of fundamental structure and is widely used in many practical applications. The FP laser, however, involves the important disadvantage that the longitudinal mode is not stable. In continuous oscillation, variations in the injection current and ambient temperature give rise to a change in or alteration of the longitudinal mode. Mode hopping is associated with a jump in the lasing wavelength and a large increase in intensity noise. Moreover, even if a laser oscillates in single longitudinal mode for continuous-wave operation, under high-speed modulation the lasing changes to multimode oscillation and the spectral width is largely broadened. These phenomena impose limitations on the applications. To solve this problem, it is necessary to implement dynamic single-mode lasers that can maintain single-mode oscillation under dynamic operation. This can be accomplished by using an optical resonator, such that effective feedback takes place only in a narrow wavelength width, to increase the threshold and substantially to prevent oscillation except for a lasing mode. Among various practical device implementation, distributed feedback (DFB) lasers and distributed Bragg reflector (DBR) lasers, utilizing a fine periodic structure, i.e., an optical grating, formed in the semiconductor waveguide, are important. The grating is formed in the active section in DFB lasers, and outside the active section in DBR lasers. Dynamic single-mode oscillation is accomplished through the wavelength-selective distributed feedback and reflection. The lasers exhibit excellent performances, including a narrow spectrum width and low noise, and therefore are practically used in applications such as long-distance optical communications. Thus the importance of DFB and DBR lasers is increasing.

In the above discussion, lasers of DH structures with an active layer of thickness around 0.1 μm (100 nm) were described. The carriers can be considered to behave as particles, as the thickness of the active layer is much larger than the wavelength of the electron wave. If the thickness is reduced to 10 nm or so, approaching the same order of magnitude as the electron wavelength, the quantum nature of the carriers as material waves appears significantly. The active layer and the surrounding cladding layers form a potential well with a narrow width, and the electrons and holes are confined in the quantum well (QW) as a wave that satisfies the Schrödinger wave equation and boundary conditions. The confinement gives rise to an increase in the effective bandgap energy and modification of the density-of-states function into a step-like function, and as a result a gain spectrum and polarization dependence unlike those of ordinary DH lasers appear. Accordingly, by appropriate design of the QW, light emission characteristics advantageous for improvement in laser performances can be obtained. Based on the use of the QW and optimization of the waveguide structure, various types of laser, i.e., single quantum well (SQW) and multiple quantum well (MQW) lasers of FP, DFB, and DBR types, have been implemented. Remarkable developments have been made in the extension of the lasing wavelength region, reduction in the threshold current, enhancement of modulation bandwidth, reduction in the noise, and improvement in the spectral purity. Although QW lasers requires advanced design and fabrication techniques, many QW lasers have already been commercialized and found many practical applications. Further development of QW lasers as an important semiconductor laser is expected.

In the lasers described above, the optical wave propagates along the optical axis parallel to the active layer and is emitted through the facet perpendicular to the axis. On the other hand, rapid progress has been made in vertical cavity surface emitting lasers (VCSELs) [21] that have a resonator with the optical axis perpendicular to the active layer and provide area emission of photons. The VCSEL, however, is outside the scope of this book.

1.2 MATERIALS FOR SEMICONDUCTOR LASERS

There are several requirements for materials for semiconductor lasers. The most important requirement is to have a bandgap of direct transition type as represented by that of GaAs. In such semiconductors, an optical transition satisfying the energy and momentum conservation rules can take place between electrons and holes at the vicinity of the band edges. Since the transition probability is large, light emission can easily be obtained. In semiconductors having a bandgap of indirect transition type, such as Si and Ge, on the other hand, the photons emitted by the transition satisfying the conservation rules are absorbed in the semiconductor itself, and the efficiency of the emission due to the transition between electrons and holes near the band edges is low because the transition requires the assistance of interaction with phonons. Therefore, implementation of lasers with this type of semiconductor is considered to be impossible or very difficult. Most III–V compound semiconductors (except for AlAs) and II–VI compound semiconductors have a direct transition bandgap and therefore can be a material for lasers.

An important requirement for heteroepitaxial fabrication of good-quality DH structures, which are considered essential for implementation of lasers for continuous oscillation, is lattice matching with an appropriate substrate crystal. Growth of an active layer crystal having a lattice constant different from that of the substrate involves a high density of defects, which causes the emission characteristics to deteriorate seriously, and therefore a practical device cannot be obtained. The maximum tolerable lattice mismatch is typically 0.1% or less. Other requirements include the possibility of producing p–n junctions by appropriate doping, and refractive indexes appropriate for waveguiding in the DH structure. For FP laser implementation, it is desirable that facet mirrors of appropriate orientation can be formed by simple cleaving.

The oscillation wavelength of a laser is determined approximately by Eq. (1.1). To implement a laser for emission of light of a given wavelength, it is therefore necessary to use a semiconductor material having an appropriate value of the bandgap energy E_g. The requirement can readily be satisfied by use of compound semiconductor alloy crystals. For example, alloy crystals of $Al_xGa_{1-x}As$ can be produced from GaAs and AlAs, E_g being a continuous function of x, and the oscillation wavelength can be arbitrarily determined in 0.7–0.9 μm range by appropriate choice of x. Figure 1.5 shows the wavelength ranges that can be covered with several

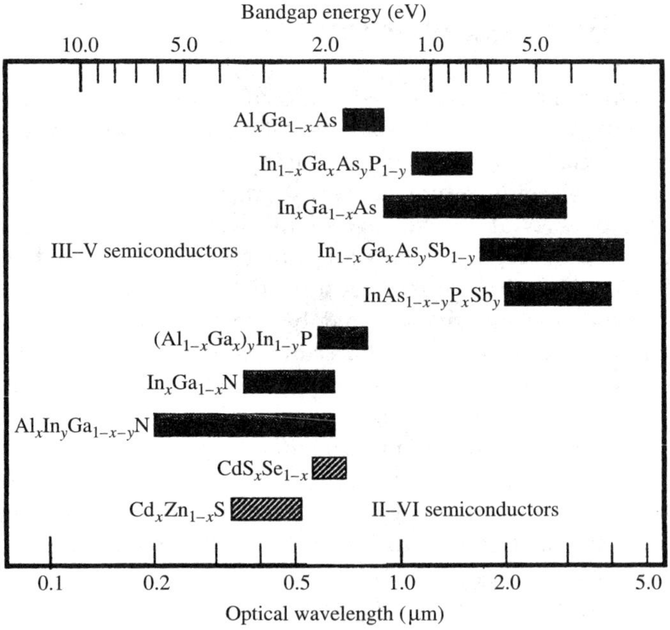

Figure 1.5 Materials and oscillation wavelength regions of semiconductor lasers.

compound semiconductor alloy crystals. The lattice constant of a ternary alloy $Al_xGa_{1-x}As$ has a very small dependence on x, and therefore the alloy of arbitrary x is lattice matched with the GaAs substrate. This situation, however, is rather exceptional; the lattice constant generally depends upon the composition ratio, and therefore for ternary alloys an arbitrary choice of E_g is not compatible with lattice matching. To solve the problem, quaternary alloys can be used. An example is $In_{1-x}Ga_xAs_yP_{1-y}$. By appropriate design of x and y, an arbitrary choice of the oscillation wavelength in the wide range 1.1–1.6 μm can be accomplished simultaneously with lattice matching with the InP substrate. $In_{1-x}Ga_xAs_yP_{1-y}$ semiconductor lasers have been widely used as lasers for optical communications in the 1.3 and 1.5 μm bands and are one of the most important semiconductor lasers.

Figure 1.5 includes other semiconductor laser materials. $In_xGa_{1-x}As$ is an important material because, although it does not lattice match with $Al_xGa_{1-x}As$, it can be combined with GaAs and $Al_xGa_{1-x}As$ to implement strained QW lasers. High-performance lasers for emission in the 0.9 μm band have been developed. $In_{1-x}Ga_xAs_ySb_{1-y}$ and $InAs_{1-x-y}P_xSb_y$, which can lattice match with the GaSb and InAs substrates, have been studied as candidates for lasers for emission in a 1.7–4.4 μm range. One of the materials for lasers for visible-light emission is $(Al_xGa_{1-x})_yIn_{1-y}P$, which can lattice match with the GaAs substrate. Lasers for red-light emission in the 0.6 μm band have been commercialized as light sources for optical disk memory such as digital versatile disk (DVD). Candidates for materials for semiconductor lasers in the green through the blue to the violet region include the II–VI compound semiconductors CdS, CdSe, ZnS, ZnSe and their alloys, and the III–V compound semiconductors GaN, InN, AlN and their alloys. A recent extensive study on lasers of this wavelength range has led to rapid and remarkable developments, as represented by accomplishments of room-temperature continuous oscillation in ZnSe QW lasers on a GaAs substrate and $In_xGa_{1-x}N$ QW lasers on an Al_2O_3 substrate. It is no doubt that $In_xGa_{1-x}N$ blue lasers will soon be employed in practical use. Not shown in Fig. 1.5 are IV–VI compound semiconductors for mid- to far-infrared lasers, such as $Pb_xSn_{1-x}Te$, $PbS_{1-x}Se_x$ and $Pb_xSn_{1-x}Se$, with which various injection lasers in the 3–34 μm wavelength range have been implemented. A unique feature of these lasers is that the oscillation wavelength can be temperature tuned over a very wide range, although they require operation at a low temperature.

Fabrication of semiconductor lasers requires growth of the fundamental multilayer DH structure with controlled composition and doping level on a substrate crystal. Such epitaxial growth is the most important technology. A technique of long history is liquid-phase epitaxy (LPE).

Development of continuous heteroepitaxy based on LPE using slide boats enabled implementation of the first DH lasers. Until now the LPE technique has been used as a method suitable for mass production. Then, the vapor-phase growth techniques, metal–organic vapor-phase epitaxy (MOVPE) and molecular-beam epitaxy (MBE), were developed. By these techniques, one can grow layers with precisely controlled composition and thickness at the level of atomic layers. They enabled fabrication of QW and strained QW structures to be carried out and have made important contributions to the development of semiconductor lasers. At present, MOVPE and MBE techniques are being employed not only in the fabrication of advanced lasers such as QW lasers but also in mass production. For monolithic integration of semiconductor lasers together with electronic devices, implementation of lasers on a Si substrate is desirable. An extensive study is being made also on superheteroepitaxy where a GaAs crystal is grown on a Si substrate having a lattice constant very different from that of GaAs.

1.3 FEATURES OF SEMICONDUCTOR INJECTION LASERS

This section considers features of semiconductor lasers from an application point of view. In comparison with other categories of lasers, semiconductor injection lasers offer the following advantages.

1. *Compactness.* Most semiconductor lasers are extremely compact and of light weight; they have a chip size of 1 mm^3 or less. Even if a heat sink and a power supply required for driving are included, the laser system can be very compact.
2. *Excitation by injection.* The laser can easily be driven by injection of a current in the milliamperes range with a low voltage (a few volts). Except for a power supply, no device or component for excitation is required. The direct conversion of electrical power into optical power ensures a high energy conversion efficiency.
3. *Room-temperature continuous oscillation.* Many semiconductor lasers can oscillate continuously at and near room temperature.
4. *Wide wavelength coverage.* By appropriate choice of the materials and design of the alloy composition ratio, lasers of arbitrary wavelengths in a very wide range, from infrared to whole visible regions, can be implemented, or at least there is possibility of the implementation.
5. *Wide gain bandwidth.* The wavelength width of the gain band of a semiconductor laser is very wide. It is possible to choose

arbitrarily the oscillation wavelength within the gain bandwidth and to implement wavelength-tunable lasers. Wide-band optical amplifiers can also be implemented.

6. *Direct modulation.* By superimposing a signal on the driving current, the intensity, the frequency and the phase of the output light can readily be modulated over a very wide (from direct current to gigahertz) modulation frequency range.
7. *High coherence.* Single-lateral-mode lasers provide an output wave of high spatial coherence. In DFB and DBR lasers, very high temporal coherence can also be obtained through stable single-longitudinal-mode oscillation of a narrow (down to submegahertz) spectrum width.
8. *Generation of ultrashort optical pulses.* It is possible to generate ultrashort optical pulses of subnanosecond to picosecond width by means of gain switching and mode locking with a simple system construction.
9. *Mass producibility.* The compact fundamental structure consisting of thin layers, along with fabrication by lithography and planar processing, is suitable for mass production.
10. *High reliability.* The device is robust and stable, since the whole laser is in a form of a chip. There is no wear-and-tear factor and, for lasers of many established materials, the fatigue problem has been solved. Thus the lasers are maintenance free, have a long lifetime, and offer high reliability.
11. *Monolithic integration.* The features 1, 2, 9, and 10 allow integration of many lasers on a substrate. It is also possible to implement optical detectors, optical modulators and electronic devices in the same semiconductor material. Monolithic integrated devices of advanced functions can be constructed.

On the other hand, semiconductor lasers involve the following drawbacks or problems.

1. *Temperature characteristics.* The performances of a laser depend largely upon temperature; the lasing wavelength, threshold current and output power change sensitively with change in ambient temperature.
2. *Noise characteristics.* The lasers utilize high-density carriers, and therefore fluctuation in the carrier density affects the refractive index of the active region. Since the lasers have a short resonator length and use facet mirrors of low reflectivity, the oscillation is affected sensitively by perturbations caused by external feedback. As a result, semiconductor lasers often involve various noise and instability problems.

3. *Divergent output beam.* The output beam is taken out through the facet in the form of a divergent beam emitted from the guided mode. An external lens is required to obtain a collimated beam.

Efforts have continued to seek improvements. Depending upon the categories of semiconductor lasers, the problems have been reduced to a practically tolerable level, or techniques to avoid the problem substantially have been developed.

1.4 APPLICATIONS OF SEMICONDUCTOR LASERS

As discussed in the previous section, semiconductor lasers exhibit many unique features in both functions and performances and also offer economical advantages. Therefore, by the development of semiconductor lasers, lasers, which had been a special instrument for scientific research and limited applications, acquired a position as a device for general and practical instruments. As will be outlined below, the applications of semiconductor lasers cover a wide area, including optical communications, optical data storage and processing, optical measurement and sensing, and optical energy applications.

One of the most important applications of semiconductor lasers is optical-fiber communications. The development of semiconductor lasers has been motivated mainly by this application. First, optical communication systems using GaAs lasers were completed, and they have been used for local area communications. Then, $In_{1-x}Ga_xAs_yP_{1-y}$ lasers which emit optical waves at wavelengths in the 1.3 μm band, where silica optical fibers exhibit the minimum group velocity dispersion, and in the 1.5 μm band, where they exhibit minimum propagation losses, were developed. Although in this application there are stringent requirements such as wide-band modulation, narrow spectral bandwidth, low noise, and high reliability, high performances have been accomplished through various improvements including developments of DFB and DBR structures and QW structures. Thus semiconductor lasers are being practically used as a completed device. Higher performances are required in the wavelength division multiplexing (WDM) communication systems and coherent communication systems. A variety of high-performance semiconductor lasers, including wavelength-tunable lasers, have been developed. Remarkable developments have been obtained also in semiconductor laser amplifiers and nonlinear-optic wavelength conversion in semiconductor lasers. There has been progress in the development of picosecond mode lock semiconductor lasers as a light source for future applications to optical-fiber soliton transmission systems. Towards development of the multimedia society using optical-fiber subscriber networks, low-cost communication semiconductor lasers are being developed. There is no doubt about the importance of semiconductor lasers which are extensively used in the communications so essential to our consumer society. Applications such as optical communications in space and local free-space optical information transmission have also been studied. Another important area of semiconductor laser applications is optical disk memories. This application requires low nose and high stability in a sense somewhat different from that in applications to communications. Important and strong requirements include a short wavelength for a high density of data recording and a low production cost. $Al_xGa_{1-x}As$ lasers were adopted as a light source for compact-disk (CD) pickup heads, and became the first laser that penetrated widely into the home. For DVD systems, $(Al_{1-x}Ga_x)_yIn_{1-y}P$ red semiconductor lasers have been adopted. Development and commercialization of semiconductor lasers of shorter wavelength are in progress, as mentioned in the previous section.

Other applications in optical information processing under practical use include laser printers, image scanners, and barcode readers. An extensive study on applications to ultrafast (picosecond) signal processing is being made. Applications in optical measurements and sensors include the use of infrared-tunable lasers in spectroscopic measurements and environment sensing, various measurements using pulse and tunable lasers, and use with optical-fiber sensors. Optical energy applications include $In_xGa_{1-x}As$ strained QW lasers as a pump source to excite fiber laser amplifiers for communication systems, broad-area lasers and arrayed lasers as a pump source to excite solid-state lasers such as yttrium aluminium garnet (YAG) lasers. These lasers are also commercially available.

Extensive research and development work is being performed to implement monolithic integrated optic devices using semiconductor lasers as a core component. Integrated devices consisting of laser and electronic circuit elements are called optoelectronic integrated circuits (OEIC) and those consisting of optoelectronic elements such as lasers, modulators, and photodetectors are called photonic integrated circuits (PIC). Many integrated devices are employed for applications to optical communication systems, optical interconnection in computer systems, and optical measurements and sensing.

As we saw above, semiconductor lasers has enabled various new applications unfeasible or difficult to accomplish with other lasers to be made. High performances and advanced functions, which had been implemented with other lasers, have been accomplished with alternative semiconductor lasers. Further developments of semiconduct or lasers are expected.

REFERENCES

1. Y. Watanabe and J. Nishizawa, Japan Patent 273217, April 22 (1957).
2. A. L. Schawlow and C.H. Townes, *Phys. Rev.*, **112**, 1940 (1958).
3. N. G. Basov, O. N. Krokhin, and Yu. M. Popov, *Sov. Phys. JETP*, **13**, 1320 (1961).
4. M. G. A. Bernard and G. Duraffourg, *Phys. Status Solidi*, **1**, 699 (1961).
5. W. P. Dumke, *Phys. Rev.*, **127**, 1559 (1962).
6. R. N. Hall, G. E. Fenner, J. D. Kingsley, T. J. Soltys, and R. O. Carlson, *Phys. Rev. Lett.*, **9**, 366 (1962).
7. M. I. Nathan, W. P. Dunke, G. Burns, F. H. Dill, Jr, and G. Lasher, *Appl. Phys. Lett.*, **1**, 62 (1962).
8. T. M. Quist, R. H. Rediker, R. J. Keyes, W. E. Krag, B. Lax, A. L. McWhorter, and H. J. Zeiger, *Appl. Phys. Lett.*, **1**, 91 (1962).
9. I. Hayashi, M. B. Panish, P. W. Foy, and S. Sumski, *Appl. Phys. Lett.*, **17**, 109 (1970).
10. Zh. I. Alferov, V. M. Andreef, D. Z. Garbuzov, Yu V. Zhilyaev, E. P. Morozov, E. Poiutnoi, and V. G. Trofim, *Fiz. Tekh. Poluprovodn.*, **4**, 1826 (1970).
11. H. Kressel and F. Z. Hawrylo, *Appl. Phys. Lett.*, **17**, 169 (1970).
12. H. Kressel and J. K. Butler, *Semiconductor Lasers and Heterojunction LEDs*, Academic Press, New York (1997).
13. H. C. Casey, Jr, and M. B. Panish, *Heterostructure Lasers*, Academic Press, New York (1978).
14. Y. Suematsu (ed.), *Semiconductor Lasers and Optical Integrated Circuits* (in Japanese), Ohmsha, Tokyo (1984).
15. Japan Society of Applied Physics (ed.), *Fundamentals of Semiconductor Lasers* (in Japanese), Ohmsha, Tokyo (1987).
16. R. Ito and M. Nakamura, *Semiconductor Lasers—Fundanentals and Applications* (in Japanese), Baifukan, Tokyo (1989).
17. G. P. Agrawal and N. K. Dutta, *Semiconductor Lasers*, second edition, Van Nostrand Reinhold, New York (1993).
18. W. W. Chow, S. W. Koch, and M. Sergeant III, *Semiconductor Laser Physics*, Springer, Berlin (1993).
19. Japan Society of Applied Physics (ed.), K. Iga (ed.), *Semiconductor Lasers* (in Japanese), Ohmsha, Tokyo (1994).
20. L. A. Coldren and S. W. Corzine, *Diode Lasers and Photonic Integrated Circuits*, John Wiley, New York (1995).
21. K. Iga and F. Koyama, *Surface Emitting Lasers* (in Japanese), Ohmsha (1990).

2

Interaction of Electrons and Photons

This chapter provides the basis for the discussion in the following chapters by summarizing the fundamental concepts and the quantum theory concerning the interaction between electrons and photons in a form that is convenient for theoretical analysis of semiconductor lasers [1–9]. First, quantization of electromagnetic fields of optical waves is outlined, and the concept of a photon is clarified. Quantum theory expressions for coherent states are also given. Then the quantum theory of electron–photon interactions and the general characteristics of optical transitions are explained. Fundamental mathematical expressions for absorption, spontaneous emission, and stimulated emission of photons are deduced, and the possibility of optical wave amplification in population-inverted states is shown.

2.1 QUANTIZATION OF OPTICAL WAVES AND PHOTONS

2.1.1 Expression of Optical Waves by Mode Expansion

The electric field $\boldsymbol{E}$ and magnetic field $\boldsymbol{H}$ of optical waves, together with the electric flux density $\boldsymbol{D}$, magnetic flux density $\boldsymbol{B}$, current density $\boldsymbol{J}$, and charge density ρ, generally satisfy the Maxwell equations

$$\nabla \times \boldsymbol{E} = -\frac{\partial \boldsymbol{B}}{\partial t} \tag{2.1a}$$

$$\nabla \cdot \boldsymbol{D} = \rho \tag{2.1b}$$

$$\nabla \times \boldsymbol{H} = \frac{\partial \boldsymbol{D}}{\partial t} + \boldsymbol{J} \tag{2.1c}$$

$$\nabla \cdot \boldsymbol{B} = 0 \tag{2.1d}$$

The electromagnetic fields can be expressed using a vector potential $\boldsymbol{A}$ and a scalar potential ϕ. For cases where there is no free charge in the medium ($\rho = 0$; $\boldsymbol{J} = \boldsymbol{0}$), in particular, we can put $\phi = 0$, and accordingly $\boldsymbol{E}$ and $\boldsymbol{H}$

can be described by using only $\boldsymbol{A}$ as

$$\boldsymbol{E} = -\frac{\partial \boldsymbol{A}}{\partial t}, \qquad \boldsymbol{H} = \frac{1}{\mu_0}\nabla \times \boldsymbol{A} \tag{2.2}$$

We express $\boldsymbol{A}$ by a superposition of sinusoidal wave components of various angular frequencies ω_m as

$$\boldsymbol{A}(\boldsymbol{r}, t) = \tfrac{1}{2}\sum_m \left[a_m(t)\boldsymbol{A}_m(\boldsymbol{r}) + a_m^*(t)\boldsymbol{A}_m^*(\boldsymbol{r})\right] \tag{2.3a}$$

$$a_m(t) = a_m \exp(-\mathrm{i}\omega_m t) \tag{2.3b}$$

Then $\boldsymbol{E}$ can be written as

$$\boldsymbol{E}(\boldsymbol{r}, t) = \tfrac{1}{2}\sum_m \left[a_m(t)\boldsymbol{E}_m(\boldsymbol{r}) + a_m^*(t)\boldsymbol{E}_m^*(\boldsymbol{r})\right] \tag{2.4a}$$

$$\boldsymbol{E}_m(\boldsymbol{r}) = \mathrm{i}\omega_m \boldsymbol{A}_m(\boldsymbol{r}) \tag{2.4b}$$

Let n_{r} be the refractive index of a medium at angular frequency ω, then the ω components of $\boldsymbol{D}$ and $\boldsymbol{E}$ are correlated by $\boldsymbol{D} = n_{\mathrm{r}}^2\varepsilon_0\boldsymbol{E}$, and $\boldsymbol{E}_m(\boldsymbol{r})$ satisfies the Helmholtz wave equation

$$\nabla^2\boldsymbol{E}_m + \left(\frac{n_{\mathrm{r}}\omega_m}{c}\right)^2\boldsymbol{E}_m = 0, \qquad \nabla\cdot\boldsymbol{E}_m = 0 \tag{2.5}$$

where $c = 1/(\varepsilon_0\mu_0)^{1/2}$ is the light velocity in vacuum. $\boldsymbol{A}_m$ also satisfies the same Helmholtz wave equation as $\boldsymbol{E}_m \cdot \boldsymbol{E}_m$ and $\boldsymbol{A}_m$ satisfying the wave equation given by Eq. (2.5) and boundary conditions constitute a mode, and the expressions in Eqs. (2.3) and (2.4) are called mode expansions. The concept of the mode expansion is illustrated in Fig. 2.1. Noting that the modes $\{\boldsymbol{E}_m(\boldsymbol{r})\}$ form an orthogonal system, we normalize them so that the energy stored in the medium of volume V satisfies

$$\int_V \frac{\varepsilon_0 n_{\mathrm{r}} n_{\mathrm{g}}}{2}\boldsymbol{E}_m(\boldsymbol{r}) \cdot \boldsymbol{E}_{m'}^*(\boldsymbol{r})\,\mathrm{d}V = \hbar\omega_m\delta_{mm'} \tag{2.6}$$

where n_{g} is the group index of refraction (see Eq. (2.10)), and $\delta_{mm'}$ is the Kronecker delta.

2.1.2 Mode Density

The solution of the wave equation given by Eq. (2.5) for the homogeneous medium occupying the space volume V can be written as

$$\boldsymbol{E}_m(\boldsymbol{r}) = \boldsymbol{E}_m \exp(\mathrm{i}\boldsymbol{k}_m \cdot \boldsymbol{r}), \qquad |\boldsymbol{k}_m| = \frac{n_{\mathrm{r}}\omega_m}{c}, \qquad \boldsymbol{E}_m \cdot \boldsymbol{k}_m = 0 \tag{2.7}$$

Each mode is a plane transverse wave propagating along the direction of a wave vector $\boldsymbol{k}_m$. Since there exist many modes with different propagation directions for a given frequency, we need the concept of mode density. As for

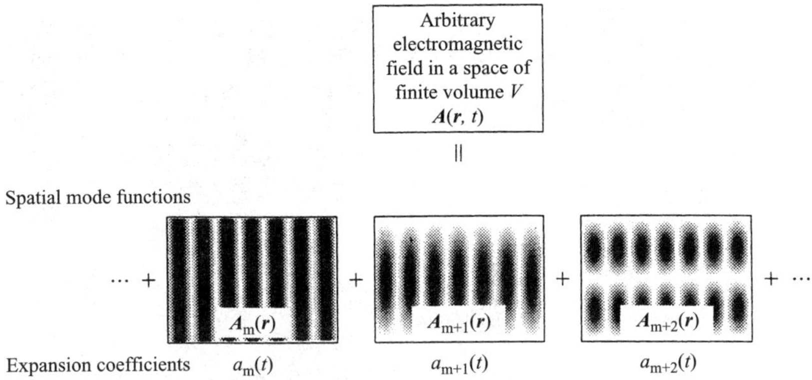

Figure 2.1 Schematic illustration of mode expansion of an optical wave in free space.

the space volume V, consider a cube of side length L much larger than the optical wavelength. Then the periodic boundary condition requires that the wave vector $\boldsymbol{k}$ $(=\boldsymbol{k}_m)$ must be in the form of

$$\boldsymbol{k} = \left(\frac{2\pi m_x}{L}, \frac{2\pi m_y}{L}, \frac{2\pi m_z}{L}\right),$$
$$m_x, m_y, m_z = \ldots, -2, -1, 0, 1, 2, \ldots \tag{2.8}$$

and, in the $\boldsymbol{k}$ space, a mode occupies a volume of $(2\pi/L)^3$. The number $\mathrm{d}N$ of modes within $\mathrm{d}k_x\,\mathrm{d}k_y\,\mathrm{d}k_z$ in the $\boldsymbol{k}$ space, per unit volume in real space, is given by

$$\begin{aligned}
\mathrm{d}N &= \frac{1}{(2\pi/L)^3}\frac{\mathrm{d}k_x\,\mathrm{d}k_y\,\mathrm{d}k_z}{L^3} \\
&= \left(\frac{1}{2\pi}\right)^3 \mathrm{d}k_x\,\mathrm{d}k_y\,\mathrm{d}k_z \\
&= \left(\frac{1}{2\pi}\right)^3 k^2 \mathrm{d}k\,\mathrm{d}\Omega \\
&= \left(\frac{1}{2\pi}\right)^3 \frac{n_\mathrm{r}^2 n_\mathrm{g}}{c^3}\omega^2 \mathrm{d}\omega\,\mathrm{d}\Omega
\end{aligned} \tag{2.9}$$

where $\mathrm{d}\Omega$ is the stereo angle for the range of propagation direction $\boldsymbol{k}$, and use has been made of

$$\frac{\mathrm{d}k}{\mathrm{d}\omega} = \frac{\mathrm{d}(n_\mathrm{r}\omega/c)}{\mathrm{d}\omega} = \frac{n_\mathrm{g}}{c}, \qquad n_\mathrm{g} \equiv n_\mathrm{r} + \frac{\omega\,\mathrm{d}n_\mathrm{r}}{\mathrm{d}\omega} \tag{2.10}$$

where n_g is the group index of refraction. There are two independent polarizations, i.e., two independent directions for $\boldsymbol{E}_m$ satisfying the third equation of Eq. (2.7). Therefore the total mode number is twice the above expression. Since the stereo angle for all directions is 4π, the mode density $\rho(\omega)$ per unit volume and per unit angular frequency width is given by

$$\rho(\omega) = \frac{2\mathrm{d}N}{\mathrm{d}\omega} = \frac{1}{\pi^2}\frac{n_r^2 n_g}{c^3}\omega^2 \tag{2.11}$$

2.1.3 Quantization of Optical Waves

The energy H stored in the medium associated with the electric field $\boldsymbol{E}$ and the magnetic field $\boldsymbol{H}$ of an optical wave can be given as follows, using the mode expansion, orthonormal relation (Eq. (2.6)) and periodic boundary condition:

$$\begin{aligned} H &= \int_V \left[\tfrac{1}{2} n_r n_g \varepsilon_0 \boldsymbol{E}^2 + \tfrac{1}{2}\mu_0 \boldsymbol{H}^2\right] \mathrm{d}V \\ &= \sum_m \frac{\hbar\omega_m}{2}\left(a_m a_m^* + a_m^* a_m\right) \end{aligned} \tag{2.12}$$

where h is the Planck constant and $\hbar = h/2\pi$. The above H can be considered as the Hamiltonian for the electromagnetic field.

In the following, we write a single mode only and omit the subscript m for simplicity. The full expression considering all modes can readily be recovered by adding the subscript m and summing to give $\sum_m$. We here define real values corresponding to the real and imaginary parts of the complex variables a and a^* by

$$q = \frac{1}{2}\left(\frac{2\hbar}{\omega}\right)^{1/2}(a + a^*) \tag{2.13a}$$

$$p = \frac{1}{2\mathrm{i}}(2\hbar\omega)^{1/2}(a - a^*) \tag{2.13b}$$

Then the Hamiltonian is written as

$$\begin{aligned} H &= \frac{\hbar\omega}{2}(aa^* + a^*a) \\ &= \tfrac{1}{2}(p^2 + \omega^2 q^2) \end{aligned} \tag{2.14}$$

and from this form we see that p and q are canonical conjugate variables.

The quantization of optical waves is accomplished by replacing a, a^* and p, q by corresponding operators. Noting that the operators p and q are canonically conjugate, we assume that the commutation relation $[q, p] = qp - pq = \mathrm{i}\hbar$ holds. Then we have the commutation relation

$$[a, a^{\dagger}] = aa^{\dagger} - a^{\dagger}a = \frac{1}{\mathrm{i}\hbar}[q, p] = 1 \tag{2.15}$$

for the operators a and $a^{\dagger}$ corresponding to a and a^*, and the Hamiltonian H is written as

$$\begin{aligned} H &= \tfrac{1}{2}(p^2 + \omega^2 q^2) \\ &= \hbar\omega(a^{\dagger}a + \tfrac{1}{2}) \\ &= \hbar\omega(N + \tfrac{1}{2}) \end{aligned} \tag{2.16}$$

$$N = a^{\dagger}a$$

Making the Heisenberg equation of motion from the above H yields

$$\begin{aligned} \frac{\mathrm{d}a}{\mathrm{d}t} &= \frac{1}{\mathrm{i}\hbar}[a, H] \\ &= -\mathrm{i}\omega a \end{aligned} \tag{2.17}$$

which corresponds to the equation for the classic amplitude a (Eq. (2.3b)).

In accordance with the replacement of the amplitude $a(t)$ by the operator a, all the electromagnetic quantities are also replaced by the corresponding operators. The operator expressions for the vector potential $\boldsymbol{A}$ and the electric field $\boldsymbol{E}$ are

$$\boldsymbol{A}(\boldsymbol{r}, t) = \tfrac{1}{2}[a\boldsymbol{A}(\boldsymbol{r}) + a^{\dagger}\boldsymbol{A}^*(\boldsymbol{r})] \tag{2.18a}$$

$$\boldsymbol{E}(\boldsymbol{r}, t) = \tfrac{1}{2}[a\boldsymbol{e}(\boldsymbol{r}) + a^{\dagger}\boldsymbol{e}^*(\boldsymbol{r})] \tag{2.18b}$$

2.1.4 Energy Eigenstates and Photons

The N defined by Eq. (2.16) is a dimensionless Hermitian operator. Let $|n\rangle$ be an eigenstate of operator N with an eigenvalue n; then we have $N|n\rangle = n|n\rangle$; using the commutation relation Eq. (2.15), we see that application of a to $|n\rangle$ results in an eigenstate of N with an eigenvalue $n-1$, and application of $a^{\dagger}$ to $|n\rangle$ results in an eigenstate of N with an eigenvalue $n+1$. From this and the normalization of the eigenstate systems, we obtain the important relations

$$a|n\rangle = n^{1/2}|n-1\rangle \tag{2.19a}$$

$$a^{\dagger}|n\rangle = (n+1)^{1/2}|n+1\rangle \tag{2.19b}$$

If the eigenvalue n is not an integer, from Eq. (2.19a) we expect the existence of eigenstates of infinitively large negative n. Since such eigenstates are not natural, the eigenvalue n should be an integer. This means that eigenstates for optical waves of a mode are discrete states of $n = 0, 1, 2, \ldots$ and, from the relation between H and N (Eq. (2.16)), the energy is given by

$$E_n = \hbar\omega\left(n + \tfrac{1}{2}\right) \qquad (n = 0, 1, 2, \ldots) \tag{2.20}$$

From Eq. (2.16), the eigenstates $|n\rangle$ of N, are energy eigenstates that satisfy

$$H|n\rangle = E_n|n\rangle \tag{2.21}$$

and form an orthonormal complete system. As Eq. (2.20) shows, the increase and decrease in energy of the optical wave of frequency ω are limited to discrete changes with $\hbar\omega$ as a unit. This implies that optical waves have a quantum nature from an energy point of view, and therefore the unit energy quantity $\hbar\omega$ is called the photon. The operator $N = a^\dagger a$ is called the photon number operator, since it gives the number of energy units, i.e., the number of photons. The amplitude operators a and $a^\dagger$, are called the annihilation operator and the creation operator, respectively, because of the characteristics of Eq. (2.19).

The energy eigenstate $|n\rangle$ plays an important role in the quantum theory treatment of optical waves. The expectation value for the energy of the optical wave in this state is

$$\langle n|H|n\rangle = E_n = \hbar\omega\left(n + \tfrac{1}{2}\right) \tag{2.22}$$

Figure 2.2 illustrates schematically the concepts of the quantization of optical wave, photons, and energy eigenstates. As the above equation shows, even the eigenstate $|0\rangle$ of the zero photon with the minimum energy is associated with a finite energy of $\hbar\omega/2$. This means that, even for the vacuum state where no photon is present, there exists a fluctuation in the electromagnetic field. The quantity $\hbar\omega/2$ is the zero-point energy, which results from fluctuations in the canonical variables following the uncertainty principle.

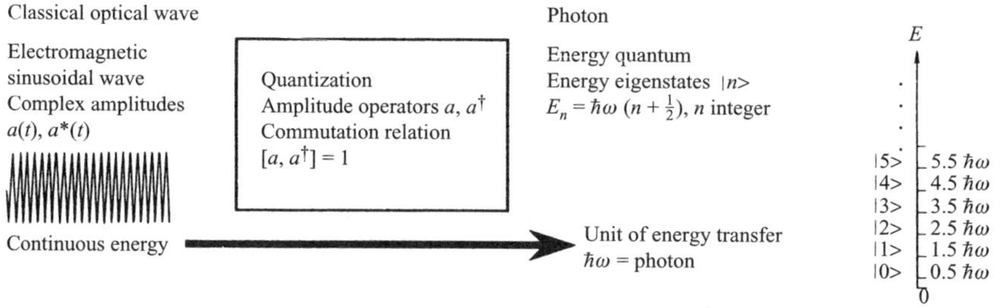

Figure 2.2 Quantization of the optical wave and the concept of a photon.

Although the energy eigenstates $|n\rangle$ of the optical wave are convenient for a discussion on the energy transfer between optical and electron systems, they are not appropriate for a discussion of the electromagnetic fields themselves. In fact, calculation of the expectation value for the electric field by using Eq. (2.18b) yields

$$\langle \boldsymbol{E} \rangle = \langle n|\boldsymbol{E}|n\rangle = 0 \tag{2.23}$$

for all instances of time, showing that, in spite of the fact that the wave has a single frequency ω, measurement of the amplitude results in fluctuations centered at zero. This is because, for an energy eigenstate with a definite photon number, the phase is completely uncertain. On the other hand, in many experiments using single-frequency optical waves such as laser light, the phase of the optical waves can be measured. The energy eigenstates are thus very unlike the ordinary state of the optical wave. It is therefore necessary to consider quantum states different from $|n\rangle$ to discuss the electromagnetic field specifically.

2.1.5 Coherent States

For a discussion of the electromagnetic field of optical waves whose amplitude can be observed as a sinusoidal wave, it is appropriate to use eigenstates of a, since the amplitude operators a and $a^\dagger$ correspond to the classic complex amplitude and its complex conjugate. Let α be an arbitrary complex value, and consider an eigenstate $|\alpha\rangle$of a with an eigenvalue α, i.e., a state satisfying

$$a|\alpha\rangle = \alpha|\alpha\rangle \tag{2.24}$$

The expectation values for amplitudes a and $a^\dagger$ at time $t=0$ are $\langle a\rangle = \alpha$ and$\langle a^\dagger\rangle = \alpha^*$, and those at time t are

$$\langle a(t)\rangle = \langle\alpha|a(t)|\alpha\rangle = \alpha\exp(-i\omega t) \tag{2.25a}$$

$$\langle a^\dagger(t)\rangle = \langle\alpha|a^\dagger(t)|\alpha\rangle = \alpha^*\exp(+i\omega t) \tag{2.25b}$$

The expectation value $\langle E\rangle$ of the electric field is given by substituting the above equations for a, $a^\dagger$ in Eq. (2.18b) and is sinusoidal. This is consistent with the well-known observations of coherent electromagnetic waves such as single-frequency radio waves and laser lights. The state $|\alpha\rangle$ is suitable for representing such electromagnetic waves and is called the coherent state.

The fluctuations in the canonical variables q, p for a coherent state $|\alpha\rangle$ are $\Delta q = \langle\Delta q^2\rangle^{1/2} = (\hbar/2\omega)^{1/2}$ and $\Delta p = \langle\Delta p^2\rangle^{1/2} = (\hbar\omega/2)^{1/2}$, respectively. They satisfy the Heisenberg uncertainty principle with the equality, and the coherent state $|\alpha\rangle$ is one of the minimum-uncertainty states. However, it should be noted that the amplitude operator a is not Hermitian, and the amplitude a with the eigenvalue α that is a complex value is not an observable physical quantity. In fact, the observable quantities are the real and imaginary parts (or combination of them) of the amplitude α. They are associated with fluctuations of amplitude 1/2, and corresponding fluctuations are inevitable in the observation. The noise caused by the fluctuations is called quantum noise.

Next, let us consider an expansion of the coherent state $|\alpha\rangle$ by the energy eigenstate systems

$$|\alpha\rangle = \sum_n c_n |n\rangle \tag{2.26}$$

The coefficients c_n can be calculated by applying $\langle n|$ to the above equation, using Eqs (2.19a) and (2.24) and normalizing so as to have $\langle\alpha|\alpha\rangle = 1$:

$$\begin{aligned} c_n &= \langle n|\alpha\rangle \\ &= \langle\alpha|n\rangle^* \\ &= \{\langle\alpha|(n!)^{-1/2} a^{\dagger n}|0\rangle\}^* \\ &= (n!)^{-1/2}\alpha^n \langle 0|\alpha\rangle \\ &= (n!)^{-1/2}\alpha^n \exp\left(-\frac{|\alpha|^2}{2}\right) \end{aligned} \tag{2.27}$$

Therefore the probability of taking each eigenstate $|n\rangle$ is given by

$$|c_n|^2 = \frac{|\alpha|^{2n}}{n!} \exp(-|\alpha|^2) \tag{2.28}$$

which is the Poissonian distribution with n as a probability variable. The coherent state is one of the Poissonian states with the Poissonian distribution given by Eq. (2.28) and is characterized by the regularity in the phases of expansion coefficients as described by Eq. (2.27).

2.2 INTERACTIONS OF ELECTRONS AND PHOTONS

2.2.1 Hamiltonian for the Photon–Electron System and the Equation of Motion

The Hamiltonian for the optical energy is obtained by taking the summation of the Hamiltonians H_m for each mode given by Eq. (2.16):

$$H = \sum_m H_m = \sum_m \hbar\omega_m\left(a_m^\dagger a_m + \tfrac{1}{2}\right) \tag{2.29}$$

The Hamiltonian for the energy of an electron in the optical electromagnetic field represented by vector potential $\boldsymbol{A}$, on the other hand, is given by

$$\begin{aligned} H &= \frac{(\boldsymbol{p} - e\boldsymbol{A})^2}{2m} + V, \qquad \boldsymbol{p} = -\mathrm{i}\hbar\boldsymbol{\nabla} \\ &= \frac{\boldsymbol{p}^2}{2m} - \frac{e}{m}\boldsymbol{A}\cdot\boldsymbol{p} + \frac{e^2}{2m}\boldsymbol{A}^2 + V \end{aligned} \tag{2.30}$$

where V describes static potential energy that may bound the electron, $\boldsymbol{p}^2/2m$ the kinetic energy of the electron, and $-(e/m)\boldsymbol{A}\cdot\boldsymbol{p}$ the energy of interaction between the electron and the optical field (photon). Usually, the term $(e^2/2m)\boldsymbol{A}^2$ is extremely small and can be neglected.

By summing Eqs (2.29) and (2.30), the Hamiltonian H for the total system of the optical wave and the electron under possible interaction is given by

$$H = H_0 + H_\mathrm{i} = H_\mathrm{e} + H_\mathrm{p} + H_\mathrm{i} \tag{2.31a}$$

$$H_\mathrm{e} = \frac{\boldsymbol{p}^2}{2m} + V \tag{2.31b}$$

$$H_\mathrm{p} = \sum_m \hbar\omega_m\left(a_m^\dagger a_m + \tfrac{1}{2}\right) \tag{2.31c}$$

$$H_\mathrm{i} = -\frac{e}{m}\boldsymbol{A}\cdot\boldsymbol{p} \tag{2.31d}$$

where H_e, H_p, and H_i are the Hamiltonians of the electron, of the field energy, and of the interaction energy, respectively, and H_0 describes a Hamiltonian for total energy under an assumption that there is no interaction. The vector potential operator $\boldsymbol{A}$ in the interaction Hamiltonian H_i is obtained by using the amplitude operators a_m and $a_m^\dagger$ for photons of each mode:

$$\boldsymbol{A} = \sum_m \tfrac{1}{2}[a_m\boldsymbol{A}_m(\boldsymbol{r}) + a_m^\dagger\boldsymbol{A}_m^*(\boldsymbol{r})] \tag{2.32}$$

Let us represent the state of a system where an electron is under interaction with optical field by $|\Psi(t)\rangle$: then the equation of motion for determining the temporal change in $|\Psi(t)\rangle$ is a Schrödinger equation with the Hamiltonian $H=H_0+H_\mathrm{i}$ of Eq. (2.31):

$$\mathrm{i}\hbar\frac{\partial}{\partial t}|\Psi(t) = (H_0 + H_\mathrm{i})|\Psi(t)\rangle \tag{2.33}$$

Then we employ an operator defined by using H_0 as

$$U_0(t) = \exp\left(-\frac{\mathrm{i}H_0 t}{\hbar}\right) \tag{2.34}$$

to convert the representation $|\Psi(t)\rangle$ of the state in the Schrödinger picture into the representation $|\Psi(t)\rangle$ in the interaction picture:

$$|\Psi(t)\rangle = U_0(t)|\Psi_{\mathrm{I}}(t)\rangle \tag{2.35}$$

We then rewrite $|\Psi_{\mathrm{I}}(t)\rangle$ as $|\Psi(t)\rangle$ to obtain an equation of motion in the interaction picture:

$$\mathrm{i}\hbar\frac{\partial}{\partial t}|\Psi(t)\rangle = H_{\mathrm{I}}(t)|\Psi(t)\rangle \tag{2.36a}$$

$$H_{\mathrm{I}}(t) = U_0^{\dagger} H_{\mathrm{i}} U_0 \tag{2.36b}$$

The solutions of the equation of motion (Eq. (2.36)) can be obtained using the expansion by energy eigenstates. The eigenstates of the Hamiltonian H_0, with interaction omitted can be written as

$$|\Psi_J\rangle = |\psi_j; n_1, n_2, \ldots, n_m, \ldots\rangle \tag{2.37}$$

using the energy eigenstates $|\psi_j\rangle$ of the electron and the eigenstates $|n_m\rangle$ of optical field of each mode m. Here J is a label for the combination of electron states and states of field modes (j and $\{n_m\}$), and $|\Psi_J\rangle$ satisfies eigenequation

$$H_0|\Psi_J\rangle = E_J|\Psi_J\rangle \tag{2.38}$$

The eigenvalue can be written as the total sum of the eigenvalues E_j of the electron and the eigenvalues of the field modes:

$$E_J = E_j + \sum_m \hbar\omega_m\left(n_m + \tfrac{1}{2}\right) \tag{2.39}$$

Then we expand the state $|\Psi(t)\rangle$ in the interaction picture as

$$|\Psi(t)\rangle = \sum_J C_J(t)|\Psi_J\rangle \tag{2.40}$$

and substitute it into Eq. (2.36a) to obtain

$$\mathrm{i}\hbar\sum_J \frac{\partial}{\partial t} C_J(t)\exp(-\mathrm{i}\Omega_J t)\,|\Psi_J\rangle = \sum_{J'} C_{J'}(t) H_{\mathrm{i}} \exp(-\mathrm{i}\Omega_{J'} t)|\Psi_{J'}\rangle \tag{2.41}$$

where

$$\begin{aligned} E_J &= \hbar\Omega_J, \qquad E_{J'} = \hbar\Omega_{J'} \\ \Omega_{JJ'} &= \Omega_J - \Omega_{J'} = \frac{E_J - E_{J'}}{\hbar} \end{aligned} \tag{2.42}$$

and use has been made of Eqs (2.34), (2.36b), and (2.39). Application of $\langle\Psi_J|$ to both sides of Eq. (2.41), with the use of the orthonormal relation of the eigenstates, yields a group of equations that determine the temporal change in the expansion coefficients C_J:

$$i\hbar\frac{\partial}{\partial t}C_J(t) = \sum_{J'} C_{J'}(t)\langle\Psi_J|H_\mathrm{i}|\Psi_{J'}\rangle\exp(\mathrm{i}\Omega_{JJ'}t) \tag{2.43}$$

The above equations are called state transition equations. The analysis of the interaction can be made by solving the equations under given initial conditions.

2.2.2 Transition Probability and Fermi's Golden Rule

Consider the transition of a state for the case where the initial condition is given by an energy eigenstate:

$$|\Psi(0)\rangle = |\Psi_i\rangle = |\psi_i; n_1, n_2, \ldots, n_m, \ldots\rangle \tag{2.44}$$

Although in general there exist several energy levels for a bound electron, we can discuss the interaction by considering only two levels for cases where only the initial state $|\psi_j\rangle$ and another state $|\psi_f\rangle$ are involved in the interaction. However, if the system does not consist of a lone electron but includes many electrons as carriers in valence and conduction bands of semiconductors, the electron states are not at discrete levels but of continuous energy, and therefore an infinite number of states must be considered. An infinite number of states must be considered also for the optical field, since it has a spectrum of continuous variable ω_m with many modes. We therefore consider the state transition of a system described by an infinite number of state transition equations. In energy eigenstate expansion, the initial condition corresponding the initial state is given by

$$C_I(0) = 1, \qquad C_F(0) = 0 \qquad (F \neq I) \tag{2.45}$$

Assuming that the state $|\Psi(t)\rangle$ does not change largely from the initial state $|\Psi(0)\rangle$ in a short time, we substitute Eq. (2.45) into the right-hand side of Eq. (2.43) to obtain an approximate equation for $C_F(t)$:

$$i\hbar\frac{\partial}{\partial t}C_F(t) = \langle\Psi_F|H_\mathrm{i}|\Psi_I\rangle\exp(\mathrm{i}\Omega_{FI}t) \tag{2.46}$$

Integration of the above equation using $C_F(0) = 0$ directly yields an expression for $C_F(t)$:

$$C_F(t) = -\mathrm{i}\langle\Psi_F|H_\mathrm{i}|\Psi_I\rangle\exp\left(\frac{\mathrm{i}\Omega_{FI}t}{2}\right)\frac{\sin(\Omega_{FI}t/2)}{\Omega_{FI}t/2}\frac{t}{\hbar} \tag{2.47}$$

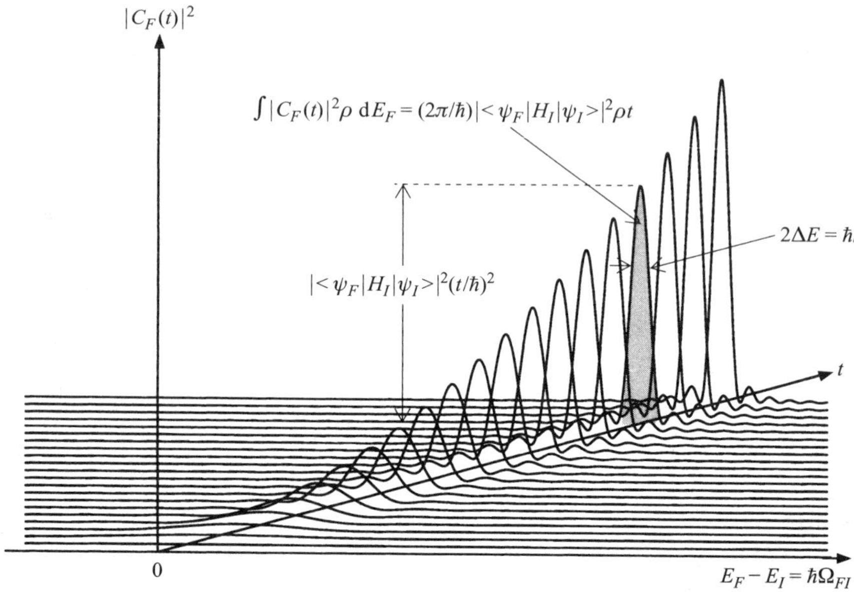

Figure 2.3 Temporal variation in the probability of the final state in the optical transition of an electron.

Accordingly, the probability that the state is $|\Psi_F\rangle$ after a short time is given by

$$|C_F(t)|^2 = |\langle \Psi_F|H_{\rm i}|\Psi_I\rangle|^2 \left(\sin \frac{(\Omega_{FI}t/2)}{\Omega_{FI}t/2} \right)^2 \left(\frac{t}{\hbar} \right)^2 \tag{2.48}$$

As shown in Fig. 2.3, the absolute value of the amplitude of the states of $\Omega_{FI}=0$, with respect to the initial state $|\Psi_I\rangle$ at $t=0$, increases with increasing t, while the amplitudes of the states of $\Omega_{FI}\neq 0$ oscillate without substantial increase. This means that the change in state is limited to such states $|\Psi_F\rangle$ that satisfy the energy conservation rule $E_I=E_F$ with respect to E_I. The change associated with the change in electron state $\psi_i \rightarrow \psi_f$ is called a transition, and $|\Psi_F\rangle$ is called the final state. Equation (2.48) indicates that some transition takes place to states $|\Psi_F\rangle$, within a region of $|E_F - E_I| < \hbar/t$, which do not exactly satisfy the energy conservation rule. This is because the energy operator $H = {\rm i}\hbar\,\partial/\partial t$ and time t are in the commutation relation of $[{\rm i}\hbar\,\partial/\partial t,\ t] = {\rm i}\hbar$, leading to an uncertainty relation $\Delta E\ \Delta t \geq \hbar/2$, and therefore the energy involves uncertainty of $\hbar/t$ or so for observation at a very short time t. In discussions on an ordinary time scale, however, the energy uncertainty is negligibly small since $\hbar$ is very small, and therefore one can consider that the energy conservation rule holds for the transition.

The probability that the system is found to be a final state $|\Psi_F\rangle$ after a time t is $|C_F(t)|^2$ given by Eq. (2.48). The final state $|\Psi_F\rangle$ of the total system must be treated as a continuous energy state. In order to calculate the probability that the electron is found to be a final state ψ_f of discrete energy, $|C_F(t)|^2$ must be multiplied by the mode density ρ of the optical field and integrated with respect to the energy E_F of the final states. This means that the probability that the electron is found in a final state ψ_f is given by $\int |C_F(t)|^2 \rho \, dE_F$ and is given by the area of the region that is surrounded by the curve and the abscissa in Fig. 2.3. The energy conservation factor $[\sin(\Omega_{FI}t/2)/(\Omega_{FI}t/2)]^2$, which is involved in the integration using Eq. (2.48), takes a small value except for the vicinity of $\Omega_{FI}=(E_F - E_I)/\hbar = 0$, and on a time scale where the energy uncertainty is not significant the factor asymptotically approaches $\pi\delta(\Omega_{FI}t/2) = (2\pi\hbar/t)\delta(E_F - E_I)$, where δ is the Dirac delta function. Accordingly, $|\langle\Psi_F|H_i|\Psi_I\rangle|^2$ and ρ can be replaced by the values at $E_F = E_I$ and put in front of the integral as follows:

$$\begin{aligned}\int |C_F(t)|^2 \rho \, dE_F &= \int |\langle\Psi_F|H_i|\Psi_I\rangle|^2 \frac{2\pi t}{\hbar} \delta(E_F - E_I) \rho \, dE_F \\ &= \frac{2\pi t}{\hbar} |\langle\Psi_F|H_i|\Psi_I\rangle|^2 \rho \end{aligned} \tag{2.49}$$

Therefore the transition probability per unit time is given by

$$w \equiv \frac{1}{t} \int |C_F(t)|^2 \rho \, dE_F$$

$$= \frac{2\pi}{\hbar} \int |\langle\Psi_F|H_i|\Psi_I\rangle|^2 \delta(E_F - E_I) \rho \, dE_F \tag{2.50a}$$

$$= \frac{2\pi}{\hbar} |\langle\Psi_F|H_i|\Psi_I\rangle|^2 \rho \tag{2.50b}$$

Equation (2.50), called Fermi's golden rule, is a very important formula that gives a simple expression for the transition probability using the matrix element of the interaction operator and the density-of-states function. For an electron in a continuous energy state, ρ in Eq. (2.50a) must be replaced by the density of electron states and the integration must be carried out with respect to the final electron energy, as will be discussed in detail in Chap. 3.

2.3 ABSORPTION AND EMISSION OF PHOTONS

2.3.1 Optical Transition and Matrix Element

From Eqs (2.31) and (2.32) the Hamiltonian representing the photon–electron interaction can be written as

$$H_{\rm i} = -\frac{e}{m}\sum_{m}\tfrac{1}{2}[a_m \boldsymbol{A}_m(\boldsymbol{r}) + a_m^{\dagger}\boldsymbol{A}_m^{*}(\boldsymbol{r})]\cdot\boldsymbol{p} \tag{2.51}$$

Application of $H_{\rm i}$ to the initial state $|\Psi_I\rangle$ given by Eq. (2.44) yields

$$\begin{aligned} H_{\rm i}|\Psi_I\rangle = -\frac{e}{m}\sum_{m}\tfrac{1}{2}[n_m^{1/2}\boldsymbol{A}_m(\boldsymbol{r})\cdot\boldsymbol{p}|\psi_i;\, n_1,\, n_2,\ldots,\, n_m-1,\ldots\rangle \\ +\,(n_m+1)^{1/2}\boldsymbol{A}_m^{*}(\boldsymbol{r})\cdot\boldsymbol{p}|\psi_i;\, n_1, n_2,\ldots, n_m+1,\ldots\rangle] \end{aligned} \tag{2.52}$$

From the above equation and the orthogonality of the energy eigenstates of the optical field, we see that the final states that give nonzero matrix element $\langle\Psi_F|H_{\rm i}|\Psi_I\rangle$ are limited to

$$|\Psi_F\rangle = |\psi_f;\, n_1, n_2,\ldots, n_m \pm 1,\ldots\rangle \tag{2.53}$$

The sign + represents an increase in the photon number or photon emission, and the sign − represents photon absorption. The energy conservation between the initial and final states, $E_F = E_I$, can also be written as

$$E_i = E_f \pm \hbar\omega_m \qquad (\omega_m > 0) \tag{2.54}$$

If the electron is initially in the upper energy level E_i ($E_i > E_f$), the electron transition $E_i \to E_f$ associated with emission of a photon of a mode of frequency $\omega_m = (E_i - E_f)/\hbar$ takes place. If the electron is initially in the lower energy level E_i ($E_i < E_f$), incidence of an optical wave of frequency $\omega_m = (E_f - E_i)/\hbar$ gives rise to absorption of a photon and excitation of the electron as $E_i \to E_f$.

From Eqs (2.52) and (2.53), the matrix element of the optical transition $\langle\Psi_F|H_{\rm i}|\Psi_I\rangle$ is given by

$$\langle\Psi_F|H_{\rm i}|\Psi_I\rangle = -\frac{e}{2m}n^{1/2}\langle\psi_f|\boldsymbol{A}(\boldsymbol{r})\cdot\boldsymbol{p}|\psi_i\rangle \tag{2.55a}$$

$$\langle\Psi_F|H_{\rm i}|\Psi_I\rangle = -\frac{e}{2m}(n+1)^{1/2}\langle\psi_f|\boldsymbol{A}^{*}(\boldsymbol{r})\cdot\boldsymbol{p}|\psi_i\rangle \tag{2.55b}$$

Equations (2.55a) and (2.55b) apply for photon absorption and photon emission, respectively. The mode label m was omitted. The spatial expanse of the electron wave function ψ is usually much smaller than the optical wavelength and hence $\boldsymbol{A}(\boldsymbol{r})$ is almost constant over the expanse of ψ.

Therefore we can put $\boldsymbol{A}(\boldsymbol{r})$ outside the angular brackets with the electron position substituted for $\boldsymbol{r}$ (dipole approximation). The matrix elements of the electron momentum $\boldsymbol{p}$ are correlated with the matrix elements of the canonically conjugate coordinate $\boldsymbol{r}$ by

$$\begin{aligned}\langle \psi_f | \boldsymbol{p} | \psi_i \rangle &= \frac{m}{\mathrm{i}\hbar}(E_i - E_f)\langle \psi_f | \boldsymbol{r} | \psi_i \rangle \\ &= \pm \frac{m}{\mathrm{i}} \omega \langle \psi_f | \boldsymbol{r} | \psi_i \rangle \end{aligned} \tag{2.56}$$

Therefore using Eq. (2.4b) the matrix elements for the photon absorption and photon emission are given by

$$\langle \Psi_F | H_{\mathrm{i}} | \Psi_I \rangle = -\tfrac{1}{2} n^{1/2} \boldsymbol{E}(\boldsymbol{r}) \langle \psi_f | e\boldsymbol{r} | \psi_i \rangle \tag{2.57a}$$

$$\langle \Psi_F | H_{\mathrm{i}} | \Psi_I \rangle = -\tfrac{1}{2} (n+1)^{1/2} \boldsymbol{E}^*(\boldsymbol{r}) \langle \psi_f | e\boldsymbol{r} | \psi_i \rangle \tag{2.57b}$$

respectively, where $\langle \psi_f | e\boldsymbol{r} | \psi_i \rangle$ on the right-hand side is called the electric dipole moment for the transition. The wave functions for an electron in an atom are even or odd functions of the displacement $\boldsymbol{r}$, since the Hamiltonian is an even function of $\boldsymbol{r}$. The dipole moment $\langle \psi_f | e\boldsymbol{r} | \psi_i \rangle$ with the odd operator $e\boldsymbol{r}$ takes nonzero values only for combinations of odd function and odd function for ψ_i and ψ_f (only for different parities). The transition takes place between states satisfying this parity selection rule.

The fundamental optical transitions discussed above are schematically illustrated in Fig. 2.4. In the following, the transition probabilities are considered for photon absorption and emission.

2.3.2 Photon Absorption

The mode function for an optical field in a medium of refractive index n_{r} given by Eq. (2.7) can be rewritten, after normalization to satisfy Eq. (2.6) for a space of volume V, as

$$\boldsymbol{E}(\boldsymbol{r}) = \left(\frac{2\hbar\omega}{\varepsilon_0 n_{\mathrm{r}} n_{\mathrm{g}} V} \right)^{1/2} \boldsymbol{e} \exp(\mathrm{i}\boldsymbol{k} \cdot \boldsymbol{r}) \tag{2.58}$$

where $\boldsymbol{e}$ is a unit vector indicating the polarization direction, and $\boldsymbol{e}$ is parallel to $\boldsymbol{E}$ and $\boldsymbol{A}$. As incident light, consider a set of modes of single polarization with the wave vectors $\boldsymbol{k}$ whose directions are within a narrow stereo angle region $\mathrm{d}\Omega$. Since from Eq. (2.9) the mode density is

$$\begin{aligned}\rho &\equiv V \frac{\mathrm{d}N}{\mathrm{d}E} \\ &= \frac{V}{\hbar} \left(\frac{1}{2c\pi} \right)^3 (n_{\mathrm{r}}^2 n_{\mathrm{g}}) \omega^2 \, \mathrm{d}\Omega \end{aligned} \tag{2.59}$$

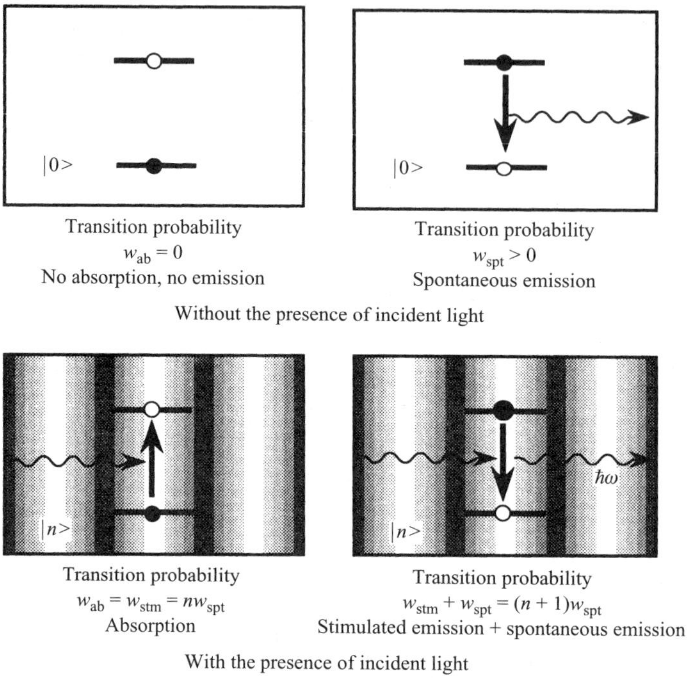

Figure 2.4 Transition probabilities for photon absorption and emission.

from Eqs (2.50b), (2.57a), (2.58), and (2.59) the transition probability for photon absorption is given by

$$\begin{aligned} w_{ab} &= \frac{\pi}{\varepsilon_0 n_r n_g V} n\omega |\boldsymbol{e} \cdot \langle \psi_f | e\boldsymbol{r} | \psi_i \rangle|^2 \rho \\ &= \frac{n_r}{8\pi^2 c^2 \hbar} \left(\frac{\mu_0}{\varepsilon_0}\right)^{1/2} n\omega^3 |\boldsymbol{e} \cdot \langle \psi_f | e\boldsymbol{r} | \psi_i \rangle|^2 \, d\Omega \end{aligned} \tag{2.60}$$

Let dI be the intensity of the components belonging to a mode set VdN of incident light; then we have $dI = v_g(n\hbar\omega/V)V\,dN$. Using the resulting relation $\rho = V(dN/dE) = (V/v_g n\hbar\omega)(dI/dE)$, we see that w_{ab} is given by

$$w_{ab} = \frac{\pi}{\hbar\varepsilon_0 c n_r} |\boldsymbol{e} \cdot \langle \psi_f | e\boldsymbol{r} | \psi_i \rangle|^2 \frac{dI}{dE} \tag{2.61}$$

Also w_{ab} is proportional to the energy density dI/dE of the incident light intensity and is proportional to the square of the component along the incident wave polarization of the dipole moment.

2.3.3 Spontaneous Emission and Stimulated Emission of Photons

From Eqs (2.50b), (2.57b), (2.58) and (2.59), the probability of photon emission is given by

$$w_{em} = w_{spt} + w_{stm} \tag{2.62a}$$

$$w_{spt} = \frac{\pi}{\varepsilon_0 n_r n_g V} \omega |\, \boldsymbol{e} \cdot \langle \psi_f |\, e\boldsymbol{r} \,| \psi_i \rangle |^2 \rho \tag{2.62b}$$

$$w_{stm} = \frac{\pi}{\varepsilon_0 n_r n_g V} n\omega |\, \boldsymbol{e} \cdot \langle \psi_f |\, e\boldsymbol{r} \,| \psi_i \rangle |^2 \rho \tag{2.62c}$$

The above equations indicate that, even if the photon number of the initial state is $n = 0$, i.e., even if there is no incident light, transition and photon emission take place with a probability w_{spt}. This is a quantum phenomena resulting from the commutation relation of the amplitude operators and is called spontaneous emission. From Eqs (2.62b) and (2.59), the probability for spontaneous emission of photons of modes with polarization direction $\boldsymbol{e}$ and wave vector direction within a stereo angle $d\Omega$ is given by

$$w_{spt} = \frac{n_r}{8\pi^2 c^2 \hbar} \left(\frac{\mu_0}{\varepsilon_0} \right)^{1/2} \omega^3 |\, \boldsymbol{e} \cdot \langle \psi_f |\, e\boldsymbol{r} \,| \psi_i \rangle |^2 \, d\Omega \tag{2.63}$$

Since the spontaneous emission takes places for all spatial modes and polarizations, it should be integrated over whole $d\Omega$. If the dipole moment is not oriented to a particular direction, using the fact that the average of $|\boldsymbol{e} \cdot \langle \psi_f | e\boldsymbol{r} | \psi_i \rangle|^2$ over many electrons is $|\langle \psi_f | e\boldsymbol{r} | \psi_i \rangle|^2/3$ and taking the two independent polarizations into account, the spontaneous emission probability w_{spt} can be calculated from

$$w_{spt} = \frac{n_r}{3\pi c^2 \hbar} \left(\frac{\mu_0}{\varepsilon_0} \right)^{1/2} \omega^3 |\langle \psi_f |\, e\boldsymbol{r} \,| \psi_i \rangle |^2 \tag{2.64}$$

An important implication of this result is that w_{spt} is proportional to the third power of the optical frequency.

If the photon number of the initial state is $n \geq 1$, i.e., if light is incident, in addition to the spontaneous emission, transition and photon emission take place with a probability w_{stm} proportional to the initial photon number n, as indicated by Eq. (2.62c). This is called stimulated emission. From Eqs (2.62c) and (2.59), the stimulated emission probability is given by

$$w_{stm} = \frac{n_r}{8\pi^2 c^2 \hbar} \left(\frac{\mu_0}{\varepsilon_0} \right)^{1/2} n\omega^3 |\, \boldsymbol{e} \cdot \langle \psi_f |\, e\boldsymbol{r} \,| \psi_i \rangle |^2 d\Omega \tag{2.65}$$

Using $\rho = (V/v_g n\hbar\omega)(\mathrm{d}I/\mathrm{d}E)$, the probability can also be rewritten as

$$w_{\mathrm{stm}} = \frac{\pi}{\hbar\varepsilon_0 c n_{\mathrm{r}}} |\boldsymbol{e} \cdot \langle\psi_f|e\boldsymbol{r}|\psi_i\rangle|^2 \, \mathrm{d}I/\mathrm{d}E \tag{2.66}$$

Equations (2.65) and (2.66) have exactly the same forms as Eqs (2.60) and (2.61), respectively. This means that the stimulated emission probability w_{stm} equals the absorption probability w_{ab}, and both probabilities are proportional to the energy density $\mathrm{d}I/\mathrm{d}E$ of the incident light intensity and to the square of the component along the incident wave polarization of the dipole moment. The spontaneous emission probability w_{spt}, which is also expressed in terms of the dipole moment, is closely related to w_{ab} and w_{stm}.

2.3.4 The Einstein Relation

Before the development of the quantum theory described above, Einstein introduced the concept of absorption, spontaneous emission and stimulated emissions of photons and proposed a mathematical relation between the transition probabilities, in order to explain the spectrum of the black-body radiation in terms of the transitions between levels of particle system. He considered an energy level 1 and another level 2 of energy higher than level 1 by $\hbar\omega$ and assumed that the probability W_{12} of the $1 \to 2$ transition and the probability W_{21} of the $2 \to 1$ transition and W_{21} in an optical field with energy density $u(E)$ per unit volume can be written as

$$W_{12} = w_{\mathrm{ab}} = B_{12}u(E) \tag{2.67a}$$

$$W_{21} = w_{\mathrm{em}} = B_{21}u(E) + A_{21} \tag{2.67b}$$

and, through a consideration based on the quantum assumption by Planck and statistical mechanics, he deduced relations between the coefficients A_{21}, B_{12}, and B_{21}:

$$A_{21} = \frac{\omega^3}{\pi^2 c^3} B_{21} \tag{2.68a}$$

$$B_{12} = B_{21} \tag{2.68b}$$

B_{12}, B_{21}, and A_{21} are the absorption, stimulated emission, and spontaneous emission coefficients respectively, and Eq. (2.68) is called the Einstein relation.

We can confirm the validity of Eqs (2.67) and (2.68), by applying the quantum theory discussed in the previous sections to the black-body radiation. The equations obtained through the quantum theory are consistent with the Einstein relation. Accordingly, the relations between the absorption, stimulated emission and spontaneous emission derived by the quantum theory are often identified with the Einstein relation. However, it should be noted that the original Einstein relation, i.e., Eq. (2.68a), does not apply for analysis of absorption and emission of coherent optical waves such as laser light.

2.4 POPULATION INVERSION AND LIGHT AMPLIFICATION

The mathematical expressions for the absorption, spontaneous emission, and stimulated emission deduced by the quantum theory in the previous section show that, in the presence of incident light, absorption and stimulated emission of photons take place with the same probability for the stimulated emission and the absorption, and, even without incident light, spontaneous photon emission takes place with a nonzero probability. Since these probabilities are for a single electron, for a system consisting of many electrons, they must be multiplied by the number of electrons in the states where absorption and emission are allowed.

In thermal equilibrium, the electron energy obeys the Fermi–Dirac distribution. An electron state of energy E is occupied by an electron with a probability

$$f = \frac{1}{\exp[(E - F)/k_B T] + 1} \tag{2.69}$$

where F is the Fermi energy, k_B the Boltzmann constant, and T the absolute temperature. According to the above equation, which indicates that the electron occupation probability for a level of higher energy is smaller than that for a level of lower energy, the number of photon emission transitions is smaller than the number of photon absorption transitions and therefore as a whole the light is absorbed. However, if a situation where the occupation of the higher level is greater than that of the lower level is produced, the larger number of stimulated photon emission transitions than that of photon absorption transitions gives rise to substantial stimulated emission. Such a situation, called population inversion, can be produced through excitation by providing the medium with an external energy. The situation of inverted population is also called the "negative temperature" state.

The fundamental principle of light amplification by stimulated emission of radiation (LASER) is based on coherent amplification of an optical wave utilizing the stimulated emission in a medium where population inversion is produced by excitation. The state of the medium with population inversion is called the laser-active state. In semiconductor injection lasers, population inversion is realized through injection of minority carriers with a high energy in the vicinity of a p–n junction by supplying a forward current. Although in this chapter the possibility of enhancement of optical power by stimulated photon emission was shown based on consideration of the states of an electron and the energy eigenstates of the optical field, this discussion does not ensure the possibility of enhancement of optical amplitude including the phase, i.e., the coherent amplification of an optical wave. The possibility of coherent amplification will be discussed on the basis of an analysis using the density matrix in Chap. 3. Spontaneous emission also takes place in parallel to the laser action. It affects significantly the laser performances; e.g., it gives rise to the laser oscillation threshold and the quantum noise in the output light.

REFERENCES

1. W. Heitler, *The Quantum Theory of Radiation*, Oxford University Press, Oxford (1954).
2. A. Yariv, *Quantum Electronics*, John Wiley, New York (1967).
3. L. I. Schiff, *Quantum Mechanics*, McGraw-Hill, New York (1968).
4. Y. Fujii, *Optical and Quantum Electronics* (in Japanese), Kyoritsu, Tokyo (1978).
5. M. Sargent III, M. O. Scully, and W. E. Lamb, Jr, *Laser Physics*, Addison-Wesley, Reading, MA (1974).
6. D. Marcuse, *Principles of Quantum Electronics*, Academic Press, New York (1980).
7. K. Shimoda, *Introduction to Laser Physics* (in Japanese), Iwanami, Tokyo (1983).
8. P. Meystre and M. Sargent III, *Elements of Quantum Optics*, Springer, Berlin (1991).
9. T. Suhara, *Quantum Electronics* (in Japanese), Ohmsha, Tokyo (1993).

3

Stimulated Emission and Optical Gain in Semiconductors

This chapter presents the basic theory and characteristics of stimulated emission and optical amplification gain in semiconductors. The former is the most important principle that enables semiconductor lasers to be implemented, and the latter is the most important parameter for analysis of the laser performances. First, stimulated emission in semiconductors is explained, and then quantum theory analysis and statistic analysis using the density matrix of the optical amplification gain are given. Stimulated emission and optical gain in semiconductor quantum well structures will be presented in the next chapter.

3.1 BAND STRUCTURE OF SEMICONDUCTORS AND STIMULATED EMISSION

3.1.1 Band Structure of Direct-Transition Bandgap Semiconductors

Semiconductor lasers utilize the interband optical transitions of carriers in a semiconductor having a direct-transition bandgap. As is well known in the electron theory of solids [1], the wave function of an electron of wave vector $\boldsymbol{k}$ (momentum $\hbar_k$) in an ideal semiconductor crystal can be written as a Bloch function

$$|\psi(\boldsymbol{r})>=|\exp(\mathrm{i}\boldsymbol{k}\cdot\boldsymbol{r})u_k(\boldsymbol{r})> \qquad (3.1)$$

where $u_k(\boldsymbol{r})$ is a periodic function with the periodicity of the crystal lattice, and $u_k(\boldsymbol{r})$ is normalized in a unit volume. The electron states form a band structure, consisting of continuous energy levels in the band. Figure 3.1 shows the band structure of GaAs [2], a representative semiconductor laser

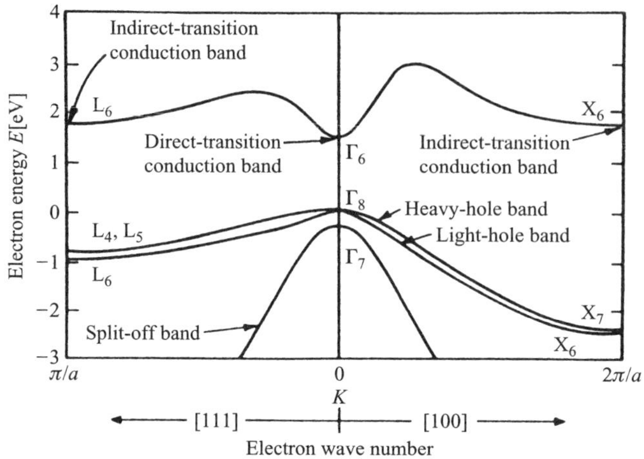

Figure 3.1 Band structure of the III–V semiconductor GaAs having a bandgap of direct transition type [2].

material. The figure shows the electron energy E dependent on $\boldsymbol{k}$ within the first Brillouin zone; the dependences on $\boldsymbol{k}$ along the [111] and [100] directions with good symmetry, in the $\boldsymbol{k}$ space, are shown in the left and right halves, respectively. Crystals of III–V compound semiconductors such as GaAs are of the zinc blende structure, and their valence and conduction bands originate from the sp^3 hybrid orbital that forms the covalent bond. The conduction band is a single band of s-like orbital, while the valence band is of p-like orbital and consists of a heavy-hole band, a light-hole band and a split-off band [3]. The upper edge of the valence band is at the Γ point, the center of $\boldsymbol{k}$ space, where the heavy-hole and light-hole bands are degenerate, and the split-off band is separated from them by the spin–orbit interaction energy Δ. The lower edge of the conduction band is at the Γ point. Thus the wave vectors for the conduction- and valence-band edges that determine the bandgap coincide with each other. This type of band edge is called a direct-transition band edge. In this situation, transitions that cause emission or absorption of photons with energy close to the bandgap energy take place with high probability, since momentum conservation is satisfied. Lasers can be implemented by using such interband transitions. Most of the III–V (GaAs, $Al_xGa_{1-x}A_s$, InP, $_cGa_xAs_yP_{1-y}$, etc.) and II–VI compound semiconductors have a band structure similar to that of Fig. 3.1 and the direct-transition bandgap. On the other hand, for group IV semiconductors such as Si and Ge, and a III–V semiconductor AlAs, the wave vectors for the conduction-band edge and the valence-band edge do not coincide with each other (indirect-transition bandgap). In these semiconductors, the transitions for emission and absorption of photons with bandgap energy require the assistance of interaction with phonons. The probability of this indirect transition is low, and therefore semiconductors of this type are not suitable for a semiconductor laser.

Using the effective-mass approximation, the energy of electrons near the conduction- and valence-band edges can be written as

$$E_c(\boldsymbol{k}) = E_c + \frac{\hbar^2}{2m_n}\boldsymbol{k}^2 \tag{3.2a}$$

$$E_v(\boldsymbol{k}) = E_v - \frac{\hbar^2}{2m_p}\boldsymbol{k}^2 \tag{3.2b}$$

where $\boldsymbol{k}$ is the electron wave vector. The band structure described by the above expressions is a parabolic band. Here, E_c and E_v are energies of the lower edge of the conduction band and the upper edge of the valence band, respectively, the difference $E_g = E_c - E_v$ is the bandgap energy, and m_n is the effective mass of electrons in the conduction band while m_p is the effective mass of holes in the valence band. The effective mass of the heavy hole is denoted by m_{ph}, and that of the light hole by m_{pl}. For GaAs, $m_n = 0.067m$, $m_{ph} = 0.45m$, and $m_{pl} = 0.082m$, where m is the mass of a free electron. The density of states (number of states per unit volume and per unit energy width) for electrons in the conduction and valence bands are calculated from Eq. (3.2) as

$$\rho_c(E) = \frac{1}{2\pi^2\hbar^3}(2m_n)^{3/2}(E - E_c)^{1/2} \qquad (E > E_c) \tag{3.3a}$$

$$\rho_v(E) = \frac{1}{2\pi^2\hbar^3}(2m_p)^{3/2}(E_v - E)^{1/2} \qquad (E_v > E) \tag{3.3b}$$

(see Appendix 2). As described by these expressions, the densities of state are given by parabolic functions with their tops at the band edge energies.

3.1.2 Condition for Stimulated Emission

Here we discuss the necessary condition for stimulated emission [4], by using the simplest model of an elementary process of the optical transition of electrons in a semiconductor. Consider an energy level E_1 in the valence band and an energy level $E_2 (> E_1)$ in the conduction band, as shown in Fig. 3.2, and consider interband transitions associated with absorption or emission of a photon of frequency ω determined by the energy conservation rule:

$$E_2 - E_1 = \hbar\omega \tag{3.4}$$

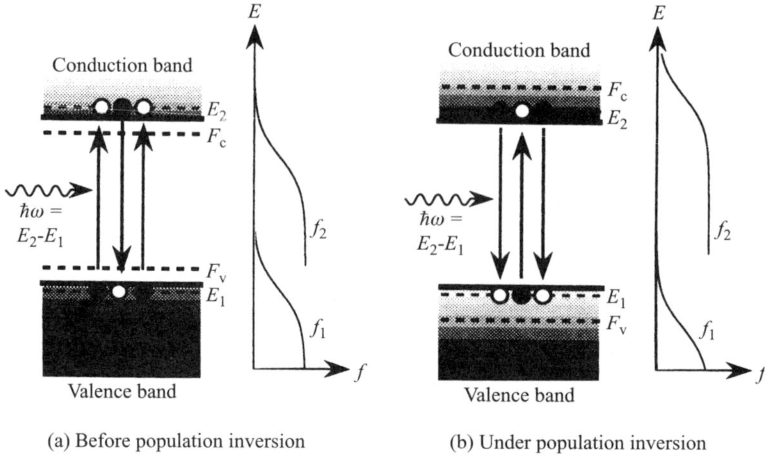

Figure 3.2 Emission and absorption of photons by electron transition in a carrier-injected semiconductor.

If the system is in thermal equilibrium, the probability of occupation of a level at energy E by an electron is generally given by the Fermi–Dirac function

$$f = \frac{1}{\exp[(E - F)/k_B T] + 1} \tag{3.5}$$

where F is the Fermi level, k_B the Boltzmann constant, and T the absolute temperature of the system. As pointed out in the previous chapter, in thermal equilibrium, substantial stimulated emission cannot be obtained. However, the population can be inverted by injecting minority carriers of energy higher than that of the majority carriers by means of current flow through a semiconductor p–n junction. In the semiconductor excited by minority-carrier injection, thermal equilibrium is violated between the conduction band and the valence band. However, one can consider that the carriers are approximately in equilibrium within each band (the quasi-thermal equilibrium approximation). Then the occupation probabilities for the energy level E_1 in the valence band and for the energy level E_2 in the conduction band can be written as

$$f_1 = \frac{1}{\exp[(E_1 - F_v)/k_B T] + 1} \tag{3.6a}$$

and

$$f_2 = \frac{1}{\exp[(E_2 - F_c)/k_B T] + 1} \tag{3.6b}$$

respectively, where F_v and F_c are quasi-Fermi levels for the valence band and the conduction band, respectively. The state of the system described by the above equation is called the quasithermal equilibrium state. When a forward voltage V is applied across a semiconductor p–n junction, an energy difference as large as eV is given between the Fermi levels for the p and n regions. Since carriers of energies represented by the Fermi levels for the p and n regions are injected into the active region, the difference $F_c - F_v$, in the quasi-Fermi levels equals eV.

In order to accomplish light amplification by stimulated emission, i.e., laser action, a substantial amount of stimulated emission more than absorption is required. Noting that the stimulated emission probability for a case where an electron is at the energy level E_2 and the energy level E_1 is vacant is given by $B_{21}u(E)$ using the Einstein coefficient, the probability of stimulated emission in the semiconductor is given by $B_{21}u(E)$ multiplied by the probability $f_2(1 - f_1)$ that the level E_2 is occupied and the level E_1 is unoccupied. Similarly, the absorption probability in the semiconductor is given by $B_{12}u(E)$ multiplied by $f_1(1 - f_2)$. Therefore, the condition for substantial spontaneous emission is

$$B_{21} f_2(1 - f_1)u(E) > B_{12} f_1(1 - f_2)u(E) \tag{3.7}$$

Using $u(E) > 0$ and the Einstein relation $B_{21} = B_{12}$, Eq. (3.7) can be rewritten as $f_2(1 - f_1) > f_1(1 - f_2)$, or

$$f_2 > f_1 \tag{3.8}$$

which implies that the population inversion is required. Substitution of Eq. (3.6) into Eq. (3.8) yields

$$E_1 - F_v > E_2 - F_c \quad \text{or} \quad F_c - F_v > E_2 - E_1 \tag{3.9}$$

Therefore, for stimulated emission to exceed absorption, it is necessary that the difference $F_c - F_v$ in the quasi-Fermi levels, i.e., the energy corresponding the applied voltage eV, exceeds the energy $E_2 - E_1 (= \hbar\omega)$ of the photon to be emitted, as shown in Fig. 3.2(b).

3.1.3 Photon Absorption and Emission, and Absorption and Gain Factors

The electron states in the band structure of a semiconductor consist of a set of many states of continuous energy. Noting this, here we analyze absorption and emission of photons, to deduce fundamental formulas useful for calculation of the gain of light amplification by laser action. Consider a case where an optical wave of a mode with an angular frequency ω is incident on a semiconductor. Let n_r and n_g be the refractive index and the group index of refraction, respectively, of the semiconductor, and n be the photon number. The modes of optical wave are normalized for volume V. In this case, the density of electron state should be used for ρ in the Fermi golden rule (Eq. (2.50a)), and then ρdE_f stands for the number of the electron states. From this equation and Eqs (2.57) and (2.58), the absorption probability and the stimulated emission probability per one set of initial and final states of electron, and per unit time, can be written as

$$\begin{aligned} w_{abs} &= w_{stm} \\ &= \frac{\pi}{\varepsilon_0 n_r n_g V} n\omega |e\langle\psi_f|er|\psi_1\rangle|^2 \delta(E_1 \pm \hbar\omega - E_f) \end{aligned} \tag{3.10}$$

where + denotes absorption and − emission. The above equation is an expression for a transition, and it must be integrated over whole combinations of electron states to yield an expression for total optical transitions.

The power absorption factor α, which is often used to describe optical absorption phenomenologically, is defined by the relative intensity attenuation per unit length as $\alpha = (-dI/d\xi)/I$, where I is the light intensity and ξ is the coordinate along the optical propagation. Since $-dI$ is the difference between energy flows per unit time for two cross sections of unit area separated by $d\xi$, $-dI/d\xi$ is the energy absorbed in unit volume per unit time, and is given by the substantial number of absorption transitions (= number of absorption transitions − number of stimulated emissions) per unit volume and unit time, multiplied by the photon energy $\hbar\omega$. On the other hand, the light intensity I, being the optical energy (except for the zero-point energy) passing across a cross section of unit area per unit time, is given by using the photon number n and the group velocity v_g of the light as

$$I = \frac{n\hbar\omega}{V} v_g = \frac{n\hbar\omega}{V} \frac{c}{n_g} \tag{3.11}$$

Therefore, a is given by substantial absorption transitions per unit volume and unit time, multiplied by $\hbar\omega/I = (V/n)(n_g/c)$. The power gain factor g, often used to describe optical gain phenomenologically, is defined by $g = (dI/d\xi)/I$ and equals the absorption factor a with the sign inverted. This is because substantial absorption (= absorption − stimulated emission) is considered for the calculation of α, while substantial emission (= stimulated emission − absorption) is considered for the calculation of g. Thus the gain factor g is given by $g = -\alpha$; the mathematical expressions for calculating α and g are essentially same. Positive α describes absorption and negative α describes amplification (the gain coefficient is $g = -\alpha = |\alpha|$).

3.2 DIRECT-TRANSITION MODEL

Consider optical transitions of electrons at the vicinity of band edges in a semiconductor having a direct-transition bandgap, as shown in Fig. 3.1, under the assumption that the semiconductor does not contain impurities and is a perfect crystal (ideal semiconductor of direct-transition type). The electric dipole moment $\langle\psi_f|e\boldsymbol{r}|\psi_\mathrm{i}\rangle$ for a combination of electrons in the valence and conduction bands can be calculated by integration using ψ_f and ψ_i in the form of Eq. (3.1). The calculation gives a nonzero value only when $\boldsymbol{k}_\mathrm{f} = \boldsymbol{k}_\mathrm{i}$ holds for the wave vectors $\boldsymbol{k}_\mathrm{f}$ and $\boldsymbol{k}_\mathrm{i}$ and for ψ_f and ψ_i. This means that only such transitions that conserve the wave (momentum) vector are allowed. In fact, the exact momentum conservation for the electron–photon system, $\boldsymbol{k}_\mathrm{f} = \boldsymbol{k}_\mathrm{i} \pm \boldsymbol{k}$, holds in the transition, but the optical wave vector $\boldsymbol{k}$ is so small in comparison with $\boldsymbol{k}_\mathrm{f}$ and $\boldsymbol{k}_\mathrm{i}$ that it can be omitted (corresponding to the dipole approximation in Section 2.3.1).

Using the periodic boundary condition for the states of electrons in a semiconductor of volume V, an electron state in each band occupies a volume of $(2\pi)^3/V$ in $\boldsymbol{k}$ space. Considering the spin, the number of states in the volume element $\mathrm{d}\boldsymbol{k} = \mathrm{d}k_x\, \mathrm{d}k_y\, \mathrm{d}k_z$ in $\boldsymbol{k}$ space, per unit volume, is therefore $(1/4\pi^3)\,\mathrm{d}\boldsymbol{k}$. The probabilities of electron occupation for energy levels E_1 and E_2 in the valence and conduction bands, respectively, are given by f_1 and f_2 in Eq. (3.6). Therefore, there are $(1/4\pi^3) f_2\, \mathrm{d}\boldsymbol{k}$ initial states for photon emission, in unit volume and in $\mathrm{d}\boldsymbol{k}$. Considering the spin, there are two final states having the same $\boldsymbol{k}$ as each initial state, and they are occupied with a probability f_1. Therefore, the number of effective combinations of initial and final states that can contribute to the emission transition, in unit volume and in $\mathrm{d}\boldsymbol{k}$, is $(1/2\pi^3) f_2(1 - f_1)\, \mathrm{d}\boldsymbol{k}$. Similarly, the number of effective combinations of initial and final states that can contribute to the absorption transition, in unit volume and in $\mathrm{d}\boldsymbol{k}$, is $(1/2\pi^3) f_1(1 - f_2)\, \mathrm{d}\boldsymbol{k}$. Since the probabilities of stimulated emission and absorption are same, as shown by Eq. (3.10), the substantial number of stimulated emission transitions per unit time, in unit volume and in $\mathrm{d}\boldsymbol{k}$, is $(1/2\pi^3)[f_2(1 - f_1) - f_1(1 - f_2)]\,\mathrm{d}\boldsymbol{k} = (1/2\pi^3)(f_2 - f_1)\,\mathrm{d}\boldsymbol{k}$ multiplied by w_stm in Eq. (3.10). Therefore, after integration with respect to $\mathrm{d}\boldsymbol{k}$, we obtain an expression for the gain factor:

$$\begin{aligned} g &= \frac{V}{n}\frac{n_\mathrm{g}}{c}\int \frac{1}{2\pi^3}(f_2 - f_1) w_\mathrm{stm}\, \mathrm{d}\boldsymbol{k} \\ &= \frac{\pi\omega}{c\varepsilon_0 n_\mathrm{r}}\frac{1}{2\pi^3}\int |e\langle\psi_2|e\boldsymbol{r}|\psi_1\rangle|^2 (f_2 - f_1)\delta(E_1 + \hbar\omega - E_2)\, \mathrm{d}\boldsymbol{k} \end{aligned} \tag{3.12}$$

Since a delta function $\delta(E_1 + \hbar\omega - E_2)$ is included in the integrand of the above equation, the $(f_2 - f_1)$ factor can be replaced by the values for E_1 and

E_2 that satisfy $E_1 + \hbar\omega = E_2$ with the same $\boldsymbol{k}$ and can be put in front of the integral. Here, E_1 and E_2 are given by

$$E_1 = E_v - \frac{m_r}{m_p}(\hbar\omega - E_g) \tag{3.13a}$$

$$E_2 = E_c + \frac{m_r}{m_n}(\hbar\omega - E_g) \tag{3.13b}$$

where

$$\frac{1}{m_r} = \frac{1}{m_n} + \frac{1}{m_p}$$

The factor $|\boldsymbol{e}\langle\psi_2|e\boldsymbol{r}|\psi_1\rangle|^2$ can be replaced by the average value for electrons near the band edges and can be put in front of the integral. From Eq. (2.56) we have

$$|\boldsymbol{e}\langle\psi_2|e\boldsymbol{r}|\psi_1\rangle|^2 = \left(\frac{e}{m\omega}\right)^2 |\boldsymbol{e}\langle\psi_2|\boldsymbol{p}|\psi_1\rangle|^2 \tag{3.14}$$

Here we denote the mean square of the momentum matrix element as

$$|M|^2 = \ll |\boldsymbol{e}\langle\psi_2|\boldsymbol{p}|\psi_1\rangle|^2 \gg \tag{3.15}$$

Then from Eq. (3.12) we obtain an expression for the gain factor g and the absorption factor α:

$$\begin{aligned} g(\hbar\omega) &= -\alpha(\hbar\omega) \\ &= \frac{\pi e^2}{n_r c\varepsilon_0 m^2\omega}|M|^2(f_2 - f_1)\rho_r(\hbar\omega) \end{aligned} \tag{3.16}$$

where $\rho_r(\hbar\omega)$ is the reduced density of states defined by

$$\rho_r(\hbar\omega) = \frac{1}{2\pi^3}\int \delta(E_1 + \hbar\omega - E_2)\,\mathrm{d}\boldsymbol{k} \tag{3.17}$$

When the $\boldsymbol{k}$ dependences of the electron energies E_2 and E_1 are given by Eq. (3.2), the reduced density of states can readily be calculated to yield

$$\rho_r(\hbar\omega) = \frac{(2m_r)^{3/2}}{\pi^2\hbar^3}(\hbar\omega - E_g)^{1/2} \tag{3.18a}$$

$$\frac{1}{m_r} = \frac{1}{m_n} + \frac{1}{m_p}, \qquad E_g = E_c - E_v \tag{3.18b}$$

where E_g is the bandgap energy and m_r is called the reduced mass. In Eq. (3.16), f_1 and f_2 are given by Eqs (3.6a) and (3.6b) with Eqs (3.13a) and (3.13b), respectively, substituted.

The gain factor g given by Eq. (3.16) takes a negative value when Eq. (3.8) holds, indicating that Eq. (3.8) is appropriate as the condition for substantial stimulated emission. As can be seen from Eq. (3.16), $g(\hbar\omega)$ is proportional to the product of the reduced density ρ_r of state and the occupation probability difference $f_2 - f_1$ representing the degree of population inversion, and therefore the gain spectrum, i.e., the ω dependence of g, is dominated by the ω dependences of ρ_r and $f_2 - f_1$. Evolution of the gain spectrum during an increase in carrier injection is illustrated in Fig. 3.3. From Eqs (3.6), (3.16), and (3.18), and Fig. 3.3, we see the following tendencies.

1. Injection of minority carriers produces amplification gain for optical wavelengths near the bandgap energy wavelength.
2. While at room temperature only a part of the carriers contributes to the gain, at low temperatures a larger part contributes to give a higher gain.
3. With increase in the carrier density, the optical frequency of maximum gain shifts to that for higher energy (band-filling effect).

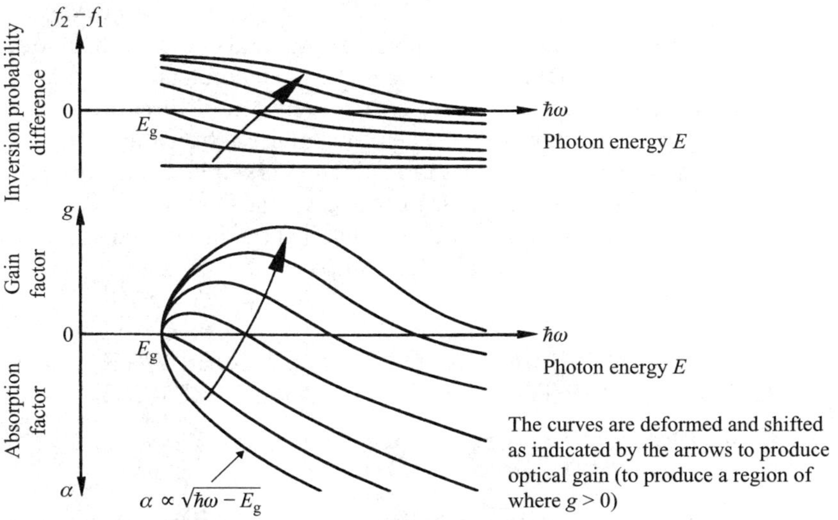

Figure 3.3 Variation in inversion occupation probability difference $f_2 - f_1$ and gain spectrum $g(\hbar\omega)$ with increasing carrier injection.

We next consider spontaneous emission. In a similar manner to Eq. (3.10), from Eqs (2.50a), (2.57), and (2.58), the probability of spontaneous emission per unit time for one set of initial and final states of electrons can be written as

$$W_{\text{spt}} = \frac{\pi}{\varepsilon_0 n_r n_g V} \omega |\boldsymbol{e}\langle\psi_f|e\boldsymbol{r}|\psi_i\rangle|^2 \delta(E_i - \hbar\omega - E_f) \tag{3.19}$$

Since the spontaneous emission radiates over all directions (stereo angle $\Omega = 4\pi$), from Eq. (2.11) the number of optical modes for spontaneous emission within the frequency range from ω to $\omega + d\omega$ in the volume V is given by $V\rho(\omega)\,d\omega = (V/\pi^2)(n_r^2 n_g/c^3)\omega^2\,d\omega$. Equation (3.19) is multiplied by this $V\rho(\omega)\,d\omega$ and also by the number of sets of electron states that can contribute to spontaneous emission in unit volume and in $d\boldsymbol{k}$, i.e., $(1/2\pi^3)f_2(1-f_1)\,d\boldsymbol{k}$, and is then integrated. Thus, in a similar way to the deduction of Eq. (3.16), the rate of spontaneous photon emission from a semiconductor of unit volume per unit time is calculated as

$$r_{\text{spt}}(\hbar\omega)\,d\omega = \frac{n_r e^2 \omega}{\pi m^2 c^3 \varepsilon_0}\,d\omega |M|^2 f_2(1-f_1)\rho_r(\hbar\omega) \tag{3.20}$$

The power of spontaneous emission within a frequency width $d\omega$ is given by the product of the spontaneous emission rate $r_{\text{spt}}\,d\omega$ and the photon energy $\hbar\omega$.

In the above discussion, a direct-transition model was used to deduce the mathematical expressions for optical absorption and emission in an ideal semiconductor of direct-transition type. The semiconductors used for implementation of semiconductor lasers, however, are of direct-transition type but are doped with impurities. Therefore indirect transitions that do not satisfy the wave vector conservation rule also take place. Accordingly, the above expressions deduced by using wave vector conservation do not exactly apply. They are not appropriate for detailed quantitative discussions of optical transitions near the band edges, in particular.

3.3 GAUSSIAN HALPERIN–LAX BAND-TAIL MODEL WITH THE STERN ENERGY-DEPENDENT MATRIX ELEMENT

3.3.1 Energy Integral Expressions for Gain and Spontaneous Emission

Optical absorption and emission including transitions without wave vector conservation can be analyzed by integration of transition probabilities with respect to energy states [5]. For an electron energy E_1 in the valence band and an electron energy E_2 in the conduction band, the densities of electron states for valence and conduction bands are given by $\rho_v(E_1)$ and $\rho_c(E_2)$ using Eq. (3.3), and the occupation probabilities for the E_1 and E_2 levels are given by f_1 and f_2, respectively, using Eq. (3.6). Considering stimulated emission transitions in unit volume, there are $\rho_c(E_2)f_2\,dE_2$ available initial states in dE_2, while there are $\rho_v(E_1)\,dE_1$ final states but they are occupied with a probability f_1. Therefore, the number of effective combinations of states that can contribute to stimulated emission in unit volume is $\rho_v(E_1)\rho_c(E_2)f_2(1-f_1)\,dE_1\,dE_2$. Similarly, the number of effective combinations of states that can contribute to absorption in unit volume is $\rho_v(E_1)\rho_c(E_2)f_1(1-f_2)\,dE_1\,dE_2$. Since the transition probabilities of stimulated emission and absorption are the same, the substantial number of stimulated emission transition in unit volume per unit time is $\rho_v(E_1)\rho_c(E_2)\times[f_2(1-f_1)-f_1(1-f_2)]\,dE_1\,dE_2=\rho_v(E_1)\rho_c(E_2)(f_2-f_1)\,dE_1\,dE_2$, multiplied by w_{stm} of Eq. (3.10). Then the gain g can be calculated through integration with respect to $dE_1\,dE_2$ as

$$\begin{aligned} g(\hbar\omega) &= \frac{V}{n}\frac{n_g}{c}\int \rho_v(E_1)\rho_c(E_2)w_{stm}(f_2-f_1)\,dE_1\,dE_2 \\ &= \frac{\pi\omega}{c\varepsilon_0 n_r}\int \rho_v(E_1)\rho_c(E_2)|\boldsymbol{e}\langle\psi_2|e\boldsymbol{r}|\psi_1\rangle|^2 \\ &\quad\times(f_2-f_1)\delta(E_1+\hbar\omega-E_2)\,dE_1\,dE_2 \\ &= \frac{\pi e^2}{c\varepsilon_0 n_r m^2\omega}\int \rho_v(E_1)\rho_c(E_2)\,|M(E_1,E_2)|^2(f_2-f_1)\,dE_1 \end{aligned} \tag{3.21}$$

$$E_2 = E_1+\hbar\omega$$

where using Eq. (3.14) we put

$$|M(E_1,E_2)|^2 = |\boldsymbol{e}\langle\psi_2|\boldsymbol{p}|\psi_1\rangle|^2 \tag{3.22}$$

Equation (3.21) can be used commonly for calculations of amplification and absorption, and the positive g describes amplification and the negative g describes absorption (absorption factor is $\alpha=-g=|g|$).

For spontaneous emission, from a similar calculation using Eq. (3.19), the rate of photon emission within frequency range from ω to $\omega+d\omega$ in the volume V per unit time is given by

$$\begin{aligned} r_{spt}(\hbar\omega)\,d\omega &= \frac{n_r e^2\omega}{\pi m^2 c^3\varepsilon_0}\,d\omega \\ &\quad\times\int \rho_v(E_1)\rho_c(E_2)|M(E_1,E_2)|^2 f_2(1-f_1)\,dE_1 \end{aligned} \tag{3.23}$$

where

$$E_2 = E_1 + \hbar\omega$$

The power of spontaneous emission within the frequency width $d\omega$ is given by the product of the above spontaneous emission rate $r_{spt}\,d\omega$ and the photon energy $\hbar\omega$. In the following sections, derivations of the densities $\rho_v(E_1)$ and $\rho_c(E_2)$ of states and the momentum matrix element $|M(E_1, E_2)|^2$, which are required for calculation of the gain factor $g = -\alpha$ and the spontaneous emission using Eqs (3.21) and (3.23), are outlined. The results on the gain and spontaneous emission will then be given in the following sections.

3.3.2 Density of Electron States in Doped Semiconductors

The direct-transition model assumes a band structure with sharp band edges described as in Eq. (3.2) and uses the parabolic density of states. Although the model applies well for pure semiconductors, such a simplified treatment is not appropriate for the impurity-doped semiconductors usually used in semiconductor lasers. The impurities doped in a semiconductor of small effective mass are easily ionized, and for cases where the doping density is large the carriers supplied by the impurities contribute to electrical conduction. This means that it is not appropriate to consider the impurity levels as localized levels. Since the ionized impurities are distributed randomly in the crystal, the periodic potential of the lattice is disturbed. It should be considered that the disturbance gives rise to modification of the densities of states for the conduction and valence bands and forms the tails of the bands. Such a model is called the band-tail model [6,7].

Kane [6] used a Gaussian distribution expression for the random potential due to the ionized impurities to analyze the density of states under the assumption that the kinetic energies of the carriers are so small that the carrier distribution is dominated by the potential. The potential produced by an ionized impurity in a semiconductor containing carriers is given by the solution of the combination of the Poisson equations and the equation for the potential-dependent carrier density. As an approximate expression, however, the potential can be written as

$$V(r) = \frac{e^2}{4\pi\varepsilon_r\varepsilon_0}\frac{1}{r}\exp\left(-\frac{r}{L_s}\right) \tag{3.24}$$

where ε_r is the relative dielectric constant. In this expression, which results from a modification of the potential produced by a point charge, the exponential factor describes the screening effect, i.e., reduction in the potential due to the distribution of carriers surrounding the ion. The screening length L_s is approximately equal to the Debye length and is written as

$$L_s = \left(\frac{k_B T \varepsilon_r \varepsilon_0}{e^2 N} \right)^{1/2} \tag{3.25}$$

where T is the temperature and N is the carrier density. Here let V_{rms} be the root-mean-square amplitude fluctuation of the potential under the assumption that the distribution of the potential fluctuation is Gaussian; then V_{rms} can be written, by using L_s, the density N_D^+ of ionized donors, and the density N_A^- of ionized acceptors as,

$$V_{rms} = \frac{e^2}{4\pi \varepsilon_r \varepsilon_0} \left[2\pi (N_D^+ + N_A^-) L_s \right]^{1/2} \tag{3.26}$$

The density of electron states for the conduction band in an n-type semiconductor having this type of potential fluctuation is calculated as

$$\rho_c(E') = \frac{m_n^{3/2}}{\pi^2 \hbar^3} (2\eta_c)^{1/2} y\left(\frac{E' - E_c}{\eta_c} \right) \tag{3.27}$$

$$\eta_c = 2^{1/2} V_{rms}$$

where $y(x)$ is the Kane function defined by

$$y(x) = \frac{1}{\pi^{1/2}} \int_{-\infty}^{x} (x - z)^{1/2} \exp(-z^2)\, dz \tag{3.28}$$

and shown in Fig. 3.4. As can be seen in this figure, ρ_c in Eq. (3.27) is approximately equal to Eq. (3.3a) and is proportional to $(E' - E_c)^{1/2}$ for large $E' - E_c$, and describes the band tail approximately proportional to a Gaussian function $\exp[-(E' - E_c)^2/\eta_c^2]$ for the low-energy region $E' < E_c$.

In the above-described Kane model, the kinetic energy of the carriers are omitted, and as a result the density of states in the band tail is overestimated. To improve the accuracy, Halperin and Lax [8] analyzed the density of states, in the band tail, taking the kinetic energy into account, and obtained the self-consistent density of states by iterative calculation. Although this method gives accurate values for the density of states in the band tail, it requires complicated and time-consuming numerical calculations

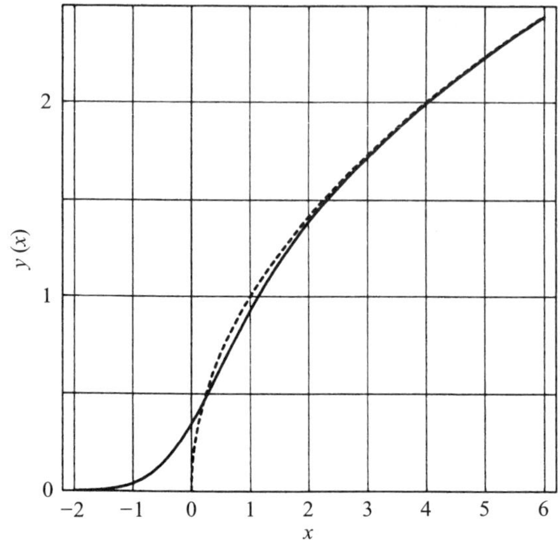

Figure 3.4 Kane function: the solid curve is Kane function $y(x)$ defined by Eq. (3.28); the dotted curve is $y = x^{1/2}$.

and is not applicable to calculations for a parabolic band with a higher energy and a band-tail transient region. As a solution to this difficulty, Stern [9] combined the two methods; in Eq. (3.27), i.e., the result of the Kane Gaussian model, η_c is considered as an unknown parameter, and the value of η_c is determined so that the value of ρ given by Eq. (3.27) may coincide with the result obtained by the method of Halperin and Lax at an appropriate junction point in the band tail. This method, called the Gaussian Halperin–Lax band-tail (GHLBT) model, offers an approximate technique very convenient for calculation of the optical transition, since it describes the density of state by the simple function in Eqs (3.27) and (3.28) over the entire region through the band edge. Although the above description is for the conduction band, the same treatment can be carried out also for the heavy holes and light holes in the valence band. The densities of states calculated by the GHLBT model for GaAs are shown in Fig. 3.5.

3.3.3 Transition Matrix Element

To calculate the momentum matrix element, given by Eqs (3.15) and (3.22), the wave functions ψ_1 and ψ_2 of electrons in the valence and conduction bands are required. The wave function of an electron of wave vector $\boldsymbol{k}$ in an ideal semiconductor crystal is given by the Bloch function in Eq. (3.1) using

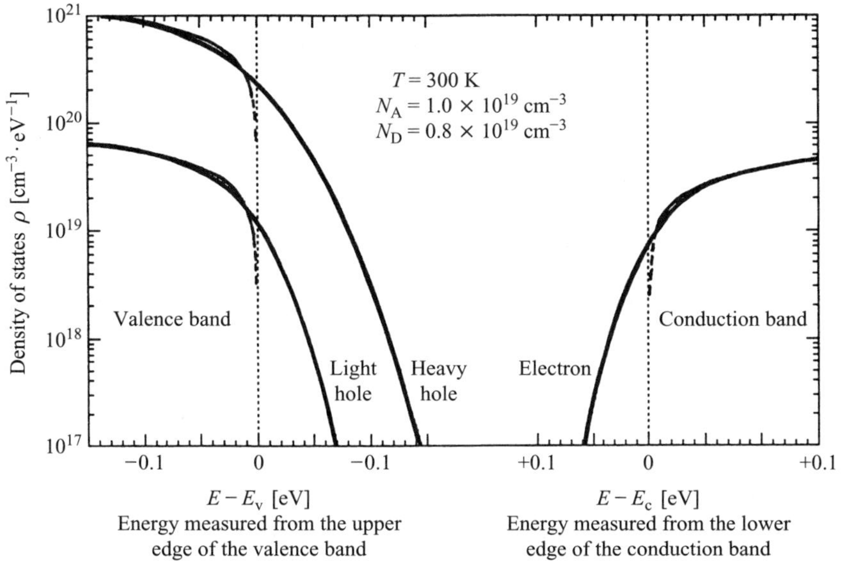

Figure 3.5 Densities of states for the conduction and valence bands of GaAs calculated upon the basis of the GHLBT model: - - -, densities of states with the band tail omitted.

$u_k(\boldsymbol{r})$. According to the ***kp*** perturbation theory developed by Kane [3] to analyze the band structure, the mean square of momentum matrix element given by Eq. (3.15) for electrons near the band edge and the effective mass m_n of electrons in the conduction band are correlated by [7]

$$|M_B|^2 = \ll |\boldsymbol{e}\langle u_2|\boldsymbol{p}|u_1\rangle|^2 \gg = \frac{m^2 P^2}{6\hbar^2} \tag{3.29a}$$

$$P^2 = \frac{\hbar^2 E_g(E_g + \Delta)}{2m_n(E_g + 2\Delta/3)} \tag{3.29b}$$

where m is the electron mass, E_g the bandgap energy, and Δ the spin–orbit interaction energy. The subscript B in the above equation indicates the value for a Bloch state in an ideal crystal, and the $\frac{1}{6}$ factor results from $\frac{1}{3}$ due to averaging $\ll\gg$ over all directions and $\frac{1}{2}$ due to the spin selection rule (the transitions are limited to those without spin inversion). For GaAs, for example, $m_n = 0.067m$, $E_g = 1.424$ eV, and $\Delta = 0.33$ eV and therefore from the above equation we have $|M_B|^2 = 1.33mE_g$. The values of $|M_B|^2$ have been determined by experimental measurements of m_n, E_g, and Δ for many materials. The data are often expressed in the form of the ratio of the value before multiplying by the $\frac{1}{6}$ averaging factor to the electron mass m, i.e., $6|M_B|^2/m$, and the data [10] include $6|M_B|^2/m = 14.4 \pm 0.1$ [eV] for GaAs, $6|M_B|^2/m = 14.9 + 1.4x$ [eV] for $Al_xGa_{1-x}As$ ($x < 0.3$), $6|M_B|^2/m = 14.4 - 3.3x$ [eV] for $In_xGa_{1-x}As$, $6|M_B|^2/m = 9.9 \pm 0.3$ [eV] for InP, and $6|M_B|^2/m = 9.9 + 2.8y$ [eV] for $In_{1-x}Ga_xAs_yP_{1-y}$ ($x = 0.47y$).

For doped semiconductors, the electrons near the band edges ($\boldsymbol{k} \sim \boldsymbol{0}$) are affected by the impurities, and therefore it should be assumed that they are localized. The wave function of such electrons can be described by the product of Eq. (3.1) and a slowly varying envelope function ψ_{env} describing the localization:

$$|\psi(\boldsymbol{r})> = |\psi_{env}(\boldsymbol{r}) \exp(i\boldsymbol{k} \cdot \boldsymbol{r})\, u_k(\boldsymbol{r})> \tag{3.30}$$

The envelope function ψ_{env}, which is the normalized wave function for the fundamental 1s state of the hydrogen atom model, is given by

$$\psi_{env}(\boldsymbol{r}) = \left(\frac{1}{\pi a^3}\right)^{1/2} \exp\left(-\frac{r}{a}\right), \qquad r = |\boldsymbol{r} - \boldsymbol{r}_i| \tag{3.31}$$

where $\boldsymbol{r}_i$ represents the center of the localization. Although in the H atom model a is the Bohr radius given by $a^* = 4\pi\varepsilon_r\varepsilon_0\hbar^2/m^*e^2$, here a is a parameter describing the localization scale and the value is determined by the method presented later. In order to calculate the matrix element M for which the electrons in the band tails are involved, the wave functions ψ_1 and ψ_2 are described in the form of the localized electron wave function (Eq. (3.30)), and the Bloch functions for the valence and conduction bands are used for $u_k(\boldsymbol{r})$. Then, using the orthogonality of the Bloch functions, the matrix element M can be written using M_B in Eq. (3.29) as

$$M = M_B M_{env} \tag{3.32}$$

$$M_{env} = \langle \psi_{envc}(\boldsymbol{r}) \exp(i\boldsymbol{k}_c \cdot \boldsymbol{r}) | \psi_{envv}(\boldsymbol{r}) \exp(i\boldsymbol{k}_v \cdot \boldsymbol{r}) \rangle \tag{3.33}$$

where the subscripts c and v denote the conduction and valence bands, respectively.

The localization character can be illustrated by assuming for the moment that the localization is only in the conduction band. Putting $\boldsymbol{k}_c \sim \boldsymbol{0}$ and $a_v \to 0$, substitution of Eq. (3.31) into Eq. (3.33) yields

$$|M_{env}|^2 = 64\pi a_c^3 (1 + a_c^2 \boldsymbol{k}_v^2)^{-4} \tag{3.34}$$

The matrix element $|M|^2$ for the transitions where wave vector conservation does not necessarily hold is given by the product of Eq. (3.29) and Eq. (3.34). Equation (3.34) indicates that $|M|^2$ is large in the region of $\boldsymbol{k}$ where $a^2\boldsymbol{k}^2$ is small.

Although Eq. (3.34) is effective for transitions between the band tail and a extended band, it does not apply for transitions between parabolic bands with larger energies and those between band tails. To solve the difficulty, Stern expressed each conduction-band electron and each valence-band electron in the form of Eqs. (3.30) and (3.31). The values of a were determined by fitting the effective energy reduction due to the localization to the effective energy reduction calculated from the density of states in the GHLBT model. The value of $|M_{\text{env}}|^2$ was calculated by using these values in Eq. (3.33) and averaging over all $\boldsymbol{k}$ directions and the localization sites $\boldsymbol{r}_i$ [11]. The result is written as

$$|M_{\text{env}}|^2 = \frac{64\pi b}{3}(t^4 - q^4)^{-5}[(b^4 - 5b^2B^2 + 5B^4)(3t^4 + q^4)(t^4 - q^4)^2 + 8b^2B^2t^2(3b^2 - 10B^2)(t^8 - q^8) + 16b^4B^4(5t^8 + 10t^4q^4 + q^8)] \tag{3.35}$$

$$B^2 = \frac{1}{a_c a_v}, \qquad b = \frac{1}{a_c} + \frac{1}{a_v}, \qquad t^2 = b^2 + k_c^2 + k_v^2, \qquad q^2 = 2k_c k_v$$

where k_c and k_v are the wave numbers of valence-band and conduction-band electrons with the effective energy reduction taken into account. The matrix element $|M|^2 = |M(E_1,\ E_2)|^2 = |M_B|^2|M_{\text{env}}|^2$ using this $|M_{\text{env}}|^2$ is called the Stern energy-dependent matrix element (SME).

Figure 3.6 shows an example of $|M_{\text{env}}|^2$ calculated by the above-described method. Considering the photon energy $\hbar\omega$ as a constant and the valence-band electron energy $E(=E_1)$ as a variable, conduction-band electrons of energy $E+\hbar\omega$ $(=E_2)$ satisfy energy conservation, and valence-band electrons of energy $E(=E_1)$ also satisfy wave vector conservation at $E = E_v - (m_r/m_p)(\hbar\omega - E_g)$ given by Eq. (3.33a). The direct transitions that exactly satisfy wave vector conservation take place only at this value of E, and this corresponds to the transition matrix element M given by a δ function that peaks at this position. In semiconductors doped with impurities, however, transitions take place even if wave vector conservation does not exactly hold, and therefore the E dependence of M is described by a peak-like function corresponding to the imperfection. Figure. 3.6 shows such an E dependence of M.

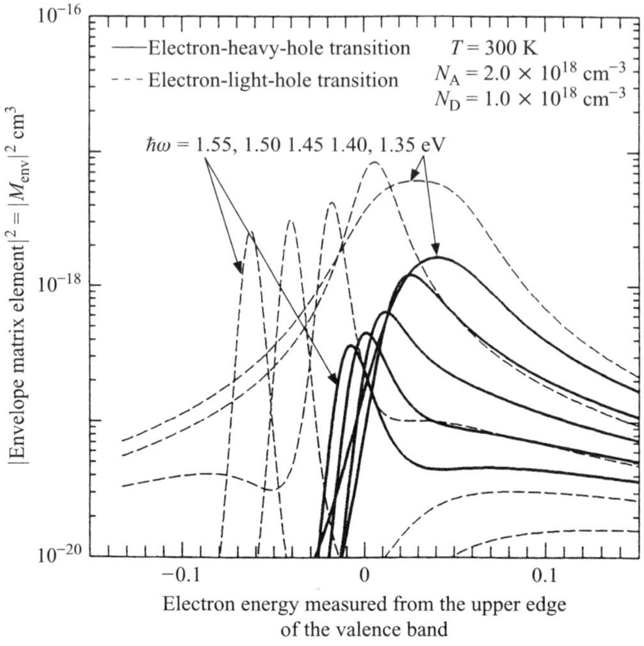

Figure 3.6 Matrix element of the interband transition in GaAs dependent upon energy (the SME).

3.4 GAIN SPECTRUM AND GAIN FACTOR

As we saw above, in the direct-transition model the gain and spontaneous emission spectra $g(\hbar\omega)$ and $r_{spn}(\hbar\omega)$ can be calculated by substituting the occupation probability (Eq. (3.6)), the density of states (Eq. (3.18)) and the matrix element (Eq. (3.29)) into Eqs (3.16) and (3.20). By this procedure the qualitative tendency of the gain spectrum as shown in Fig. 3.3 has been clarified. Accurate evaluation can be accomplished in the GHLBT–SME model where $g(\hbar\omega)$ and $r_{spn}(\hbar\omega)$ are calculated by substituting the occupation probability (Eq. (3.6)), the GHLBT density of states (Eqs (3.27) and (3.28)), and the SME (Eqs (3.32) and (3.35)) into Eqs (3.21) and (3.23) [12, 13].

3.4.1 Determination of the Quasi-Fermi Levels

The expressions for the occupation probabilities f_1 and f_2 (Eqs (3.6)) are in the form of an energy distribution dependent on the quasi-Fermi levels F_v

and F_c. To clarify the dependences of f_1 and f_2 on the carrier injection, it is therefore necessary to correlate the minority-carrier density and the quasi-Fermi levels. The densities of electrons and holes in the semiconductor excited by carrier injection can be written by using the densities ρ_v and ρ_c of states in Eq. (3.3) and the occupation probabilities f_1 and f_2 in Eqs (3.6) as

$$n = \int \rho_c(E_2) f_2 \, dE_2 \tag{3.36a}$$

$$p = \int \rho_v(E_1)(1 - f_1) \, dE_1 \tag{3.36b}$$

Here the band tails are omitted, since it is known that the band tails do not affect significantly the quasi-Fermi levels. Since the above integration cannot be evaluated analytically, various approximate but analytical expressions for the Fermi levels dependent on the carrier density have been developed [10]. One of them is the Joyce–Dixon [14] expression

$$\frac{F_c - E_c}{k_B T} = \ln\left(\frac{n}{N_c}\right) + \sum_{i=1}^{4} A_i \left(\frac{n}{N_c}\right)^i \tag{3.37a}$$

$$N_c = 2\left(2\pi k_B \frac{T}{h^2}\right)^{3/2} (m_n)^{3/2}$$

$$\frac{E_v - F_v}{k_B T} = \ln\left(\frac{p}{N_v}\right) + \sum_{i=1}^{4} A_i \left(\frac{p}{N_v}\right)^i \tag{3.37b}$$

$$N_v = 2\left(2\pi k_B \frac{T}{h^2}\right)^{3/2} \left(m_{ph}^{3/2} + m_{pl}^{3/2}\right)$$

$$A_1 = +3.53553 \times 10^{-1}, \qquad A_2 = -4.95009 \times 10^{-3}$$

$$A_3 = +1.48386 \times 10^{-4}, \qquad A_4 = -4.42563 \times 10^{-6}$$

which is a good approximation for a wide range of carrier densities. The first term on the right-hand side is the major term, and it corresponds to the result obtained by approximating the Fermi–Dirac distribution by the

Boltzmann distribution. Generally in the semiconductor the electrical neutrality condition

$$n + N_A^- = p + N_D^+ \tag{3.38}$$

is maintained, where N_A^- and N_D^+ are the densities of ionized acceptors and donors, respectively. Although N_A^- and N_D^+ are given by the acceptor and donor densities N_A and N_D multiplied by the Fermi–Dirac coefficients, we can approximate them as $N_A^- \approx N_A$ and $N_D^+ \approx N_D$ except for the low-temperature range. When the minority carriers are injected, the majority-carrier density also increases according to Eq. (3.37). For a sufficiently high injection level, N_A^-, $N_D^+ \ll n$, p and $n \approx p$. When under carrier injection the minority-carrier density is given as an independent variable, one of the quasi-Fermi levels F_v and F_c is determined by one of Eqs (3.36), and the other quasi-Fermi level is determined by Eq. (3.38) and the other equation of Eqs (3.36). Thus, once the minority carrier density is given, the occupation probabilities f_1 and f_2 are determined. Examples of the energy distribution and the quasi-Fermi levels calculated for carriers in excited GaAs are shown in Fig. 3.7.

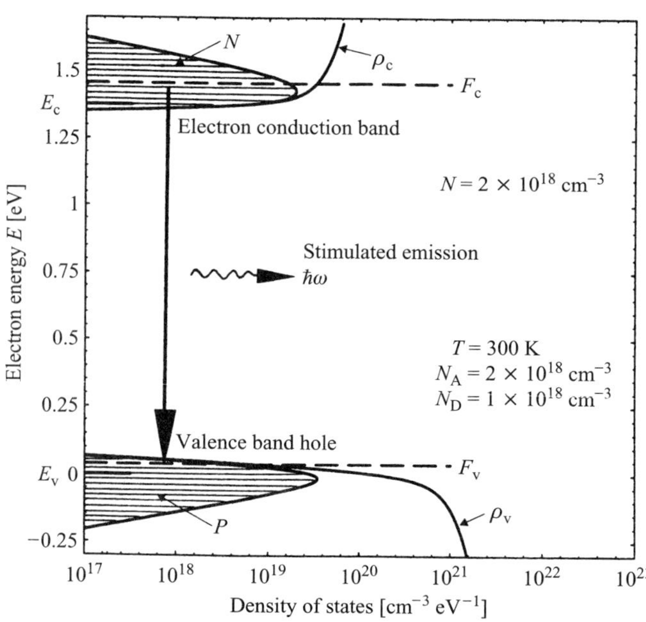

Figure 3.7 Energy distribution of the carrier density in an excited semiconductor.

3.4.2 Bandgap Shrinkage

To analyze precisely the dependences of the gain spectrum and spontaneous emission spectrum on the carrier injection density, the dependences of the band edges E_c and E_v on the carrier density must be taken into account. The band structure and the position of the band edges are calculated for a single electron. When the carrier density increases, the band edges shift owing to many-body effects. The existence of high-density movable carriers gives rise to screening of the potential, which reduces the repulsion between the electrons and the localization of electron waves, and as a result the energy bandwidths of each band increase. Therefore, the bandgap energy E_g decreases with increasing carrier density. This phenomena is called bandgap shrinkage. The magnitude of the shrinkage ΔE_g may be assumed to be inversely proportional to the distance between adjacent carriers and therefore proportional to the cubic root of the carrier density. The effect has been theoretically analyzed as a many-body problem. For GaAs, from experimental measurement of the dependence of the absorption spectrum on the carrier density, the following expression has been obtained [12]:

$$E_g = 1.424 + \Delta E_g \tag{3.39}$$

$$\Delta E_g = -1.6 \times 10^{-8}(n^{1/3} + p^{1/3})$$

where E_g is in [eV], and the carrier densities n and p are in [cm^{-3}]. The above expression for ΔE_g is often used not only for GaAs but also for other materials. Although actually the bandgap shrinkage deforms the gain spectrum, the deformation is not significant. In many cases, satisfactory approximate results are obtained by simply shifting the spectrum by ΔE_g.

3.4.3 Gain Spectrum

The validity of the GHLBT–SME model has been confirmed by the good agreement of the absorption and spontaneous emission spectra calculated by the above-described procedure with experimental results. An example of the gain (absorption) spectra and the dependence on the minority-carrier density, calculated for GaAs by this method, is shown in Fig. 3.8. The tendencies explained by the direct-transition model can be confirmed in this result. In addition, the GHLBT–SME model shows that the gain (absorption) spectra are associated with a gradient tail at or near the

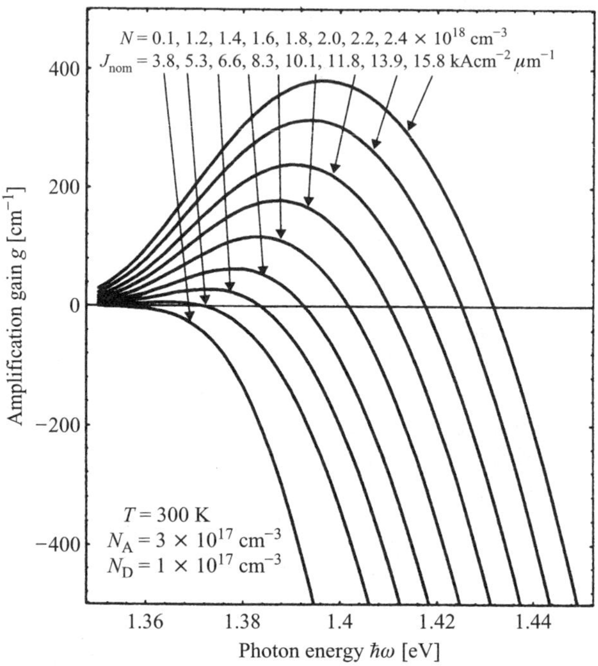

Figure 3.8 Gain (absorption) spectra of GaAs dependent upon the minority-carrier density calculated upon the basis of the GHLBT–SME model.

bandgap energy, unlike the result of the direct-transition model, which gives a sharp gain (absorption) edge. This is because band tails exist and they contribute to the gain (absorption). With increasing carrier density, the band-filling effect shifts the gain peak towards a higher energy. The gain peak wavelength shift is dominated by the band-filling effect, although it is partially cancelled out by the band-shrinkage effect.

The α-dependence of the gain on the impurity densities can be analyzed by the GHLBT–SME model, since it allows quantitative evaluation of the gain in semiconductors doped with impurities. The calculation shows that, with increases in the acceptor density N_A and donor density N_D, the minimum injection carrier density required to obtain a gain (transparency carrier density) is reduced, and the gain is enhanced. This is because the shift in the Fermi level due to the doping affects the situation favorably, and the population inversion condition (Eq. (3.9)) is satisfied with a smaller minority-carrier injection. The doping effect is significant with n-type doping rather than with p-type doping. This is attributed to the fact that the effective mass

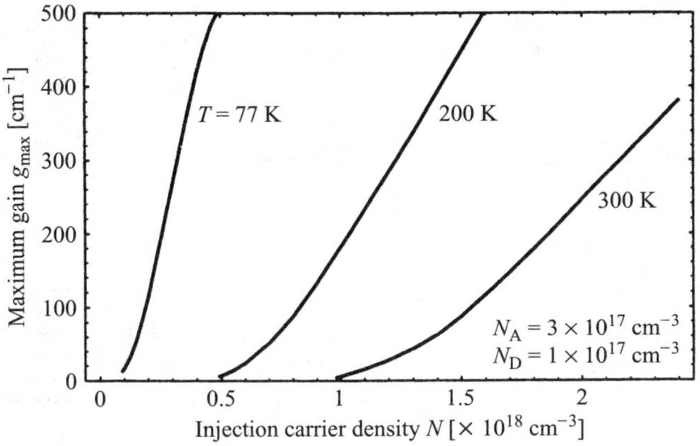

Figure 3.9 Injection carrier density dependence of maximum amplification gain.

of the conduction-band electrons is smaller than that of the valence-band electrons.

3.4.4 Carrier Density Dependence of the Maximum Gain

The data of the gain spectrum shown in Fig. 3.8 can be rewritten as curves describing the carrier density dependence of the gain at each optical wavelength. The envelope of the curves describes the dependence of the maximum gain on the injection carrier density. The envelope gives the carrier density dependence of the maximum in the gain spectrum, as shown in Fig. 3.9. The figure shows that, for lower temperatures, the carrier density at transparency is smaller and the gain is larger. We also see that, at room temperature, a gain as large as several hundred reciprocal centimeters is obtained with carrier densities of the order of magnitude of $10^{18}\,\text{cm}^{-3}$. As the graph shows, the carrier density dependence of the maximum gain g_{max} at a temperature can be expressed by a linear function

$$g_{max} = A_g(N - N_0) \tag{3.40}$$

Here A_g is a constant and N_0 denotes the transparency carrier density. The values depend upon the material, composition, impurity densities and temperature; for GaAs at room temperature, A_g is in the range $(3.0\text{–}7.0) \times 10^{-16}\,\text{cm}^2$ and N_0 is $(0.6\text{–}1.4) \times 10^{18}\,\text{cm}^{-3}$; for $In_{1-x}Ga_xAs_yP_{1-y}$ at room temperature, A_g is in the range $(1.2\text{–}2.5) \times 10^{-16}\,\text{cm}^2$ and N_0 is $(0.9\text{–}1.5) \times 10^{18}\,\text{cm}^{-3}$ [15]. The above expression together with these values are often used in the analysis of semiconductor laser characteristics as a simple and convenient phenomenological expression for the gain factor.

3.5 SPONTANEOUS EMISSION AND INJECTION CURRENT DENSITY

3.5.1 Spontaneous Emission Spectrum

The spontaneous emission spectrum can be determined by calculating $r_{spn}(\hbar\omega)$ in Eq. (3.23) using the GHLBT–SME model in a similar manner to the calculation of the gain spectrum. As shown in Fig. 3.10, the spontaneous emission exhibits a spectrum having a peak near the bandgap energy E_g; with increasing minority-carrier density the peak is enhanced and shifted towards the short-wavelength side and the width is broadened. When the impurity density is increased, the band-tail effect becomes more significant, the peak shifts to the long-wavelength side and the width is broadened. The spontaneous emission rate R_{spt} per unit volume and per unit time is given by the spontaneous emission rate $r_{spt}\, d\omega$ integrated by the photon frequency:

$$R_{sp} = \int r_{spt}(\hbar\omega)\, d\omega \tag{3.41}$$

This is indicated by the area of the region below the spontaneous emission spectrum curves in Fig. 3.10.

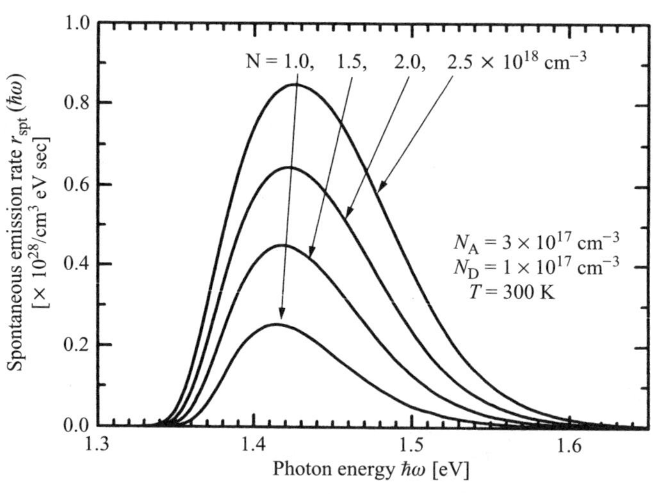

Figure 3.10 Spontaneous emission spectrum of GaAs dependent upon the carrier density.

3.5.2 Carrier Density and Injection Current Density

Now we consider the relation between the minority-carrier density and the injection current density. The injected minority carriers are consumed by the recombination with the majority carriers associated with the stimulated and spontaneous emissions. They are also consumed by the nonradiative recombination described later. For an injection level below the laser oscillation threshold or for the case where the optical amplitude is small, recombination by stimulated emission is negligible, and most of the carriers are consumed by spontaneous emission recombination and nonradiative recombination. The spontaneous emission rate is calculated by Eq. (3.41). On the other hand, under the assumption that spontaneous emission transitions take place also with electrons and holes regardless of the wave vector conservation rule, the recombination rate is considered to be proportional to the product of the carrier densities n and p. Then, the substantial spontaneous emission rate R_{sp} with the thermal excitation subtracted can be written as

$$R_{sp} = B_r(np - n_0 p_0) = B_r(\Delta n\, p_0 + \Delta p\, n_0 + \Delta n\, \Delta p) \tag{3.42}$$

$$\Delta n = n - n_0, \qquad \Delta p = p - p_0$$

where n_0 and p_0 are the carrier densities under thermal equilibrium and B_r is a constant. Considering a case where electrons as the minority carriers are injected into a p-type semiconductor, we have $n \gg n_0$, $p_0 \gg n_0$, $\Delta p \approx \Delta n \approx n$, and therefore

$$R_{sp} \approx B_r(p_0 n + n^2) \tag{3.43}$$

A similar relation holds also for an n-type semiconductor. The spontaneous emission rate R_{sp} calculated by Eq. (3.41) exhibits a carrier density dependence approximately similar to that given by Eq. (3.43). Let η_{sp} be the ratio of the spontaneous emission recombination to the total carrier recombination (internal quantum efficiency for the spontaneous emission), s and d be the area and the thickness respectively, of the active layer, and q be the elementary electric charge; then R_{sp} is correlated with the injection current I by $sdR_{sp} = \eta_{sp}(I/q)$. We call the injection current density required for obtaining a carrier density under the assumption that $\eta_{sp} = 1$ and $d = 1\,\mu\mathrm{m}$ the nominal current density and denote it by J_{nom}. Then the current density required for obtaining the same carrier density in an active layer of thickness $d\,\mu\mathrm{m}$ is written as

$$J = \frac{I}{s} = \frac{J_{nom} d}{\eta_{sp}} \tag{3.44a}$$

$$J_{nom} = {}_qR_{sp}\ \mathrm{A/cm^2\mu m} \tag{3.44b}$$

For a given minority carrier density, R_{sp} is calculated by Eq. (3.41). Using this result and Eq. (3.44b), the carrier density is correlated with J_{nom}.

3.5.3 Injection Current Density Dependence of the Maximum Gain

Using the relation between the carrier density and J_{nom}, the carrier density dependence of the gain spectrum $g(\hbar\omega)$ can readily be rewritten into the dependence on the nominal current density J_{nom}. In Fig. 3.8, the values of the corresponding J_{nom} determined in this manner are also shown. Furthermore, the dependence of the maximum value g_{max} of the gain g on J_{nom} can be obtained. The results are shown in Fig. 3.11. Because of the approximate relation described by Eq. (3.43), the shapes of the graphs for the current density dependence are somewhat different from those for the carrier density dependence shown in Fig. 3.9; the slopes of the graphs for the current density dependence become smaller for the region of high injection currents. The dependence can still be approximated by a linear function similar to Eq. (3.40):

$$g_{max} = B_g(J_{nom} - J_0) \tag{3.45}$$

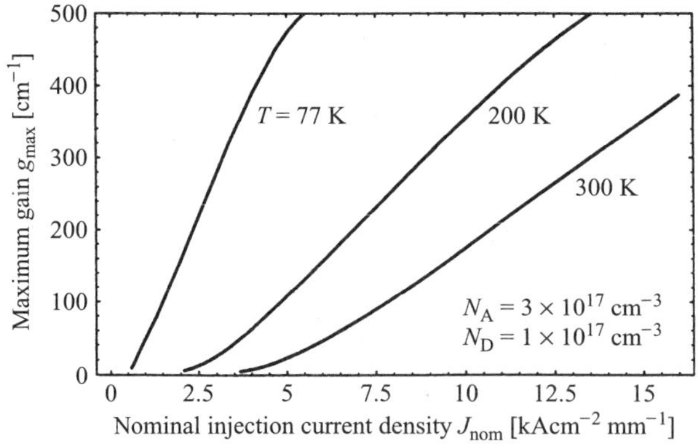

Figure 3.11 Dependence of the maximum amplification gain upon the injection current density.

where B_g is a constant and J_0 is the nominal current density at transparency.

3.5.4 Nonradiative Recombination and Carrier Overflow

Some of the injected minority carriers are consumed in the active layer by the nonradiative recombination without photon emission. This gives rise to a factor that causes the internal quantum efficiency η_{sp} for spontaneous emission in Eq. (3.44a) to deteriorate. Although the nonradiative recombination includes recombination due to lattice defects and recombination due to impurities, and also recombination at the interfaces of the heterostructure, it is negligibly small for ordinary lasers that use high-quality crystals. More important and essential nonradiative recombination is that caused by the Auger processes [15,16]. Collision of two electrons and one hole resulting in recombination of one electron and the hole, and excitation of another electron with the energy released by the recombination up to a higher level in the conduction band (CCCH process), and collision of two heavy holes with one electron resulting in similar recombination and excitation (CHHS and CHHL processes), as shown in Fig. 3.12, are called Auger processes. The excited carriers give energy to the crystal lattice in the form of heat and return to the unexcited level. Since both energy conservation and momentum conservation must hold also for the Auger process, recombination occurs with carriers at energies apart from the band edge, as shown in Fig. 3.12. Accordingly, the recombination rate R_A exhibits a remarkable temperature dependence and bandgap energy dependence; R_A is larger for a narrow bandgap and a high temperature.

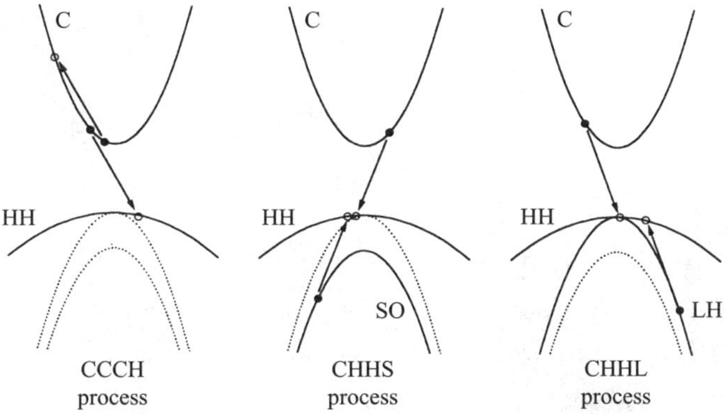

Figure 3.12 Interband Auger recombination processes: C, conduction band; HH, heavy-hole band; LH, light-hole band; SO, split-off band.

Since the process is a collision of three particles, the carrier density dependence of R_A can be phenomenologically written as

$$R_A \approx C_n n^2 p + C_p n p^2 \approx C_A n^3 \tag{3.46}$$

where C_n, C_p, and C_A are constants and use has been made of $n \approx p$.

The values of C_A at room temperature, determined experimentally by the carrier lifetime measurement, are of the order of $10^{-30}\,cm^6/s$ for GaAs, while they are larger, namely, $(2\text{–}3) \times 10^{-29}\,cm^6/s$, for $In_{1-x}\,Ga_xAs_yP_{1-y}$ for emission in the 1.3 μm band, and $(7\text{–}9) \times 10^{-29}\,cm^6/s$ for $In_{1-x}Ga_xAs_yP_{1-y}$ for emission in the 1.5 μm band. This means that, while the Auger recombination is usually negligibly small for lasers for short-wavelength emission, it is an important factor that significantly affects η_{sp} in $In_{1-x}\,Ga_xAs_y\,P_{1-y}$ lasers for long-wavelength emission. The Auger recombination can be reduced by using (strained) quantum well structures.

Another important factor that causes the emission efficiency to deteriorate is the overflow of minority carriers injected in the active layer into the region opposite to the injection side, which gives rise to additional current without contribution to the laser action. As shown in Fig. 3.13, the minority carriers are injected and confined in the potential barriers, and same of them having energies higher than the barrier overflow into the opposite side. The magnitude of the overflow current can be evaluated by calculating the carrier density at the boundary between the active and barrier regions under the assumption that the quasi-Fermi level is continuous across the boundary, and solving the carrier diffusion equation in the barrier region with the use of the boundary value. The overflow is

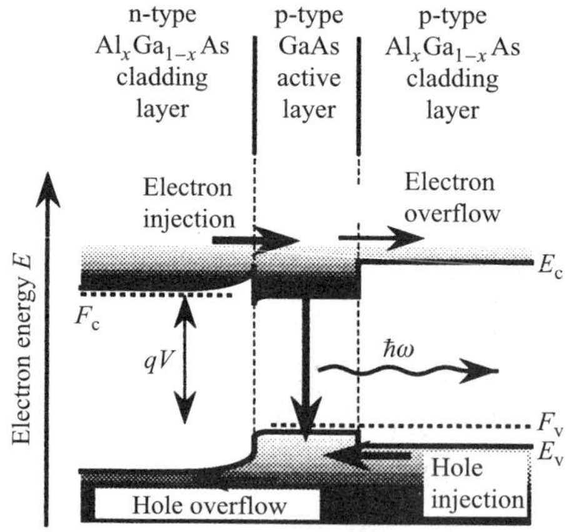

Figure 3.13 Leakage of carriers by overflow in a double heterostructure.

larger for lower barrier heights and higher temperatures. The current consists of diffusion current and drift current with the ratio dependent on the thickness and the resistivity of the barrier region. In many lasers the overflow can be suppressed to a negligible value by appropriate design. For short-wavelength lasers where it is difficult to assure sufficient barrier height, and for laser operation at elevated temperatures, the overflow can be an important factor that limits the performances.

3.6 DENSITY MATRIX ANALYSIS

In the previous section, the GHLBT–SME model and the quasithermal equilibrium approximation were used to discuss the interaction of optical waves and carriers in a semiconductor and to clarify the gain factor and the spontaneous emission. The analysis, however, requires complicated computations of the density of states and the matrix element. Another problem is that the quasithermal equilibrium approximation is not appropriate for accurate analysis of the characteristics of lasers oscillating or amplifying an optical wave with a large amplitude, since the carrier state deviates significantly from quasithermal equilibrium by the strong interaction of the carriers with the optical wave. This means that the results obtained in the previous section are valid only for injection levels below the threshold of laser oscillation and in the vicinity of it.

Interaction of many carriers and optical waves can be analyzed, with the deviation from quasithermal equilibrium taken into account, by using the density matrix [17, 18]. The density matrix enables a statistical analysis of a system consisting of many electrons to be made based on the quantum theory for an electron. The fundamental concept of the density operator and an outline of the density matrix analysis, in general, are given in Appendix 1.

3.6.1 Polarization and Gain Factor

In density matrix analysis, it is possible to treat both electrons and optical waves by the quantum theory. However, if the optical wave is represented by an energy eigenstate, the phase of the optical wave cannot be properly treated. Although representation by a coherent state can be used to solve the problem, the analysis becomes much more complicated. It is practical to employ semiclassical theory where the electrons are treated by quantum theory and the optical wave by classical theory. Although the semiclassical theory cannot deal with spontaneous emission, it can be included phenomenologically by using the results obtained in the last section. Then the semiclassical theory can give results to a good approximation, except for cases where the laser output is very small.

Let us use the complex expressions $\boldsymbol{E}\exp(-\mathrm{i}\omega t)$ and $\boldsymbol{H}\exp(-\mathrm{i}\omega t)$, with time dependence $\exp(-\mathrm{i}\omega t)$, for the electromagnetic field of an optical wave in a semiconductor. Then $\boldsymbol{E}$ and $\boldsymbol{H}$ satisfy the Maxwell equations

$$\nabla \times \boldsymbol{E} = +\mathrm{i}\omega \boldsymbol{B}, \qquad \nabla \times \boldsymbol{H} = -\mathrm{i}\omega \boldsymbol{D} \tag{3.47a}$$

$$\nabla \cdot \boldsymbol{D} = 0, \qquad \nabla \cdot \boldsymbol{B} = 0 \tag{3.47b}$$

Let the refractive index n_r represent the effects of polarizations due to the electrons at energy levels lower than the deep part of the valence band and those due to the carrier plasma motion. The effects of the optical transitions of the carriers can phenomenologically be represented by the polarization $\boldsymbol{P}$ at the optical frequency. Assuming that $\boldsymbol{P}$ is proportional to the optical electric field, we can write the electric flux density $\boldsymbol{D}$ as

$$\boldsymbol{D} = \varepsilon_0 n_\mathrm{r}^2 \boldsymbol{E} + \boldsymbol{P}, \qquad \boldsymbol{P} = \varepsilon_0 \chi \boldsymbol{E} \tag{3.48}$$

where $\boldsymbol{P}$ is the electron polarization representing the transition and χ is a complex electric susceptibility. The magnetic flux density is given by $\boldsymbol{B} = \mu_0 \boldsymbol{H}$. From these equations we obtain a wave equation for the electric field $\boldsymbol{E}$:

$$\nabla^2 \boldsymbol{E} + \left[(n_\mathrm{r}^2 + \chi)\left(\frac{\omega}{c}\right)^2 \right] \boldsymbol{E} = \boldsymbol{0} \tag{3.49}$$

Consider a plane wave propagating along the $+z$ direction, and let $\tilde{\beta}$ be the complex propagation constant. Substituting $\boldsymbol{E} = \boldsymbol{E}_0 \exp(i\tilde{\beta}z)$ into the above equation yields

$$\tilde{\beta} = \frac{n_\mathrm{r}\omega}{c} + \frac{\omega}{2n_\mathrm{r}c}\mathrm{Re}\{\chi\} + i\frac{\omega}{2n_\mathrm{r}c}\mathrm{Im}\{\chi\}) \tag{3.50}$$

where use has been made of $n_\mathrm{r}^2 \gg |\chi|$. From this equation, we see that the transition electron polarization represented by χ is equivalent to a change Δn in refractive index, and amplification (or absorption) represented by a gain factor g (or a absorption factor α) given by

$$\Delta n = \frac{1}{2n_\mathrm{r}}\mathrm{Re}\{\chi\} \tag{3.51a}$$

$$g = -\alpha = \frac{1}{|\boldsymbol{E}|^2}\frac{\mathrm{d}|\boldsymbol{E}|^2}{\mathrm{d}z} = -\frac{\omega}{n_\mathrm{r}c}\,\mathrm{Im}\{\chi\} \tag{3.51b}$$

This means that the gain (absorption) factor can be given by using the imaginary part of the phenomenological complex electric susceptibility χ.

Although the above expression gives the rate of the amplification per unit length of propagation, the factor can be converted into a temporal gain factor representing the rate per unit time:

$$G = \frac{1}{|\boldsymbol{E}|^2}\frac{\mathrm{d}|\boldsymbol{E}|^2}{\mathrm{d}t} = \frac{1}{|\boldsymbol{E}|^2}\left(\frac{\mathrm{d}|\boldsymbol{E}|^2}{\mathrm{d}z}\right)v_\mathrm{g} = v_\mathrm{g}g \tag{3.52}$$

where $v_\mathrm{g} = c/n_\mathrm{g}$ is the group velocity and n_g is the group index of refraction.

3.6.2 Density Matrix Expression for Polarization

As shown by Eq. (A1.10) in Appendix 1, the equation of motion for the density operator ρ (here ρ_I is simply written as ρ) is given by

$$\frac{\mathrm{d}}{\mathrm{d}t}\rho(t) = \frac{1}{\mathrm{i}\hbar}[H_\mathrm{I}(t), \rho(t)] \tag{3.53}$$

where H_I is the interaction Hamiltonian. Consider an optical wave of a mode with frequency ω, and let $\boldsymbol{E}(\boldsymbol{r})\exp(-\mathrm{i}\omega t)$ be the complex expression for the electric field. Then the interaction Hamiltonian H_I in the semiclassical theory is given by

$$H_\mathrm{I}(t) = \exp\left(\frac{\mathrm{i}H_0 t}{\hbar}\right)H_\mathrm{i}\exp\left(\frac{-\mathrm{i}H_0 t}{\hbar}\right) \tag{3.54a}$$

$$\begin{aligned} H_\mathrm{i} &= -\frac{e}{m}\boldsymbol{A}\cdot\boldsymbol{p} \\ &= -\frac{e}{2\mathrm{i}m\omega}[\exp(-\mathrm{i}\omega t)\boldsymbol{E}(\boldsymbol{r}) - \exp(+(\mathrm{i}\omega t)\boldsymbol{E}^*(\boldsymbol{r})]\cdot\boldsymbol{p} \end{aligned} \tag{3.54b}$$

Here H_0 is the electron Hamiltonian with the interaction omitted, given by Eq. (2.31b), $\boldsymbol{p}$ is the electron momentum operator, and use has been made of Eq. (2.31d) and the expression for the optical wave in Eq. (2.3). In order to deal with the interband transition, we use $|n\rangle$ to represent the electron states in the conduction band and $|m\rangle$ to represent those in the valence band. Considering the combinations of these, the transition matrix element can be written as

$$\begin{aligned} H_{nm} &= \langle n|H_\mathrm{I}|m\rangle \\ &= \langle n|H_\mathrm{i}|m\rangle\exp(\mathrm{i}\omega_{nm}t) \\ &= \frac{-e}{\mathrm{i}m\omega}\langle n|\boldsymbol{p}|m\rangle \\ &\quad\times\frac{1}{2}\{\boldsymbol{E}(\boldsymbol{r})\exp[\mathrm{i}(\omega_{nm}-\omega)t] - \boldsymbol{E}^*(\boldsymbol{r})\exp[\mathrm{i}(\omega_{nm}+\omega)t]\} \end{aligned} \tag{3.55}$$

where

$$\hbar\omega_{nm} = E_n - E_m$$

The second term on the right-hand side of the above expression oscillates rapidly and does not substantially contribute to the change in $\rho(t)$. Therefore it can be omitted (rotating-wave approximation). Using Eq. (2.56), the above equation is rewritten as

$$H_{nm} = H_{nm}^* \\ = -\tfrac{1}{2}\boldsymbol{P}_{nm} \cdot \boldsymbol{E}(\boldsymbol{r}) \exp[\mathrm{i}(\omega_{nm} - \omega)t] \tag{3.56a}$$

$$\boldsymbol{P}_{nm} = \langle n|\boldsymbol{P}|m\rangle = \langle n|e\boldsymbol{r}|m\rangle, \tag{3.56b}$$

where $\boldsymbol{P}_{nm}$ is the matrix element of polarization.

The density matrix ρ in the interaction picture is correlated with the density matrix ρ^{s} in the Schrödinger picture by Eq. (A1.11) in Appendix 1, and the electron polarization is represented by an operator $\boldsymbol{P} = e\boldsymbol{r}$. As shown by Eq. (A1.12), the expectation value for the operator $\boldsymbol{P}$ can be calculated by taking the trace of $\rho^{\mathrm{s}}\boldsymbol{P}$. The total polarization representing the interband transitions in a semiconductor is given by

$$N_{\mathrm{vc}}\langle \boldsymbol{P}\rangle = N_{\mathrm{vc}}\,\mathrm{Tr}(\rho^{\mathrm{s}}\boldsymbol{P}) \\ = N_{\mathrm{vc}} \sum\nolimits_{jj'} \rho_{jj'} \exp(-\mathrm{i}\omega_{jj'}t)\boldsymbol{P}_{j'j} \\ = N_{\mathrm{vc}} \sum_{nm} [\rho_{nm}\boldsymbol{P}_{mn} \exp(-i\omega_{nm}t) + \rho_{mn}\boldsymbol{P}_{nm} \exp(+\mathrm{i}\omega_{nm}t)] \tag{3.57}$$

where N_{vc} is the total number of electrons in the conduction and valence bands. Writing the complex expression for the polarization as $\boldsymbol{P}(\boldsymbol{r})$ $\exp(-\mathrm{i}\omega t)$, the real part of this expression coincides with the above equation. Thus $\boldsymbol{P}(\boldsymbol{r})$ is given by

$$\boldsymbol{P}(\boldsymbol{r}) = 2N_{\mathrm{vc}} \sum_{nm} \rho_{nm}\boldsymbol{P}_{mn} \exp[-\mathrm{i}(\omega_{nm} - \omega)t]. \tag{3.58}$$

The complex susceptibility χ, defined by the second equation of Eq. (3.48), can be calculated by taking the ratio of $\boldsymbol{P}$ to $\varepsilon_0\boldsymbol{E}$. Then the gain factors g, G and the carrier-induced refractive index change Δn can readily be obtained by using Eqs (3.51) and (3.52).

3.6.3 Relaxation Effects

The large number of carriers in the conduction and valence bands in a semiconductor undergo complex and random interactions such as collision between carriers, collision with impurities, and collision with phonons (crystal lattice vibration). These interactions give rise to a reduction in the interaction with the optical wave for the laser action and shift the states of each band towards the quasithermal equilibrium. It is known that, in the density matrix analysis, a temporal change in the density operator due to such relaxation effects can be described by

$$\frac{d}{dt}\rho = -\tfrac{1}{2}[\Gamma(\rho - \rho^{QE}) + (\rho - \rho^{QE})\Gamma] \tag{3.59}$$

where Γ is an operator representing the relaxation; Γ can be expressed by a diagonal matrix having elements given by

$$\Gamma_{jj'} = \frac{\delta_{jj'}}{\tau_j} \tag{3.60}$$

where $\delta_{jj'}$ is the Kronecker delta, and τ_j denotes the relaxation time for the eigenstate $|j\rangle$. Here ρ^{QE} is the density operator representing the quasithermal equilibrium state and can be expressed by a diagonal matrix with elements given by

$$\rho_{jj'}^{QE} = \delta_{jj'}\rho_j^{QE} \tag{3.61}$$

where ρ_j^{QE} is the probability that the eigenstate $|j\rangle$ of the system is in quasithermal equilibrium. Equation (3.59) indicates that, provided that there is no interaction except for the relaxation phenomenon, the system approaches quasithermal equilibrium with a time constant τ_j and eventually reaches steady-state quasithermal equilibrium. The intraband relaxation times τ_j have been experimentally determined, and it is known that they are very short, of the order of 0.1 ps.

3.6.4 Density Matrix Equations

Here we formulate and solve the equation of motion for the density operator describing the behavior of electrons in a semiconductor injected with carriers. The semiclassical theory, in principle, cannot describe the effects of spontaneous emission. Spontaneous emission, however, is a process in which transitions of electrons from the conduction band to the valence band occur with emission of photons regardless of the incident optical wave and shift the system towards thermal equilibrium of the whole semiconductor; therefore it can be considered as a relaxation process for electrons. This means that spontaneous emission can phenomenologically be interpreted in terms of interband relaxation. Thus, combining the interaction with the optical wave giving rise to absorption and stimulated emission, intraband relaxation due to collisions of electrons with carriers and impurities, interband relaxation due to the spontaneous emission, and the effect of carrier injection, the equation of motion for the density operator can be formulated as

$$\begin{aligned}\frac{\mathrm{d}}{\mathrm{d}t}\rho = &\frac{1}{\mathrm{i}\hbar}[H_{\mathrm{I}}(t),\ \rho]\\ &-\tfrac{1}{2}[\Gamma_{\mathrm{r}}(\rho-\rho^{\mathrm{QE}})+(\rho-\rho^{\mathrm{QE}})\Gamma_{\mathrm{r}}]\\ &-\tfrac{1}{2}[\Gamma_{\mathrm{s}}(\rho-\rho^{\mathrm{TE}})+(\rho-\rho^{\mathrm{TE}})\Gamma_{\mathrm{s}}]\\ &+\Lambda\end{aligned} \tag{3.62}$$

where H_{I} is the interaction Hamiltonian for optical absorption and stimulated emission. The operators Γ_{r} and Γ_{s}, representing intraband relaxation and spontaneous emission, respectively, can be described in the form of a diagonal matrix as in Eq. (3.60). The operators ρ^{QE} and ρ^{TE} represent quasithermal equilibrium and thermal equilibrium, respectively, and are given by diagonal matrixes in the form of Eq. (3.61). The operator Λ describes the carrier injection and is given by a diagonal matrix with elements $\Lambda_{jj'} = \delta_{jj'}\Lambda_j$. The electron energy distribution for cases including laser oscillation is given by the diagonal elements ρ_{jj} of the solution ρ of Eq. (3.62) multiplied by the total electron number N_{vc} per unit volume. The electron energy distribution $N_{\mathrm{vc}}\,\rho_{nn}^{\mathrm{QE}}$ in the quasithermal equilibrium state, and the electron energy distribution $N_{\mathrm{vc}}\rho_{nn}$ under the laser oscillation are schematically illustrated in Fig. 3.14. The electron energy distribution $N_{\mathrm{vc}}\,\rho_{nn}^{\mathrm{TE}}$, in the thermal equilibrium state is usually much smaller than $N_{\mathrm{vc}}\,\rho_{nn}^{\mathrm{QE}}$ and $N_{\mathrm{vc}}\,\rho_{nn}$.

The electron distributions described by the density matrix are dominated not only by stimulated emission and absorption of photons but also by relaxation and injection. As a result, the solution ρ of Eq. (3.62) exhibits a nonlinear dependence on the amplitude of the incident optical wave. Therefore, to solve Eq. (3.62), we expand the solution ρ in a polynomial series with respect to the amplitude of the optical wave:

$$\rho = \rho^{(0)} + \rho^{(1)} + \rho^{(2)} + \cdots + \rho^{(l)} + \cdots \tag{3.63}$$

where $\rho^{(l)}$ is the term proportional to the lth power of the optical amplitude. Noting that H_{I} in Eq. (3.62) is proportional to the optical amplitude, we equate each term of the same power on the left- and right-hand sides of Eq. (3.62), in which Eq. (3.63) has been substituted, to obtain

$$\begin{aligned}\frac{\mathrm{d}}{\mathrm{d}t}\rho^{(0)} &= -\tfrac{1}{2}[\Gamma_{\mathrm{r}}(\rho^{(0)}-\rho^{\mathrm{QE}})+(\rho^{(0)}-\rho^{\mathrm{QE}})\Gamma_{\mathrm{r}}]\\ &\quad -\tfrac{1}{2}[\Gamma_{\mathrm{s}}(\rho^{(0)}-\rho^{\mathrm{TE}})+(\rho^{(0)}-\rho^{\mathrm{TE}})\Gamma_{\mathrm{s}}]+\Lambda\\ \frac{\mathrm{d}}{\mathrm{d}t}\rho^{(l)} &= \frac{1}{\mathrm{i}\hbar}[H_{\mathrm{I}},\rho^{(l-1)}]\\ &\quad -\tfrac{1}{2}[(\Gamma_{\mathrm{r}}+\Gamma_{\mathrm{s}})\rho^{(l)}+\rho^{(l)}(\Gamma_{\mathrm{r}}+\Gamma_{\mathrm{s}})] \qquad (l=1,2,3,\ldots)\end{aligned} \tag{3.64}$$

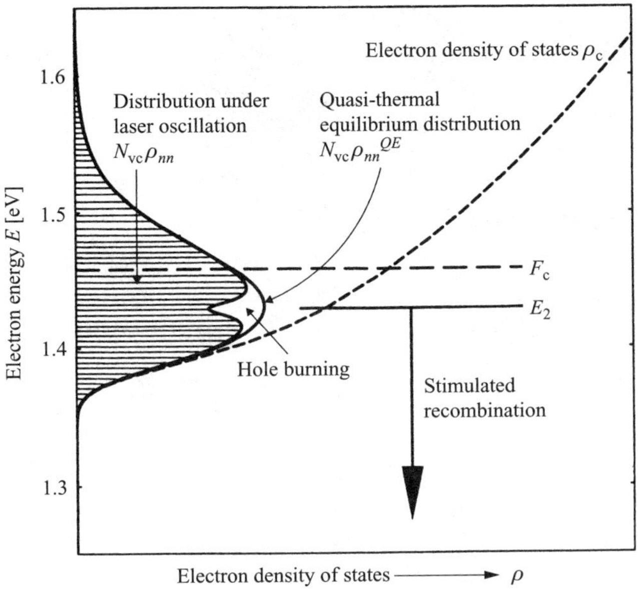

Figure 3.14 Electron energy distribution in the conduction band of an excited semiconductor (under laser oscillation).

Converting these equations into matrix expressions, noting that H_{I} is given by Eq. (3.56) and the off-diagonal elements are zero, we see that ρ can be described by diagonal elements for even power only and off-diagonal elements for odd power only according to

$$\rho_{nn} = \rho_{nn}^{(0)} + \rho_{nn}^{(2)} + \cdots + \rho_{nn}^{(2l)} + \cdots \tag{3.65a}$$

$$\rho_{nm} = \rho_{nm}^{(1)} + \rho_{nm}^{(3)} + \cdots + \rho_{nm}^{(2l+1)} + \cdots \tag{3.65b}$$

and that Eq. (3.64) is rewritten as

$$\frac{\mathrm{d}}{\mathrm{d}t}\rho_{nn}^{(0)} = -\frac{\rho_{nn}^{(0)} - \rho_{nn}^{\mathrm{QE}}}{\tau_{\mathrm{c}}} - \frac{\rho_{nn}^{(0)} - \rho_{nn}^{\mathrm{TE}}}{\tau_{\mathrm{s}}} + \Lambda_n \tag{3.66a}$$

$$\begin{aligned}\frac{\mathrm{d}}{\mathrm{d}t}\rho_{nn}^{(2l)} &= \frac{1}{\mathrm{i}\hbar}\left(H_{nm}\rho_{mn}^{(2l-1)} - \rho_{nm}^{(2l-1)}H_{mn}\right)\\ &\quad - \rho_{nn}^{(2l)}\left(\frac{1}{\tau_{\mathrm{c}}} + \frac{1}{\tau_{\mathrm{s}}}\right) \qquad (l = 1, 2, 3, \ldots)\end{aligned} \tag{3.66b}$$

$$\frac{\mathrm{d}}{\mathrm{d}t}\rho_{nm}^{(2l+1)} = \frac{1}{\mathrm{i}\hbar}\left(\rho_{mm}^{(2l)} - \rho_{nn}^{(2l)}\right)H_{nm} - \frac{\rho_{nm}^{(2l+1)}}{\tau_{\mathrm{r}}} \qquad (l = 0, 1, 2, 3, \ldots)$$

$$\frac{1}{\tau_{\mathrm{r}}} = \frac{1}{\tau_{\mathrm{c}}} + \frac{1}{\tau_{\mathrm{v}}} \tag{3.66c}$$

Here τ_c is the relaxation time for the conduction-band electrons, and τ_v is that for the valence-band electrons. The expressions for ρ_{mm} are given by replacing n by m, and c by v, in Eqs (3.66a) and (3.66b), and the expressions for ρ_{mn} are given by complex conjugate of Eq. (3.66c).

3.6.5 Linear Gain and Refractive Index Change

Since, as shown by Eq. (3.58), the polarization $\boldsymbol{P}$ is calculated from the off-diagonal elements ρ_{nm} of the density matrix, we first derive the expression for the first power term $\rho_{nm}^{(1)}$. From Eqs (3.56) and (3.66c), we have

$$\frac{\mathrm{d}}{\mathrm{d}t}\rho_{nm}^{(1)} = \left(\frac{1}{2\mathrm{i}\hbar}\right)(\rho_{nn}^{(0)} - \rho_{mm}^{(0)})\boldsymbol{P}_{nm} \cdot \boldsymbol{E}\exp[\mathrm{i}(\omega_{nm} - \omega)t] - \frac{\rho_{nm}^{(1)}}{\tau_r} \tag{3.67}$$

Solving the above equation for a steady oscillation state, we obtain

$$\rho_{nm}^{(1)} = \frac{1}{2\mathrm{i}\hbar}(\rho_{nn}^{0} - \rho_{mm}^{(0)}) \times \frac{\boldsymbol{P}_{nm} \cdot \boldsymbol{E}\exp[\mathrm{i}(\omega_{nm} - \omega)t]}{\mathrm{i}(\omega_{nm} - \omega) + 1/\tau_r} \tag{3.68}$$

Substitution of Eq. (3.68) into Eq. (3.58) yields

$$\boldsymbol{P}^{(1)} = \frac{N_{vc}}{\mathrm{i}\hbar}\sum_{nm}(\rho_{nn}^{(0)} - \rho_{mm}^{(0)}) \times \frac{|\boldsymbol{P}_{mn}|^2\boldsymbol{E}}{\mathrm{i}(\omega_{nm} - \omega) + 1/\tau_r} \tag{3.69}$$

Therefore, from Eqs (3.48) and (3.69) the linear complex susceptibility is calculated as

$$\chi^{(1)}(\hbar\omega) = \frac{N_{vc}}{\varepsilon_0\hbar}\sum_{nm}(\rho_{nn}^{(0)} - \rho_{mm}^{(0)}) \times \frac{|\boldsymbol{P}_{mn}|^2}{(\omega - \omega_{nm}) + \mathrm{i}/\tau_r} \tag{3.70}$$

The above expression contains $\rho_{nn}^{(0)}$ and $\rho_{mm}^{(0)}$, which are solutions of Eq. (3.66a). Considering the fact that the intraband relaxation times τ_c and τ_v are very short, we can approximate them by the quasithermal equilibrium states as $\rho_{nn}^{(0)} \approx \rho_{nn}^{\mathrm{QE}}$, $\rho_{mm}^{(0)} \approx \rho_{\mathrm{mm}}^{\mathrm{QE}}$, and ρ_{nn}^{QE} and ρ_{mm}^{QE} can be replaced by the Fermi–Dirac functions f_2 and f_1 at the quasi-Fermi levels for the conduction and valence bands, respectively. On the other hand, $\boldsymbol{P}_{mn}$ vanishes to zero unless the wave vectors $\boldsymbol{k}$ for the states m and n are same. Therefore, $N_{\mathrm{vc}} \sum_{nm}$ in the above expression can be rewritten as integration with respect to the density of states for the electron wave vector $\mathrm{d}\boldsymbol{k}/2\pi^2$ and then as energy integration:

$$\begin{aligned}
\chi^{(1)}(\hbar\omega) &= \frac{1}{\varepsilon_0 \hbar} \int \frac{(f_2 - f_1)\mathrm{d}\boldsymbol{k}}{2\pi^3} \\
&\quad \times \frac{|\boldsymbol{P}_{mn}|^2}{(\omega - \omega_{nm}) + \mathrm{i}/\tau_{\mathrm{r}}} \\
&= \frac{1}{\varepsilon_0 \hbar} \int (f_2 - f_1) \left(\frac{1}{2\pi^3} \int \delta(E_m + \hbar\omega_{nm} - E_n)\, \mathrm{d}\boldsymbol{k} \right) \mathrm{d}(\hbar\omega_{nm}) \\
&\quad \times \frac{|\boldsymbol{P}_{mn}|^2}{(\omega - \omega_{nm}) + \mathrm{i}/\tau_{\mathrm{r}}} \\
&= \frac{1}{\varepsilon_0} \int (f_2 - f_1) \rho_{\mathrm{r}}(\hbar\omega_{nm})\, \mathrm{d}\omega_{nm} \\
&\quad \times \frac{|\boldsymbol{P}_{mn}|^2}{(\omega - \omega_{nm}) + \mathrm{i}/\tau_r}
\end{aligned} \tag{3.71}$$

where $\rho_{\mathrm{r}}(\hbar\omega_{nm})$ is the reduced density of states defined by Eq. (3.18) with $\hbar\omega$ substituted by $\hbar\omega_{nm}$, and f_1 and f_2 are given by Eq. (3.6). The fact that $\boldsymbol{P}^{(1)}$ is proportional to electric field amplitude $\boldsymbol{E}$ with a coefficient $\varepsilon_0 \chi^{(1)}$ having an imaginary part implies that the optical wave is coherently amplified. The real and imaginary parts of $\chi^{(1)}(\hbar\omega)$ given by Eq. (3.71) satisfy the Kramers–Kronig relation (see Appendix 3).

From Eqs (3.51) and (3.71), we obtain expressions for the induced linear index change and the linear gain:

$$\begin{aligned}
\Delta n(\hbar\omega) &= \frac{\pi e^2}{2n_{\mathrm{r}}\varepsilon_0 m^2 \omega^2} \int (f_2 - f_1) \rho_{\mathrm{r}}(\hbar\omega_{nm}) |M_{mn}|^2 \\
&\quad \times [\tau_{\mathrm{r}}(\omega - \omega_{nm})] L_{nm}(\omega - \omega_{nm})\, \mathrm{d}\omega_{nm}
\end{aligned} \tag{3.72}$$

$$\begin{aligned}
g^{(1)}(\hbar\omega) &= \frac{\pi e^2}{n_{\mathrm{r}} c \varepsilon_0 m^2 \omega} \\
&\quad \times \int (f_2 - f_1) \rho_{\mathrm{r}}(\hbar\omega_{nm}) |M_{mn}|^2 L_{nm}(\omega - \omega_{nm})\, \mathrm{d}\omega_{nm}
\end{aligned} \tag{3.73}$$

where we put $|P_{mn}|^2 = (e/m\omega)^2 |M_{mn}|^2$ from Eqs (3.14), (3.15), and (3.56b). In the above expressions, $L_{nm}(\omega - \omega_{nm})$ is a function defined by

$$L_{nm}(\omega - \omega_{nm}) = \frac{1}{\pi\tau_{\mathrm{r}}[(\omega - \omega_{nm})^2 + 1/\tau_{\mathrm{r}}^2]} \tag{3.74}$$

Equation (3.73) is in the form of a convolution of the result in Eq. (3.16) obtained in the direct-transition model using the wave vector conservation rule and L_{nm} given by Eq. (3.74). In the limiting case when $\tau_r \rightarrow \infty$ the function $L_{nm}(\omega - \omega_{nm})$ approaches a delta function $\delta(\omega - \omega_{nm})$ representing the energy conservation rule. Therefore, for cases where the relaxation is negligible, Eq. (3.73) coincides with the result of the direct-transition model. Equation (3.74) implies that with the relaxation present the state of each electron changes in a short relaxation time τ_r, which allows transitions with deviation from the exact energy conservation, and combinations of carriers with ω_{nm} such that $|\omega - \omega_{nm}| < 1/\tau_r$ contributes to the amplification and absorption of the optical wave of frequency ω. The function $L_{nm}(\omega - \omega_{nm})$ describes the spectrum of stimulated emission (absorption) by combinations of carriers of energy difference $\hbar\omega_{nm}$ only and is called the line-shape function.

As shown above, by representing the intraband relaxation in terms of the relaxation time τ_r, the gain factor can be calculated simply by Eq. (3.73), which is an expansion of the direct-transition model result. The band-tail effect can also be taken into account by identifying it as an intraband relaxation and including it in τ_r. The refractive index change induced by the carriers can also be calculated in a similar manner. The transition matrix element $|M_{mn}|^2$ on the right-hand side of these expressions can be approximated by the value at the bandgap described by Eq. (3.29) and put outside the integral. The relaxation time τ_r can be calculated by Eq. (3.66c) from the relaxation times τ_c and τ_v for each band. Although τ_c and τ_v actually depend upon m and n, they can be represented by a single value for each, which can be determined by experimental measurements on the carriers in each band. In this approximation, the line-shape function of Eq. (3.74) reduces to $L(\omega - \omega_{nm})$. This procedure has been used to calculate amplification gains for many III–V compound alloy semiconductors [18]. As an example, the result of the calculation for GaAs is shown in Fig. 3.15. It has been confirmed for GaAs that this procedure gives results approximately the same as those obtained by the GHLBT–SME model described in the previous section. However, it has also been pointed out that the gain spectrum calculated by using the Lorenzian line-shape function with single relaxation time τ_r is not sufficiently accurate [19]. One of the problems is that, as we can see in Fig. 3.15, slight absorption appears for photon energy $\hbar\omega$ smaller than the bandgap energy E_g. This is because in the above analysis the carrier relaxation was described simply by a single τ_r in spite of the fact that the relaxation for carriers away from the band edge is not so fast as that for carriers near the band edge. The relaxation effect has been examined in detail through more careful density matrix analysis, and, based on the

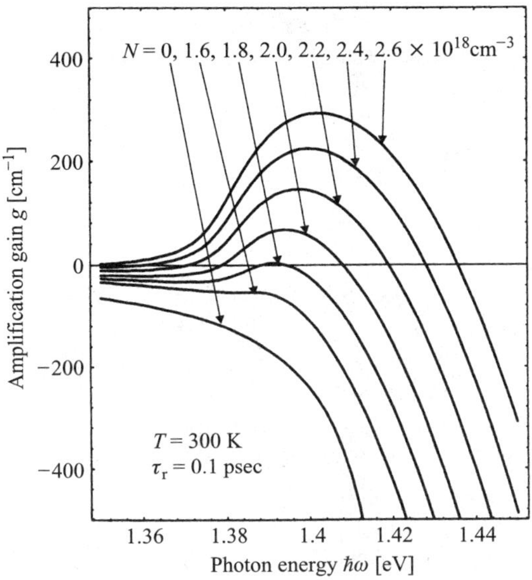

Figure 3.15 Linear amplification gain spectrum of GaAs calculated using Eq. (3.73) based on density matrix analysis.

results, line-shape functions modified for improvement have been proposed [19]. One of these is a hyperbolic secant function

$$L(\omega - \omega_{nm}) = \frac{\tau_r}{\pi} \operatorname{sech}[\tau_r(\omega - \omega_{nm})] \tag{3.75}$$

which decays for large $|\omega - \omega_{nm}|$ more quickly than the Lorenzian function. Use of this line-shape function eliminates the above-mentioned problem [20]. The carrier-induced refractive index change can be calculated by Kramers–Kronig transformation of the gain spectrum (see Appendix 3). While the above-described line-shape functions are expressions in the frequency (energy) domain, expressions in the time domain can also be used to calculate the gain spectrum in a form of a product of Fourier transformations instead of the convolution of Eq. (3.73) [10].

3.6.6 Saturation Effect

The terms $\rho_{nn}^{(2)}$ and $\rho_{nm}^{(3)}$, and those for higher powers, in the series expansion of the density matrix in Eq. (3.65) describe effects that are nonlinear with respect to the electric field $\boldsymbol{E}$. Firstly, solving Eq. (3.66b) with $l=1$, and Eqs (3.56) and (3.68) substituted, for the steady-state oscillation, yields

$$\rho_{nn}^{(2)} = -\frac{\pi}{2\hbar^2}\sum_m\left(\frac{1}{\tau_c}+\frac{1}{\tau_s}\right)^{-1}\left(\rho_{nn}^{(0)}-\rho_{mm}^{(0)}\right) \times |\boldsymbol{P}_{mn}|^2\,|\boldsymbol{E}|^2\,L_{nm}(\omega-\omega_{nm}) \tag{3.76}$$

The expression for $\rho_{nn}^{(2)}$ is given by replacing m by n and c by v in the above equation. These expressions describe the fact that, with the presence of a strong optical field $\boldsymbol{E}$ produced by laser oscillation, the carrier state distribution deviates from quasithermal equilibrium in such a manner that the population inversion of ω_{nm} close to the optical frequency ω decreases by an amount proportional to the oscillation intensity $|\boldsymbol{E}|^2$. This means that, when a strong optical wave is incident or when laser oscillation takes place, the densities of electrons and holes contribute by the energy relation to the optical amplification decrease. The decrease in the electron density is illustrated in Fig. 3.14. This effect is called hole burning, since it is associated with a deep depression of the carrier energy distribution curve at $\omega_{nm}\approx\omega$.

We next substitute Eqs (3.56) and (3.76) into Eq. (3.66c) for $l=1$ and solve for steady oscillation state to obtain

$$\rho_{nm}^{(3)} = -\frac{\pi}{4\hbar^3}\left[\left(\frac{1}{\tau_v}+\frac{1}{\tau_s}\right)^{-1}+\left(\frac{1}{\tau_c}+\frac{1}{\tau_s}\right)^{-1}\right] \times \left(\rho_{nn}^{(0)}-\rho_{mm}^{(0)}\right)|\boldsymbol{P}_{mn}|^2\,|\boldsymbol{E}|^2\,\boldsymbol{P}_{nm}\cdot\boldsymbol{E}\,\exp[\mathrm{i}(\omega_{nm}-\omega)t] \times \frac{L_{nm}(\omega-\omega_{nm})}{(\omega-\omega_{nm})+\mathrm{i}/\tau_r} \tag{3.77}$$

Substituting Eq. (3.77) into Eq. (3.58) yields an expression for nonlinear polarization:

$$\boldsymbol{P}^{(3)} = -\frac{\pi}{4\hbar^3}\left[\left(\frac{1}{\tau_v}+\frac{1}{\tau_s}\right)^{-1}+\left(\frac{1}{\tau_c}+\frac{1}{\tau_s}\right)^{-1}\right] \times 2\hbar\int (f_2-f_1)\,\rho_r(\hbar\omega_{nm})\,\mathrm{d}\omega_{nm}|\boldsymbol{P}_{mn}|^4\,|\boldsymbol{E}|^2\,\boldsymbol{E} \times \frac{L_{nm}(\omega-\omega_{nm})}{(\omega-\omega_{nm})+\mathrm{i}/\tau_r} \tag{3.78}$$

From Eqs (3.48), (3.51), and (3.78) the nonlinear susceptibility and nonlinear gain factor are calculated as

$$\chi^{(3)}(\hbar\omega) = -\frac{\pi}{2\hbar^2\varepsilon_0}\left(\frac{e}{m\omega}\right)^4\left[\left(\frac{1}{\tau_\mathrm{v}}+\frac{1}{\tau_\mathrm{s}}\right)^{-1}+\left(\frac{1}{\tau_\mathrm{c}}+\frac{1}{\tau_\mathrm{s}}\right)^{-1}\right]$$
$$\times |\boldsymbol{E}|^2\int (f_2-f_1)\,\rho_\mathrm{r}(\hbar\omega_{nm})\,\mathrm{d}\omega_{nm}\,|M_{mn}|^4$$
$$\times\frac{L_{nm}(\omega-\omega_{nm})}{(\omega-\omega_{nm})+\mathrm{i}/\tau_\mathrm{r}} \tag{3.79}$$

$$g^{(3)}(\hbar\omega) = -\frac{\pi^2\omega}{2\hbar^2 n_\mathrm{r} c\varepsilon_0}\left(\frac{e}{m\omega}\right)^4$$
$$\times\left[\left(\frac{1}{\tau_\mathrm{v}}+\frac{1}{\tau_\mathrm{s}}\right)^{-1}+\left(\frac{1}{\tau_\mathrm{c}}+\frac{1}{\tau_\mathrm{s}}\right)^{-1}\right]|\boldsymbol{E}|^2$$
$$\times\int (f_2-f_1)\,\rho_\mathrm{r}(\hbar\omega_{nm})\,\mathrm{d}\omega_{nm}\,|M_{mn}|^4\,[L_{nm}(\omega-\omega_{nm})]^2 \tag{3.80}$$

In these expressions, $N_\mathrm{vc}\sum_{nm}$ was replaced by the integration with respect to the energy using the reduced density of states in the same manner as described above. Equation (3.80) shows that, with increase in the optical field $\boldsymbol{E}$, the gain in the frequency in the vicinity of ω is reduced from the value calculated with the quasithermal equilibrium approximation by an amount proportional to $|\boldsymbol{E}|^2$. This phenomenon is called gain saturation. In a laser oscillator where the saturation effect is taken into account, with increase in the injection current above the threshold, the gain is kept constant to maintain the oscillation condition, while the carrier density continues to increase slightly to compensate for the gain saturation.

3.6.7 Extension to Multimode Expressions

Thus far the optical wave has been assumed to be of a single mode. For general cases including cases where the optical field consists of components of several frequencies, a similar analysis can be made by using a mode expansion expression $\sum_q \boldsymbol{E}_q(\boldsymbol{r})\exp(-\mathrm{i}\omega_q t)$ for the complex electric field. Then, since $\rho_{nm}^{(1)}$ and $\boldsymbol{P}^{(1)}$ are linear functions of the complex electric field, the linear susceptibility $\chi^{(1)}(\hbar\omega_q)$ and the linear gain factor $g(\hbar\omega_q)$ for each frequency component q are given by the same expressions as those for single-frequency case, with ω replaced by ω_q. On the other hand, since $\rho_{nn}^{(2)}$ and $\rho_{nm}^{(3)}$ are terms proportional to the second and third powers, respectively, of the electric field, we must take summations of the forms $\sum_{qq'}\mathrm{Re}\{[\boldsymbol{E}_q(\boldsymbol{r})\ \exp(-\mathrm{i}\omega_q t)][\boldsymbol{E}_{q'}(\boldsymbol{r})\exp(-\mathrm{i}\omega_{q'}t)]\}$ and $\sum_{qq'q''}\mathrm{Re}\{[\boldsymbol{E}_q(\boldsymbol{r})'\exp(-\mathrm{i}\omega_q t)]\times[\boldsymbol{E}_{q'}(\boldsymbol{r})\exp(-\mathrm{i}\omega_{q'}t)][\boldsymbol{E}_{q''}(\boldsymbol{r})\exp(-\mathrm{i}\omega_{q''}t)]\}$, respectively. Therefore, the nonlinear gain factor $g^{(3)}(\hbar\omega_q)$ for a component of a frequency ω_q is given by summation of the contributions by all frequency components:

$$g^{(3)}(\hbar\omega_p) = -\sum_{p'}\zeta_{pp'}|\boldsymbol{E}_{p'}|^2 \quad (0 < \zeta_{pp'}) \tag{3.81}$$

The self-contribution on the right-hand side of the above expression, $-\zeta_{pp}|\boldsymbol{E}_p|^2$, is given by the right-hand side of Eq. (3.80) with ω and $|\boldsymbol{E}|^2$ replaced by ω_p and $|\boldsymbol{E}_p|^2$. The cross contribution by the frequency component $p'(\neq p)$ on the right-hand side of the above expression is $-\zeta_{pp'}|\boldsymbol{E}_{p'}|^2$. We can now compare the magnitudes of the coefficients ζ_{pp} and $\zeta_{pp'}\,(p'\neq p)$. In $\sum_{qq'q''}$ of $\rho^{(3)}_{nm}$, only two terms contribute to the self-saturation coefficient ζ_{pp} describing the saturation of gain for the frequency component p due to the same frequency component p. On the other hand, there are four terms that contribute to the cross saturation coefficient $\zeta_{pp'}$ describing the saturation of gain for the frequency component p due to the

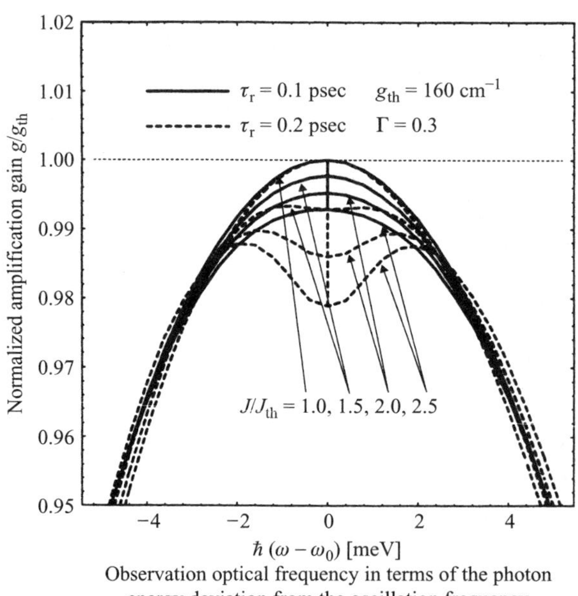

Figure 3.16 Example of calculated spectrum hole burning by gain saturation in GaAs.

other frequency component p'. This suggests that $\zeta_{pp'}$ is roughly twice ζ_{pp}. It has been shown that $\zeta_{pp} < \zeta_{pp'}(p' \neq p)$ although the difference in the spatial distributions of the fields reduces the magnitude of ζ_{pp}.

Consider next a case where a laser is oscillating in a single mode of frequency ω_L corresponding to the peak of the small signal gain spectrum $g^{(1)}(\hbar\omega)$, and let $|\boldsymbol{E}|^2$ be the square of the electric field amplitude of the oscillating mode. Then, without taking the summation $\sum$ in Eq. (3.81), we simply have $g^{(3)}(\hbar\omega_L) = -\zeta|\boldsymbol{E}|^2$ for $\omega = \omega_L$, $g^{(3)}(\hbar\omega) = -\zeta'|\boldsymbol{E}|^2$ for $\omega \neq \omega_L$, and $0 < \zeta < \zeta'$. The gain spectrum for the frequency ω at and near the oscillating frequency ω_L is given by $g(\hbar\omega) = g^{(1)}(\hbar\omega) + g^{(3)}(\hbar\omega)$. For the oscillating frequency ω_L, the gain is reduced to $g^{(1)}(\hbar\omega_L) - \zeta|\boldsymbol{E}|^2$ owing to the saturation, and this reduced value is kept at the threshold gain g_{th} to maintain the oscillation. In this situation, the gain for frequency is reduced even more to $g^{(1)}(\hbar\omega) - \zeta'|\boldsymbol{E}|^2$. Accordingly, the gain spectrum is such that the gain equals the threshold gain g_{th} at the lasing frequency and is lower than g_{th} in the vicinity of it, as illustrated by the result of the calculation shown in Fig. 3.16. This means that hole burning is caused in the gain spectrum. For cases where a laser is oscillating in several modes, the nonlinear gain is given by Eq. (3.81) and $\zeta_{pp} < \zeta_{pp'}$ $(p' \neq p)$. From this, we see that, once a mode starts to oscillate, the oscillating mode gives rise to larger saturation of gain for other modes than for the oscillation mode itself. The multimode oscillation characteristics affected by the interaction between modes will be discussed in Chap. 6.

REFERENCES

1. C. Kittel, *Introduction to Solid State Physics*, John Wiley, New York (1986).
2. F. H. Pollak, C. W. Higginbotham, and M. Cardona *J. Phys. Soc. Japan* **21**, Suppl., 20 (1966).
3. E. O. Kane, *J. Phys. Chem. Solids*, **1**, 249 (1957).
4. M. G. A. Bernard and G. Duraffourg, *Phys. Status Solidi*, **1**, 699 (1961).
5. G. Lasher and F. Stern, *Phys. Rev.*, **133**, A553 (1964).
6. E. O. Kane, *Phys. Rev.*, **131**, 79 (1963).
7. H. C. Casey, Jr, and M. B. Panish, *Heterostructive Lasers*, Academic Press, New York (1978).
8. B. I. Halperin and M. Lax, *Phys. Rev.*, **148**, 722 (1966).
9. F. Stern, *Phys. Rev.*, **148**, 186 (1966).
10. L. A. Coldren and S. W. Carzine, *Diode Lasers and Photonic Integrated Circuits*, John Wiley, New York (1995).
11. F. Stern, *Phys. Rev. B* **B3**, 2636 (1971).
12. F. Stern, *J. Appl. Phys.*, **47**, 5382 (1976).
13. N. K. Dutta, *J. Appl. Phys.*, **52**, 55 (1981).

14. W. B. Joyce and R. W. Dixon, *Appl. Phys. Lett.*, **31**, 354 (1977).
15. N. K. Dutta and R. J. Nelson, *J. Appl. Phys.*, **53**, 74 (1982).
16. G. P. Agrawal and N. K. Dutta, *Semiconductor Lasers*, second edition, Van Nostrand Reihold, New York (1993).
17. M. Yamada and Y. Suematsu, *J. Appl. Phys.*, **52**, 2653 (1981).
18. Y. Suematsu (ed.), *Semiconductor Lasers and Optical Integrated Circuits* (in Japanese), Ohmsha, Tokyo (1987).
19. M. Yamanishi and Y. Lee, *IEEE J. Quantum Electron.*, **QE-23**, 367 (1987).
20. W. W. Chow, S. W. Koch, and M. Sergeant III, *Semiconductor Laser Physics*, Springer, Berlin (1993).

4

Stimulated Emission in Quantum Well Structures

In semiconductor microstructures of nanometer size, the behavior of electrons is strongly affected by the quantum nature of the electron and exhibits a remarkable dependence on the parameters specifying the structure. Therefore, by appropriate design of the structure parameters, one can implement artificial novel electronic properties unlike the intrinsic characteristics of the bulk materials [1–4]. Recent advanced techniques for crystal growth have enabled precise fabrication of such the quantum structures, and quantum structures have offered very effective and attractive possibilities for improvement in semiconductor laser performances [5,6]. This chapter presents the fundamental theory of the quantum well (QW) as the most important quantum structure and the optical amplification by stimulated emission in it.

4.1 ELECTRON STATE IN QUANTUM WELL STRUCTURES

4.1.1 Formation of a Quantum Well Structure

The de Broglie wavelength of electrons in a very thin (less than a few tens of nanometers) semiconductor film structure is comparable with the thickness of the film. In such a structure, electrons exhibit interesting electric and optical characteristics dissimilar to those in bulk semiconductors and ordinary double heterostructures (DHs). The most fundamental semiconductor quantum heterostructure is a single quantum well (SQW), which consists of a very thin layer of a semiconductor sandwiched between two layers of a semiconductor having a bandgap energy larger than that of the thin layer. The conduction- and valence-band edges of this structure form potential wells, as shown in Fig. 4.1. The region of the thin layer is called a well, and the regions of the side layers are called barriers. Let E_{gW} and E_{gB}

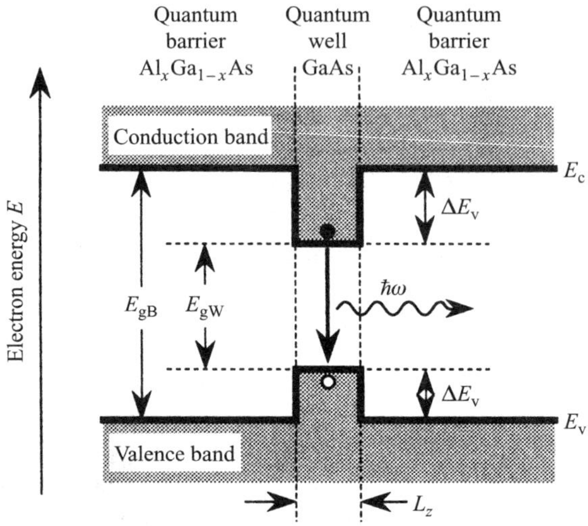

Figure 4.1 Potential energy of conduction and valence bands in a quantum well structure.

be the bandgap energies of the well and barrier regions, respectively, and $\Delta E_g = E_{gB} - E_{gW}$ be the difference between them. The potential well is formed in such a manner that ΔE_g is divided into the potential discontinuity (band offset) for the conduction-band edge ΔE_c and that for the valence-band edge ΔE_v, as shown in Fig. 4.1. The ratio of the band offset has been experimentally determined [7,8] as

$$\frac{\Delta E_c}{\Delta E_g} = 0.67 \pm 0.01 \tag{4.1a}$$

$$\frac{\Delta E_v}{\Delta E_g} = 0.33 \pm 0.01 \tag{4.1b}$$

for GaAs–$Al_xGa_{1-x}As$ quantum wells, and

$$\frac{\Delta E_c}{\Delta E_g} = 0.39 \pm 0.01 \tag{4.2a}$$

$$\frac{\Delta E_v}{\Delta E_g} = 0.61 \pm 0.01 \tag{4.2b}$$

for InP–$In_{1-x}Ga_xAs_yP_{1-y}$ quantum wells. The electrons and holes are confined in the well (in the thin layer) by the potential distribution. A structure consisting of several quantum wells stacked with thick barriers, where coupling between wells is negligible, is called a multiple-quantum-well (MQW) structure. On the other hand, a structure consisting of many alternatively stacked thin wells and thin barriers, where the wells are coupled each other, is called a superlattice (SL).

4.1.2 Quantum Confinement Effect

Consider a SQW structure, consisting of a thin layer of a semiconductor crystal W sandwiched between two regions of semiconductor crystal B having a bandgap energy larger than that of W. We take the z coordinate along the direction perpendicular to the layer, in such a manner that the potential well of width (thickness of the layer) L_z is positioned in the region $-L_z/2 < z < L_z/2$. Let $V_c(z)$ be the energy level of the conduction band-edge. Using the effective-mass approximation, the wave function ψ of an electron in a semiconductor crystal can be written in an expansion form by the Wannier function (Fourier transform of Bloch function) $W(\boldsymbol{r}-\boldsymbol{r}_j)$:

$$\psi(\boldsymbol{r}) = \sum_j F(\boldsymbol{r}_j) W(\boldsymbol{r}-\boldsymbol{r}_j) \tag{4.3}$$

where $F(\boldsymbol{r})$ is an envelope function for the electron wave function. The function $F(\boldsymbol{r})$ satisfies an effective-mass equation in the same form as the Schrödinger equation:

$$\left(-\frac{\hbar^2}{2m^*(z)}\nabla^2 + V_c(z)\right)F(\boldsymbol{r}) = EF(\boldsymbol{r}) \tag{4.4}$$

where $\boldsymbol{r}_j$ denotes each lattice point, $m^*(z)$ describes the effective mass of electron in each region, and E denotes the electron energy. Noting that m^* and V_c are functions of z only, we put

$$F(\boldsymbol{r}) = \exp(\mathrm{i}\boldsymbol{k}_{xy}\cdot\boldsymbol{r})\,\chi(z) \tag{4.5}$$

using a wave vector $\boldsymbol{k}_{xy}$ in the plane of the well (x–y plane). Then, from Eqs (4.4) and (4.5) we obtain an equation for $\chi(z)$:

$$\left(-\frac{\hbar^2}{2m^*(z)}\frac{\mathrm{d}^2}{\mathrm{d}z^2} + V_c(z)\right)\chi(z) = \left(E - \frac{\hbar^2 k^2}{2m^*(z)}\right)\chi(z) \tag{4.6}$$

where

$$k = |\boldsymbol{k}_{xy}|$$

The electron wave function ψ, given by Eqs (4.3) and (4.5) with the solution of the above equation $\chi(z)$, can be rewritten as an approximate expression using the Bloch function:

$$\psi(\boldsymbol{r}) = \exp(\mathrm{i}\boldsymbol{k}_{xy}\cdot\boldsymbol{r})\,u_0(\boldsymbol{r})\chi(z) \tag{4.7}$$

Thus the wave function ψ is given by the Bloch functions in each region with their amplitudes modulated by the envelope function $\chi(z)$ describing the distribution along the direction perpendicular to the plane of the well.

We consider electrons in and around the quantum well such that $|\chi(z)|$ is large for $|z| < L_z/2$ and is small for $|z| > L_z/2$. Assuming that the difference between the effective mass m^*_{W} in the well and the effective mass m^*_{B} in the barrier and is small, we can approximate $m^*(z)$ on the right-hand side of Eq. (4.6) as m^*_{W}. Thus Eq. (4.6) is rewritten as

$$\left(-\frac{\hbar^2}{2m^*(z)}\frac{\mathrm{d}^2}{\mathrm{d}z^2} + V_{\mathrm{c}}(z)\right)\chi(z) = \varepsilon\chi(z) \tag{4.8a}$$

$$E = \frac{\hbar^2 k^2}{2m^*_{\mathrm{W}}} + \varepsilon \qquad (k = |\boldsymbol{k}_{xy}|) \tag{4.8b}$$

Since the functions $m^*(z)$ and $V_{\mathrm{c}}(z)$ on the left hand side of the above equation have discontinuity at $z = \pm L_z/2$, the solution $\chi(z)$ must be obtained in each region and be connected. From the continuity of the wave function ψ, continuity of $\chi(z)$ is required at the boundary. Since the electric charge flow across the boundary is continuous, the z component $J_z = e\,\mathrm{Re}\{\chi^*(-\mathrm{i}\hbar/m^*)\,\mathrm{d}\chi/\mathrm{d}z\}$ of the charge flow density vector $J = e\,\mathrm{Re}\{\psi^* \boldsymbol{v}\psi\}$, must be continuous, and therefore continuity of $(1/m^*)\,\mathrm{d}\chi/\mathrm{d}z$ is required.

We next derive solutions of the eigenequation in Eq. (4.8a). A SQW of depth ΔE can be described by

$$\mathrm{V_c}(z) = \begin{cases} 0 & \left(|z| < \dfrac{L_z}{2}\right) \\ \Delta E & \left(|z| > \dfrac{L_z}{2}\right) \end{cases} \tag{4.9}$$

The energy is described with reference to the bottom of the well. Then, from the symmetry of $m^*(z)$ and $V_{\mathrm{c}}(z)$, the solutions $\chi(z)$ are even or odd functions of z, and they can be written as

$$\chi(z) = \begin{cases} A\cos\left(\dfrac{\kappa L_z}{2}\right)\exp\left[-\gamma\left(z - \dfrac{L_z}{2}\right)\right] & \left(z > +\dfrac{L_z}{2}\right) \\ A\cos(\kappa z) & \left(|z| < \dfrac{L_z}{2}\right) \\ A\cos\left(\dfrac{\kappa L_z}{2}\right)\exp\left[+\gamma\left(z + \dfrac{L_z}{2}\right)\right] & \left(z < -\dfrac{L_z}{2}\right) \end{cases} \tag{4.10a}$$

or

$$\chi(z) = \begin{cases} A\sin\left(\frac{\kappa L_z}{2}\right)\exp\left[-\gamma\left(z-\frac{L_z}{2}\right)\right] & \left(z>+\frac{L_z}{2}\right) \\ A\sin(\kappa z) & \left(|z|<\frac{L_z}{2}\right) \\ -A\sin\left(\frac{\kappa L_z}{2}\right)\exp\left[+\gamma\left(z+\frac{L_z}{2}\right)\right] & \left(z<-\frac{L_z}{2}\right) \end{cases} \tag{4.10b}$$

where $\kappa=(2m_{\mathrm{W}}^{*}\varepsilon)^{1/2}/\hbar$ and $g=[2m_{\mathrm{B}}^{*}(\Delta E-\varepsilon)]^{1/2}/\hbar$. The above equations show that the envelope function $\chi(z)$ exhibits oscillatory variation in the well and decays monotonically with increasing distance from the well in the barriers; i.e., the electron wave is confined in and around the well as shown in Fig. 4.2. From the boundary conditions, we obtain the characteristic equations

$$\kappa\tan\left(\frac{\kappa L_z}{2}\right)=\frac{m_{\mathrm{W}}^{*}}{m_{\mathrm{B}}^{*}}\gamma \tag{4.11a}$$

$$-\kappa\cot\left(\frac{\kappa L_z}{2}\right)=\frac{m_{\mathrm{W}}^{*}}{m_{\mathrm{B}}^{*}}\gamma \tag{4.11b}$$

for determining the eigenvalue ε. These equations can be graphically solved as shown in Fig. 4.3. Putting $\nu=\kappa L_z/2=(2m_{\mathrm{W}}^{*}\varepsilon)^{1/2}L_z/2\hbar$, the above equations are rewritten as

$$\nu\tan\nu=(m_{\mathrm{W}}^{*}/m_{\mathrm{B}}^{*})^{1/2}\left\{(m_{\mathrm{W}}^{*}\Delta E L_z^2/2\hbar^2)-\nu^2\right\}^{1/2} \tag{4.12a}$$

$$-\nu\cot\nu=(m_{\mathrm{W}}^{*}/m_{\mathrm{B}}^{*})^{1/2}\left\{(m_{\mathrm{W}}^{*}\Delta E L_z^2/2\hbar^2)-\nu^2\right\}^{1/2} \tag{4.12b}$$

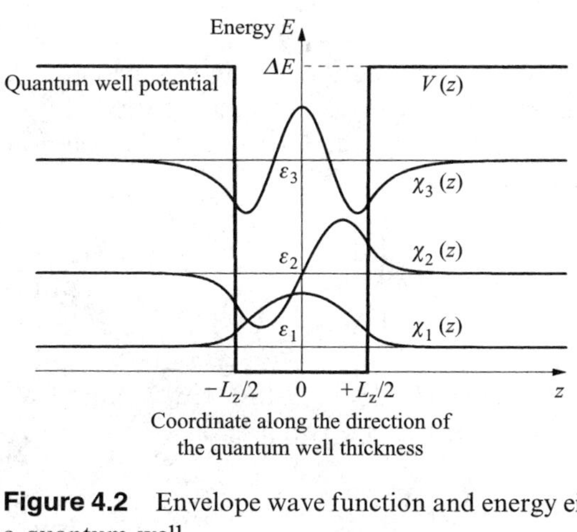

Figure 4.2 Envelope wave function and energy eigenvalues of electrons confined in a quantum well.

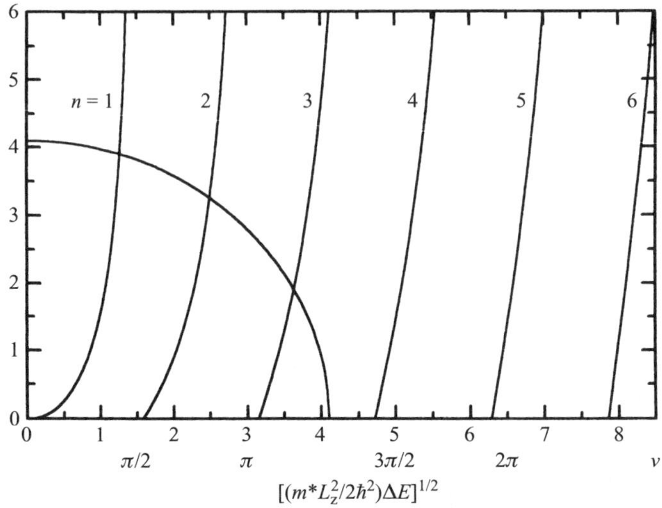

Figure 4.3 Graphic solution of the eigenvalue equation in Eq. (4.11).

By plotting graphs for the right- and left-hand sides of the above equations as functions of ν, the roots of ν are given by the values of the abscissa of the cross points. The graph shows that the number of the roots is $1+[(2m_{\mathrm{W}}^{*}\,\Delta E\,L_z^2/\pi^2\hbar^2)^{1/2}]$, and there is at least one root. Denoting the roots as $\nu_n\,(n=1,2,\ldots)$, the eigenvalues ε are given by

$$\varepsilon=\varepsilon_n=\frac{\hbar^2}{2m_{\mathrm{W}}^{*}}\left(\frac{2\nu_n}{L_z}\right)^2\qquad(n=1,2,\ldots)\tag{4.13}$$

The relations between the depth of the well and the energy eigenvalues are plotted in Fig. 4.4 using normalized parameters, where the approximation $m_{\mathrm{B}}^{*}\approx m_{\mathrm{W}}^{*}=m^{*}$ is used. For an infinitely deep well ($\Delta E\to\infty$), the solution of the eigenequation in Eq. (4.8a) is given by

$$\chi(z)=\begin{cases}A\cos(\kappa z)\\0\end{cases}\quad\text{or}\quad\begin{cases}A\,\sin(\kappa z)&\left(|z|<\dfrac{L_z}{2}\right)\\[2ex]0&\left(|z|>\dfrac{L_z}{2}\right)\end{cases}\tag{4.14}$$

From the boundary conditions we have $\kappa L_z/2=n\pi/2\ (n=1,2,\ldots)$, and we obtain a simple analytical expression for the eigenvalues:

$$\varepsilon=\varepsilon_n=\frac{\hbar^2}{2m_{\mathrm{W}}^{*}}\left(\frac{n\pi}{L_z}\right)^2\qquad(n=1,2,\ldots)\tag{4.15}$$

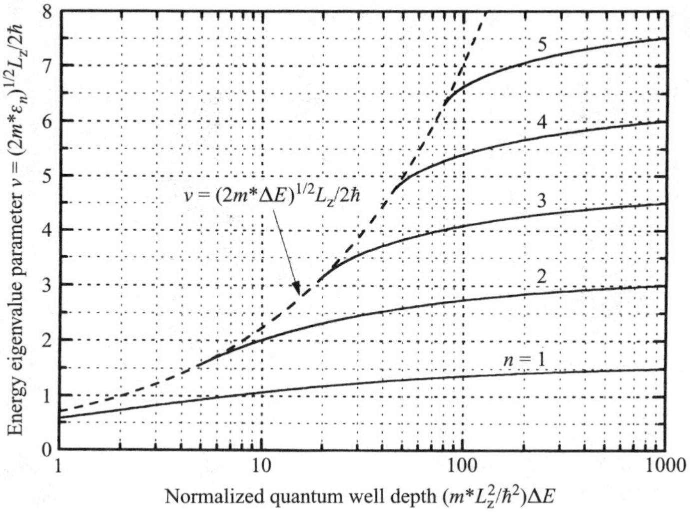

Figure 4.4 Relations between quantum well depth (or quantum well width) and energy eigenvalues.

4.1.3 Two-Dimensional Electron Gas

Although electrons in a quantum well can move with arbitrary momentum along the direction parallel to the plane of the well, they are confined in the well without freedom of motion along the direction perpendicular to the well. Unlike electrons in a bulk semiconductor (including a thick DH) which have three-dimensional freedom, electrons in a quantum well have only two-dimensional freedom. These electrons of two-dimensional freedom are called a two-dimensional electron gas (2DEG). There is similarity between the electron wave confinement in a quantum well and optical wave confinement in a dielectric thin-film optical waveguide. A quantum well may be considered as a waveguide for an electron wave.

Equations (4.8b) and (4.13) show that the energy of the 2DEG can be written in a form of the sum of the discrete eigenvalue ε_n concerning the confinement in the well and the kinetic energy concerning the motion along the plane of the well:

$$\begin{aligned} E &= \varepsilon_n + \frac{\hbar^2 k^2}{2m_W^*} \\ &= \varepsilon_n + \frac{\hbar^2}{2m_W^*}(k_x^2 + k_y^2) \end{aligned} \tag{4.16}$$

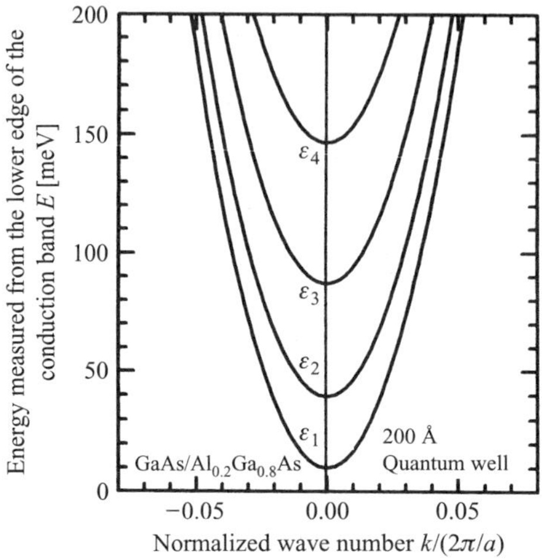

Figure 4.5 Subband structure (energy dispersion) of conduction-band electrons in a quantum well.

The conduction band can be described by the energy dispersion concerning the momentum in the plane of the well for each quantum number n concerning the confinement. Accordingly, the conduction band is divided into states described by dispersion curves for each quantum number n (Fig. 4.5). Each set of electron states labeled by a quantum number is called a subband.

4.1.4 Holes in a Quantum Well

In a similar manner to electrons in the conduction band, holes in the valence band are also confined in the well and subbands are formed. The valence band, however, consist of a heavy-hole (HH) band, a light-hole (LH) band, and a split-off band as shown in Fig. 3.1, and it cannot be described by a single effective mass and a single dispersion. Accordingly, the subband structure for holes in a quantum well is even more complicated. An example of calculated results is shown in Fig. 4.6. Consider first the upper edge of the valence band ($\boldsymbol{k} = \boldsymbol{0}$). Whereas in a bulk semiconductor, HHs and LHs are degenerate and have the same energy E_v, in a quantum well the degeneracy is removed by the perturbation caused by the well potential and the effective-mass difference. When the well potential is approximated by a potential with infinitely high barriers, the confinement

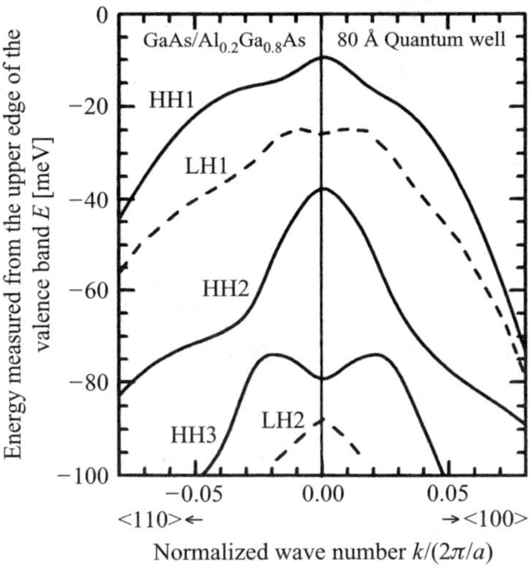

Figure 4.6 Subband structure (energy dispersion) of valence-band holes in a quantum well [9].

energy eigenvalues for heavy holes and light holes at the upper edge of the valence band ($\boldsymbol{k} = \boldsymbol{0}$) are given by

$$\varepsilon_{n,\mathrm{hh}} = \frac{\hbar^2}{2m_{\mathrm{hh}}^{*z}}\left(\frac{\pi n}{L_z}\right)^2 \tag{4.17a}$$

$$\varepsilon_{n,\mathrm{lh}} = \frac{\hbar^2}{2m_{\mathrm{lh}}^{*z}}\left(\frac{\pi n}{L_z}\right)^2 \tag{4.17b}$$

which are in the same form as Eq. (4.15). The envelope wave function $\chi(z)$ is also given by a formula similar to Eq. (4.14). The effective masses m_{hh}^{*z} and m_{lh}^{*z} for the z direction, however, take values different from those of the bulk HH effective mass m_{hh}^{*} and the bulk LH effective mass, although $m_{\mathrm{hh}}^{*z} > m_{\mathrm{lh}}^{*z}$ holds. Although holes of eigenvalues $\varepsilon_{n,\mathrm{hh}}$ and $\varepsilon_{n,\mathrm{lh}}$ in a quantum well are also called heavy holes and light holes, respectively, this is only for convenience and their characters are different from those of HHs and LHs in bulk semiconductor. Consider next the dispersion with respect to the momentum along the plane of quantum well (wave vector $\boldsymbol{k}_{xy}$). The dispersion is affected by the band mixing owing to the perturbation caused by the well potential and exhibits singular behavior significantly different from that in a bulk semiconductor. Although the dispersion in the region where $|\boldsymbol{k}_{xy}|$ is small can be written in the same form as Eq. (4.16), i.e.,

$$E_{\mathrm{hh}} = -\varepsilon_{n,\mathrm{hh}} - \frac{\hbar^2 k^2}{2m^{*xy}_{n,\mathrm{hh}}} \qquad (k = |\boldsymbol{k}_{xy}|) \tag{4.18a}$$

$$E_{\mathrm{lh}} = -\varepsilon_{n,\mathrm{lh}} - \frac{\hbar^2 k^2}{2m^{*xy}_{n,\mathrm{lh}}} \qquad (k = |\boldsymbol{k}_{xy}|) \tag{4.18b}$$

Not only are the values of m^{*xy}_{hh} and m^{*xy}_{lh} different from those of m^{*z}_{hh} and m^{*z}_{lh} (anisotropy of effective mass), but also the order of magnitude is inverted as $m^{*xy}_{\mathrm{hh}} < m^{*xy}_{\mathrm{lh}}$. In addition, in the region where $|\boldsymbol{k}_{xy}|$ is large, the dispersion deviates significantly from the parabolic function (nonparabolic band). The singularity also include a negative effective mass along the plane of the well for some holes and the associated electron-like behavior.

4.1.5 Density of States

An important difference between electrons in a quantum well and electrons in a bulk semiconductor appears not only in the confinement of the wave function in the well but also in the density of states.

As is well known, the (three-dimensional) density $\rho_{3\mathrm{D}}(E)$ of states per unit volume for electrons in a bulk semiconductor is given by a parabolic function:

$$\rho_{3\mathrm{D}}(E) = \frac{1}{2\pi^2\hbar^3}(2m^*)^{3/2}E^{1/2} \tag{4.19}$$

where the energy E is measured with reference to the lower edge of the conduction band. Here we derive expressions for the density of states for electrons in a quantum well. Consider first a subband (the nth subband) and neglect the nonparabolicity. Then the dispersion is given by

$$E = \varepsilon_n + \frac{\hbar^2 k^2}{2m^*} \qquad (k = |\boldsymbol{k}_{xy}|) \tag{4.20}$$

and therefore $E > \varepsilon_n$, and the density of states is $\rho_{2\mathrm{D}}(E) = 0$ for $E < \varepsilon_n$. Considering a square region of side length L in the plane of the quantum well (x–y plane), and applying the periodic boundary condition, the wave vector in the plane is limited to values given by $\boldsymbol{k}_{xy} = (2\pi n_x/L,\ 2\pi n_y/L)$ (n_x, $n_y = 0, \pm 1, \pm 2, \ldots$), and therefore each electron state occupies an area of $(2\pi/L)^2$ in the two-dimensional $\boldsymbol{k}_{xy}$ plane. Therefore, for $E > \varepsilon_n$ the number of states of energy between E and $E + \mathrm{d}E$ per unit area is given by

$$\begin{aligned}\rho_{2D}(E)\,\mathrm{d}E &= \frac{2[2\pi k\,\mathrm{d}k/(2\pi/L)^2]}{L^2} \\ &= \frac{k\,\mathrm{d}k}{\pi} \\ &= \frac{m^*}{\pi\hbar^2}\,\mathrm{d}E \qquad (E > \varepsilon_n)\end{aligned} \tag{4.21}$$

where spin is taken into account. This result shows that the density $\rho_{2D}(E)$ of states per unit area, does not depend upon E and takes a constant value $m^*/\pi\hbar^2$. Since the total density of states is given by the sum of the density of states for each subband, it can be written by using the unit step function $u(\xi)$ ($u(\xi)=0$ $(\xi<0)$; $u(\xi)=1$ $(0<\xi)$) as

$$\rho_{2D}(E) = \frac{m^*}{\pi\hbar^2}\sum_n u(E - \varepsilon_n) \tag{4.22}$$

Thus the density of states is given by a multistep function having steps at positions of eigenvalues ε_n, as shown in Fig. 4.7. When the quantum well is approximated by a well with infinitely high barriers, the eigenvalues ε_n are given by Eq. (4.15). Then the values of $\rho_{2D}(E)$ in Eq. (4.22) at the subband edges $E=\varepsilon_n$ coincide with the values of the three-dimensional density $\rho_{3D}(E)$ of states given by Eq. (4.19), multiplied by the well thickness L_z to convert it into the density $\rho_{3D}(E)L_z$ per unit area.

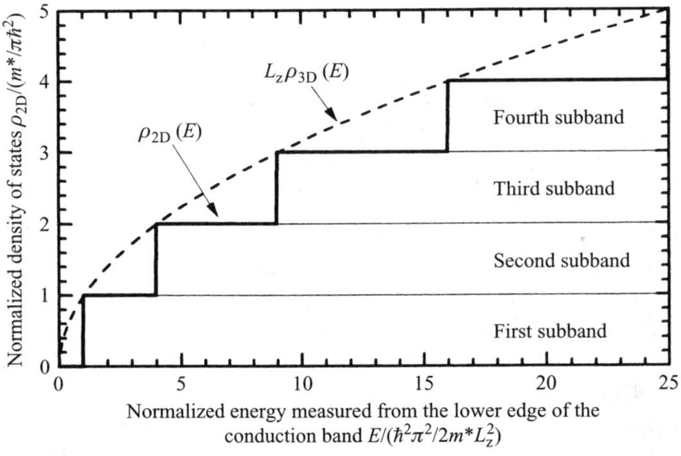

Figure 4.7 Density of states of two-dimensional electrons compared with that of three-dimensional electrons.

Now let us compare the density $\rho_{2D}(E)$of states for a quantum well with the density $\rho_{3D}(E)$ of states for the bulk. While the bulk band edge is at $E=0$, the band edge for a quantum well is at $E=\varepsilon_1$ raised by the minimum eigenvalue ε_1. While the bulk density of states vanishes as $\rho_{3D}(E)\to 0$ for the band edge $E\to +0$, the density of states of the quantum well does not vanish but remains as $\rho_{2D}(E)\to m^*/\pi\hbar^2$ for the band edge $E\to \varepsilon_1+0$. These differences in the density of states are fundamental and important characteristics that explain the properties of a quantum well.

Electrons in a quantum well are confined by a one-dimensional potential well and have two-dimensional freedom. When electrons are confined in a two-dimensional potential well, the freedom of the electrons is reduced to one-dimensional freedom. In a three-dimensional potential well, electrons lose spatial freedom (zero-dimensional freedom). Structures realizing such two- and three-dimensional potential wells are called a quantum wire and a quantum dot, respectively. The densities of states for one- and zero-dimensional electrons confined in a quantum wire or a quantum dot are given by

$$\rho_{1D}(E)=\frac{2}{\pi\hbar}\left(\frac{m^*}{2}\right)^{1/2}\sum_{nn'}(E-\varepsilon_{nn'})^{-1/2} \tag{4.23}$$

and

$$\rho_{0D}(E)=2\sum_{nn'n''}\delta(E-\varepsilon_{nn'n''}) \tag{4.24}$$

respectively (see Appendix 2). These densities are peak-like functions, and the lower-dimensional electrons exhibit a behavior even more singular than that of two-dimensional electrons.

4.2 DIRECT-TRANSITION MODEL

4.2.1 Expressions for Amplification Gain and Spontaneous Emission

The optical amplification and absorption by interband transition in a quantum well can be analyzed by using the direct-transition model in a similar manner to that described in Section 3.2. Consider first transitions between a subband in the valence band and a subband in the conduction band. In a similar manner to Eq. (3.16), the factors of amplification and absorption by this transition can be written as

$$\begin{aligned} g(\hbar\omega) &= -\alpha(\hbar\omega) \\ &= \frac{\pi e^2}{n_r c\varepsilon_0 m^2\omega}|M|^2\frac{(f_2-f_1)\rho_{2Dr}(\hbar\omega)}{L_z} \end{aligned} \tag{4.25}$$

where $\rho_{2\mathrm{Dr}}(\hbar\omega)$ is the two-dimensional reduced density of states defined by

$$\rho_{2\mathrm{Dr}}(\hbar\omega) = \frac{1}{\pi^2}\int \delta(E_1 + \hbar\omega - E_2)\,\mathrm{d}\boldsymbol{k}_{xy} \tag{4.26}$$

In order to convert the reduced density $\rho_{2\mathrm{Dr}}$ of states per unit area into the density per unit volume, the well width L_z is inserted at the end of the left-hand side of Eq. (4.25). An important difference between the expressions for a bulk semiconductor (Eqs (3.16) and (3.17)) and the expressions for a quantum well (Eqs (4.25) and (4.26)) is that the former are given by a volume integral with respect to a three-dimensional wave vector $\boldsymbol{k}$, while the latter are given by a area integral with respect to a two-dimensional wave vector $\boldsymbol{k}_{xy}$.

Consider also spontaneous emission due to the direct transition of electrons from a subband in the conduction band to a subband in the valence band. Then, in a similar manner as Eq. (3.20), the rate of spontaneous emission, per unit volume per unit time, of photons of frequency within the range from ω to $\omega + \mathrm{d}\omega$ can be written as

$$\gamma_{\mathrm{spt}}(\hbar\omega)\,\mathrm{d}\omega = \frac{n_{\mathrm{r}}e^2\omega}{\pi m^2 c^3 \varepsilon_0}\,\mathrm{d}\omega |M|^2 \frac{f_2(1-f_1)\rho_{2\mathrm{Dr}}(\hbar\omega)}{L_z} \tag{4.27}$$

4.2.2 Transition Matrix Element

The matrix element for an interband transition in a quantum well can be calculated by using a definition similar to Eq. (3.15), namely,

$$\begin{aligned} |M|^2 &= |\boldsymbol{e}\langle\psi_2|\boldsymbol{p}|\psi_1\rangle|^2 \\ &= \left(\frac{m\omega}{e}\right)^2 |\boldsymbol{e}\langle\psi_2|e\boldsymbol{r}|\psi_1\rangle|^2 \end{aligned} \tag{4.28}$$

and substituting the electron wave function ψ_2 and hole wave function ψ_1 into Eq. (4.28). Since the wave functions ψ in the well can be written as Eq. (4.7) using the normalized envelope wave function $\chi(z)$, the above expression is written as

$$\begin{aligned} |M|^2 &= (m\omega)^2 \left|\boldsymbol{e}\langle u^{\mathrm{e}}(\boldsymbol{r})\chi^{\mathrm{e}}(z)|\boldsymbol{r}|u^{\mathrm{h}}(\boldsymbol{r})\chi^{\mathrm{h}}(z)\rangle\right|^2 \\ &\approx (m\omega)^2 \left|\boldsymbol{e}\langle u^{\mathrm{e}}(\boldsymbol{r})|\boldsymbol{r}|u^{\mathrm{h}}(\boldsymbol{r})\rangle \int \chi^{\mathrm{e}*}(z)\chi^{\mathrm{h}}(z)\mathrm{d}z\right|^2 \end{aligned} \tag{4.29}$$

where the superscripts e and h denote electrons and holes, respectively. In the above expression, the in-plane wave vector for electrons and holes are assumed to be equal ($\boldsymbol{k}_{xy}^{\mathrm{e}} = \boldsymbol{k}_{xy}^{\mathrm{h}}$), since $|M|^2 = 0$ for $\boldsymbol{k}_{xy}^{\mathrm{e}} \neq \boldsymbol{k}_{xy}^{\mathrm{h}}$. Because of the orthonormality of the envelope wave function (although the orthogonality holds rigorously for a well of infinite barrier height, it is an approximate relation for a realistic well of finite barrier height), the factor $\int \chi^{\mathrm{e}*}(z)\chi^{\mathrm{h}}(z)\,\mathrm{d}z$ on the right-hand side of the above expression equals 1 for the case where $n^{\mathrm{e}} = n^{\mathrm{h}}$ and equals approximately 0 for the cases where $n^{\mathrm{e}} \neq n^{\mathrm{h}}$, where n^{e} and n^{h} are the quantum numbers of the electron and hole confinements respectively, with respect to the z direction. Therefore, only transitions which satisfy the in-plane wave vector conservation rule $\boldsymbol{k}_{xy}^{\mathrm{e}} = \boldsymbol{k}_{xy}^{\mathrm{h}}$ and the quantum number conservation rule ($\Delta n = 0$ selection rule) occur, and transitions with different quantum numbers are inhibited.

In Eq. (4.29), $\boldsymbol{e}$ is a unit vector describing the direction of the optical electric field vector (polarization), and $\langle u^{\mathrm{e}}(\boldsymbol{r})|\boldsymbol{r}|u^{\mathrm{h}}(\boldsymbol{r})\rangle$ is a matrix element which can be calculated by using the Bloch functions for electrons and holes in the well. For a bulk semiconductor, $|M|^2$ is polarization independent and is given by the averaged Bloch state matrix element (Eq. (3.29)), which is the value derived from the $\boldsymbol{kp}$ perturbation theory

$$|M|^2 = \frac{m^2 E_{\mathrm{g}}(E_{\mathrm{g}} + \Delta)}{2m_n(E_{\mathrm{g}} + 2\Delta/3)} \tag{4.30}$$

multiplied by the factor $\frac{1}{3}$ resulting from the averaging $\ll\gg$ over all directions for the hole wave function and the factor $\frac{1}{2}$ resulting from the spin selection rule. For a quantum well, however, such averaging does not apply because of the anisotropy of $u_{\mathrm{h}}(\boldsymbol{r})$. It has been shown through a detailed analysis that $|M|^2$ of a quantum well depends not only on polarization but also on $\boldsymbol{k}_{xy}$ because of the hole band mixing. The values of $|M|^2$ at and near the band edge $\boldsymbol{k}_{xy} \approx \boldsymbol{0}$ are as follows [10]:

For polarization parallel to the plane of well ($e \| \mathrm{QW}$), i.e., TE wave:

1/2 times for electron – heavy hole transition
1/6 times for electron – light hole transition

For polarization perpendicular to the plane of well ($\boldsymbol{e} \perp \mathrm{QW}$), i.e., TM wave:

0 times times for electron – heavy hole transition
2/3 times for electron – light hole transition

of the value given by Eq. (4.30). They must be multiplied by a factor $\frac{1}{2}$ resulting from the spin selection rule.

4.2.3 Reduced Density of States

Under the assumption that the parabolic approximation applies for the electron and hole dispersions as in Eqs (4.16) and (4.18) and they can be written as

$$E_e = E_2 = E_{cb} + \varepsilon_{n,e} + \frac{\hbar^2 k^2}{2m_e} \qquad (k = |\boldsymbol{k}_{xy}|) \tag{4.31a}$$

$$E_h = E_1 = E_{vb} - \varepsilon_{n,h} - \frac{\hbar^2 k^2}{2m_h} \qquad (k = |\boldsymbol{k}_{xy}|) \tag{4.31b}$$

the electron and hole energies, satisfying the in-plane wave vector ($\boldsymbol{k}_{xy}$) conservation rule, the quantum number ($\Delta n = 0$) selection rule, and the energy conservation rule $E_1 + \hbar\omega - E_2 = 0$ simultaneously, are

$$E_2 = E_{cb} + \varepsilon_{n,e} + \frac{m_r}{m_e}(\hbar\omega - E_{g,n}) \tag{4.32a}$$

$$E_1 = E_{vb} - \varepsilon_{n,h} - \frac{m_r}{m_h}(\hbar\omega - E_{g,n}) \tag{4.32b}$$

$$\frac{1}{m_r} = \frac{1}{m_e} + \frac{1}{m_h} \tag{4.33}$$

$$E_{g,n} = E_{gb} + \varepsilon_{n,e} + \varepsilon_{n,h} = (E_{cb} - E_{vb}) + \varepsilon_{n,e} + \varepsilon_{n,h} \tag{4.34}$$

where m_r is the reduced effective mass, E_{gb} the bulk bandgap energy, and $E_{g,n}$ the gap energy for the nth subbands of a quantum well. Then the reduced density $\rho_{2Drn}(\hbar\omega)$ of states for the nth subbands can readily be calculated from Eqs (4.26) and (4.31) to yield

$$\rho_{2Drn}(\hbar\omega) = \frac{2m_r}{\pi\hbar^2} u(\hbar\omega - E_{g,n}) \tag{4.35}$$

where u denotes the unit-step function.

For the Fermi–Dirac factors f_2 and f_1 in the expressions for the gain factor and the spontaneous emission (Eqs (4.25) and (4.27)), E_2 and E_1 of Eq. (4.32) should be used for substitution into Eqs (3.6).

4.3 GAIN SPECTRUM AND GAIN FACTOR

4.3.1 Structures of Quantum Well Lasers

Use of quantum well(s) as an active layer allows implementation of quantum well lasers offering performances better than those of ordinary DH lasers. However, as discussed in detail in Chap. 5, mere reduction in the active layer thickness of a DH laser to form a SQW or MQW structure does not lead to realization of high performances, since the resultant quantum well (QW) is too thin in comparison with the optical wavelength to ensure strong optical confinement, and therefore a high effective gain for the guided wave cannot be attained. Another problem is that it is not easy to attain a high efficiency of carrier injection in MQW structures having many heterojunctions, and the carriers injected in the thin QW may leak, thereby reducing the effective carrier injection efficiency. To overcome these problems and to implement high-performance quantum well lasers, various improved structures as shown in Fig. 4.8 have been developed. They are separate-confinement heterostructures (SCHs), consisting of the QW active layer for carrier confinement and a refractive index structure for optical confinement outside it, and their modifications.

4.3.2 Determination of Quasi-Fermi Level

When the nonparabolicity of the band can be omitted, the density (per unit volume) of electrons and holes in a carrier-injected quantum well of width L_z can be written as

$$n = \int \frac{\rho_{2\mathrm{Dc}}(E_2) f_2 \,\mathrm{d}E_2}{L_z} \tag{4.36a}$$

$$p = \int \frac{\rho_{2\mathrm{Dv}}(E_1)(1 - f_1) \,\mathrm{d}E_1}{L_z} \tag{4.36b}$$

where $\rho_{2\mathrm{D}}$ is the two-dimensional density (per unit area) of states in the form of Eq. (4.22), and f_1 and f_2 are the Fermi–Dirac factors of Eqs (3.6) with the quasi-Fermi levels F_c and F_v. Since $\rho_{2\mathrm{D}}$ is constant in this case, the above integration can be carried out analytically to yield

$$n = \frac{m_\mathrm{e} k_\mathrm{B} T}{\pi \hbar^2 L_z} \sum_n \ln\left[\exp\left(\frac{F_\mathrm{c} - E_{\mathrm{c}n}}{k_\mathrm{B} T}\right) + 1\right] \tag{4.37a}$$

$$p = \frac{m_\mathrm{h} k_\mathrm{B} T}{\pi^2 \hbar^2 L_z} \sum_n \ln\left[\exp\left(\frac{E_{\mathrm{v}n} - F_\mathrm{v}}{k_\mathrm{B} T}\right) + 1\right] \tag{4.37b}$$

For electrons, assuming that the contributions by subbands except for the first subbands can be neglected, from the above equation we obtain an expression for F_c:

$$\frac{F_\mathrm{c} - E_{\mathrm{c}1}}{k_\mathrm{B} T} = \ln\left[\exp\left(\frac{\pi \hbar^2 n L_z}{m_\mathrm{e} k_\mathrm{B} T}\right) - 1\right] \tag{4.38}$$

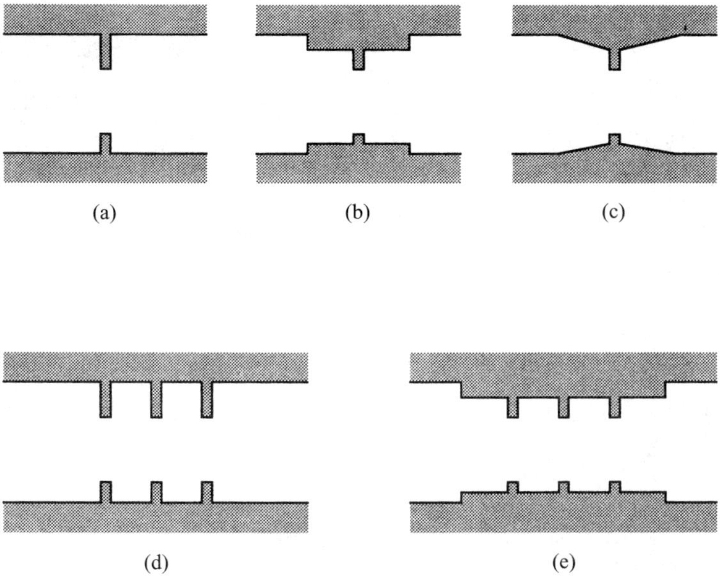

Figure 4.8 Various quantum heterostructures for quantum well lasers: (a) SQW; (b) modified SQW; (c) graded-index separate-confined heterostructure single quantum well (GRIN-SCH-SQW); (d) MQW; (e) modified MQW. The ordinate represents electron energy, and the abscissa is along the direction perpendicular to the junction plane. The upper and lower shaded areas indicate the conduction band and the valence band, respectively.

For holes, Eq. (4.37b) should be used to obtain F_v, since generally contributions by both the heavy hole bands and light hole bands must be considered. When the minority-carrier density is given as independent, the quasi-Fermi levels F_c and F_v can be determined by these expressions and the electric neutrality condition Eq. (3.38). Although the procedure described here applies with good approximation for cases where the carrier density is not very high, calculation for accurate analysis for cases including those of high carrier density requires consideration of nonparabolicity of the bands and contribution of carriers with energy above the quantum well barrier potential.

4.3.3 Characteristics of the Gain Spectrum of Quantum Well

The laser amplification gain under carrier injection into a quantum well structure results from interband transitions satisfying the quantum number

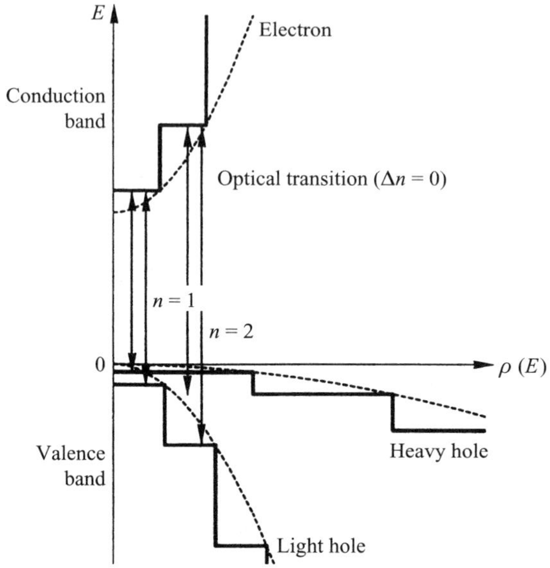

Figure 4.9 Interband optical transition in a quantum well structure.

conservation rule, as shown in Fig. 4.9. Summing the contributions by each subband given by Eq. (4.25), and using the reduced density of states for the parabolic bands, the gain (absorption) spectrum is given by

$$g(\hbar\omega) = -\alpha(\hbar\omega)$$
$$= \frac{\pi e^2}{n_r c \varepsilon_0 m^2 \omega} \times \sum_n |M|^2 (f_2 - f_1) \frac{2m_r}{\pi \hbar^2 L_z} u(\hbar\omega - E_{g,n}) \tag{4.39}$$

where m_r is the reduced effective mass, $E_{g,n}$ is the gap energy of the subbands, u is a unit step function, and Σ_n indicates summation with respect to subbands of confinement quantum number n. For the occupation probabilities f_2 and f_1 for the conduction and valence bands, respectively, the probabilities for the carrier injected state given by Eq. (3.6) with the quasi-Fermi levels and Eq. (4.32) should be used. The quasi-Fermi levels are correlated with the carrier density as described in the previous Section 4.3.2.

Equation (4.39) shows that the laser gain spectrum is formed as a total effect of contributions from all the subbands. Therefore, the gain or absorption resulting from the interband transitions in a quantum well exhibits a spectrum in the form of the product of the step-like reduced density of states rising $E_{g,1}$, with a deviation from the bulk bandgap E_{gb} by

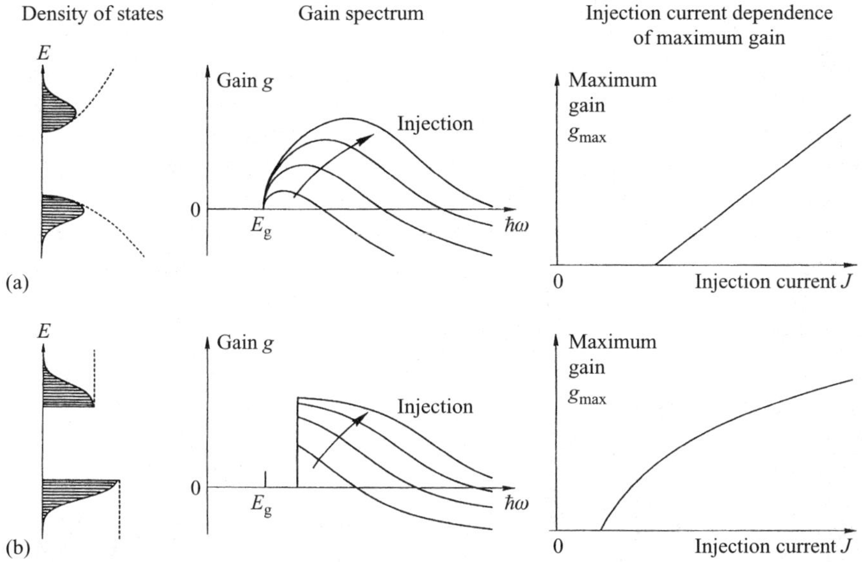

Figure 4.10 Comparison between laser gains in (a) bulk DHs and (b) quantum well structures.

the fundamental eigenvalues $\varepsilon_{1,e} + \varepsilon_{1,h}$, and the population inversion function $f_2 - f_1$. This means that the gain has a peak-like spectrum with a comparatively narrow width rising rapidly at the effective bandgap energy $E_{g,1}$. The gain spectrum of a quantum well is illustrated in Fig. 4.10 in comparison with that of bulk semiconductor.

Figure 4.10 shows that a quantum well offers many advantages in laser device implementation. Firstly, a quantum well laser oscillates at a wavelength shorter, by a value corresponding to the quantum confinement eigenvalues, than that for a bulk DH laser using an active layer of the same composition as the quantum well. Although the oscillation wavelength of a DH laser can be made shorter by changing the composition of the active layer to enlarge the bandgap energy, laser action cannot be obtained when the composition goes into the indirect-transition region. A quantum well laser enables shortening of the wavelength to be obtained outside this limitation. The second point is that, whereas in a bulk semiconductor the gain peak wavelength shifts with increasing injection level through band filling, in a quantum well the gain peak wavelength hardly changes. Another advantage of quantum wells is that because of the step-like density of states the gain is less sensitive to a change in temperature (the characteristic temperature T_0 for the threshold current is high).

4.3.4 Polarization Dependence

As we see from Eqs (4.17) and (4.22) and Fig. 4.9, the heavy holes have a smaller confinement eigenvalues and a larger density of state in comparison with light holes. Accordingly, the portion of heavy hole density of the hole density p is larger than that of the light hole density. This means that the major optical transitions are electron–heavy hole transitions, which contribute to amplification of the TE waves but not to amplification of the TM waves as we see from the description of the matrix element given in Section 4.2.2. Thus the gain in a quantum well exhibits a remarkable polarization dependence, and the gain is larger for TE waves than for TM waves.

4.3.5 Exciton Effects

The optical absorption due to interband transitions in a quantum well at thermal equilibrium, which is described by Eqs (4.39) with f_2 and f_1 substituted by the Fermi–Dirac factors with the Fermi level at thermal equilibrium, exhibits a step-like spectrum. Consideration of the absorption spectrum of a quantum well structure, however, must take the existence of excitons into account [4]. An exciton is a pair consisting of an electron and a hole mutually bound by the Coulomb attraction force and moving together. It can be represented by a H-atom-like wave function. The excitons form absorption peaks at a position shifted from the (sub)bandgap energy towards lower energy by the exciton binding energies. In bulk semiconductors, the exciton absorption peak is observed only at low temperatures, since the binding energy is around 4 meV which is considerably smaller than the thermal energy for room temperature ($k_B T \approx 25$ meV). In quantum wells, on the other hand, excitons collapse, becoming a flat two-dimensional nature, the extent of the wave function shrinks, and the binding energy increases to up to four times the value for the three-dimensional exciton in the bulk. Moreover, the oscillator strength per unit volume (proportional to the matrix element $|M|^2$) for the exciton transition in a quantum well is enhanced to several times the value in the bulk owing to shrinkage of the wave function. For these reasons, the exciton absorption peak is clearly observed even at room temperature. Thus the optical absorption spectrum of a quantum well is dominated by an exciton absorption peak superimposed on the absorption due to the interband transition. In a quantum well where carriers are injected with a density sufficient for obtaining amplification gain, however, excitons do not contribute substantially to optical transitions, as they are ionized immediately into electrons and holes by collision with carriers. We can thus omit the exciton effects in the gain analysis.

4.3.6 Effects of the Band Tail and Relaxation

Using the results of the density matrix analysis, described in Chap. 3, to take the band-tail and intraband relaxation effects into account, the gain spectrum for a quantum well can be written in a form of convolution of Eq. (4.25), derived from the analysis using the direct-transition model, and the line-shape function:

$$g(\hbar\omega) = \frac{\pi e^2}{n_r c \varepsilon_0 m^2 \omega} \times \sum_n \int |M|^2 (f_2 - f_1) \frac{\rho_{2Drn}(\hbar\omega')}{L_z} L(\omega - \omega')\, d\omega' \tag{4.40}$$

The line-shape function $L(\omega{-}\omega')$ is given in a form such as Eq. (3.74) or Eq. (3.75). As in a bulk semiconductor, the band-tail and relaxation effects modify the gain spectrum in such a manner that the rising step is rounded and the gain peak is somewhat reduced and broadened.

As in a bulk semiconductor, the many-body effects of carriers in a quantum well give rise to bandgap shrinkage. The bandgap shrinkage for completely two-dimensional states is expected to be proportional to the carrier density to the power 1/2. However, it has been shown experimentally that the dependence is actually close to the 1/3 power dependence of a bulk semiconductor. Therefore, ΔE_g of Eq. (3.39) for bulk semiconductors can be used also as an approximate expression for the carrier density dependence of $E_{g,n}$ of a quantum well. It should also be noted that, for accurate calculation of the above expression, the reduced density of states derived by taking the nonparabolicity of the valence band into account must be used for $\rho_{2Drn}(\hbar\omega')$. Then, $\Sigma_n \rho_{2Drn}$ is not exactly a multistep function as described by the sum in Eq. (4.35) but is a function with each step deviating from a flat horizontal step. The gain spectrum in the vicinity of the effective bandgap, however, does not deviate greatly.

Using the procedure described above, the gain characteristics have been calculated for quantum wells and strained quantum wells of various materials [11]. The gain spectrum for TE polarization calculated from Eq. (4.40) for a GaAs/Al$_x$G$_{1-x}$As quantum well is shown in Fig. 4.11 [11].

4.3.7 Dependence of the Maximum Gain on the Carrier Density

The dependence of the maximum gain for TE polarization calculated from Eq. (4.40) for a GaAs/Al$_x$G$_{1-x}$As quantum well on the area density ($L_z N$) of the carriers is shown in Fig. 4.12 [11]. In the same figure, the maximum gain for a bulk semiconductor is shown for comparison, with the volume density

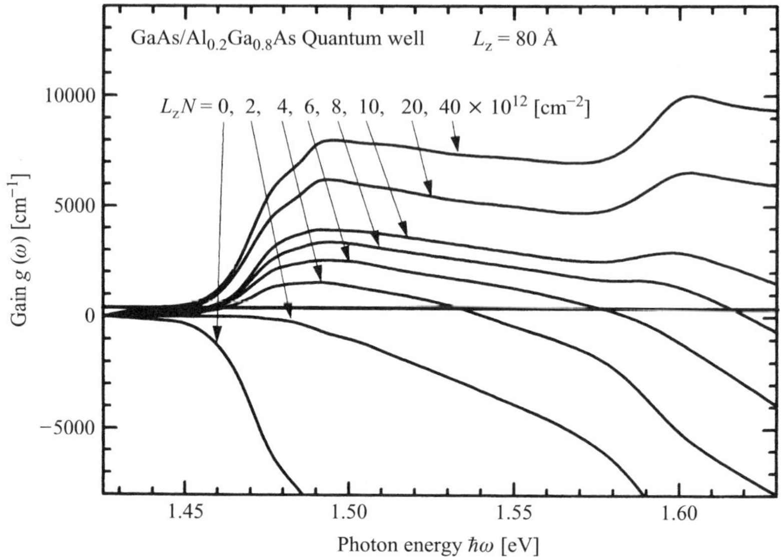

Figure 4.11 Example of the dependence of the calculated gain spectrum of a quantum well on the carrier density [11].

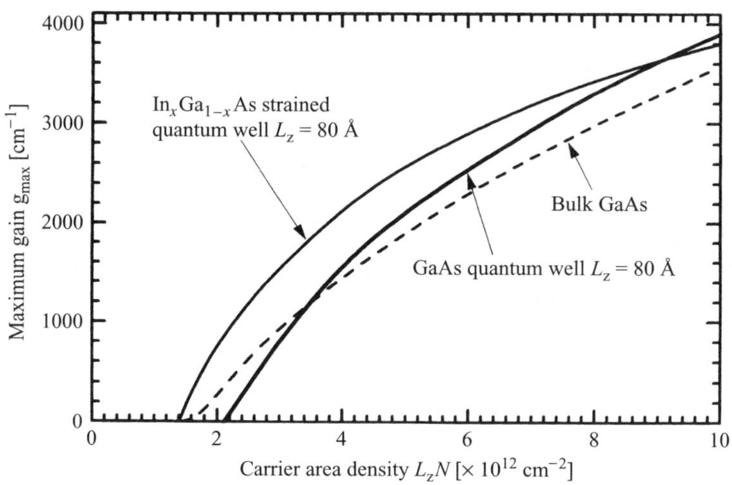

Figure 4.12 Dependence of the maximum gain of the quantum well on the carrier density [11]. (For the curve for bulk GaAs shown for comparison, the abscissa is the carrier volume density N multiplied by 80 Å.)

N of the carriers multiplied by $L_z(=80\,\text{Å})$ as the abscissa. As we see from these results, the carrier densities at transparency and the gains for a quantum well and a bulk semiconductor do not differ greatly from each other, provided that a comparison is made with carrier densities for the same thickness. However, a quantum well, being very thin, reaches transparency at a carrier density per unit area of the active layer much smaller than that required in the bulk. The carrier density per unit area of active layer required in a quantum well to obtain a given gain is also smaller than that required in the bulk. It should also be noted that the gain of a quantum well increases with increasing carrier density after transparency. This is because the density of states has a sharp step-like rise up at the subband edge; therefore the carriers contribute efficiently to the maximum gain, and the value of the matrix element is large. However, since the density of states takes a constant value, a further increase in carrier density approaching perfect inversion is associated with saturation of the gain increase. Although the carrier density dependence of the maximum gain can be approximately described by

$$g_{\max} = A_g(N - N_0) \tag{4.41a}$$

An expressions such as

$$g_{\max} = g_0 \ln\left(\frac{N + N_s}{N_0 + N_s}\right) \tag{4.41b}$$

$$g_{\max} = g_0 \ln\left(\frac{N}{N_0}\right) \tag{4.41c}$$

are also used to describe phenomenologically the sublinear dependence due to saturation. Here N is the injection carrier density, N_0 is the carrier density at transparency, and A_g and g_0 are constants. In Eq. (4.14b), N_s is a constant representing the degree of saturation; Eq. (4.14b) coincides with Eq. (4.41a) for $N_s \to \infty$ and coincides with Eq. (4.41c) for $N_s = 0$.

4.3.8 Absorption Spectrum and Maximum Gain Wavelength

The absorption spectrum for the thermal equilibrium state without carrier injection and the gain spectrum under carrier injection for a quantum well, in comparison with those for a bulk semiconductor, are shown in Fig. 4.13 [12]. A quantum well exhibits absorption and gain spectra having a sharp rise, and the bandgap shrinks on carrier injection. Accordingly, the absorption in a quantum well without injection at the wavelength of the gain maximum for an injected quantum well is much smaller than the

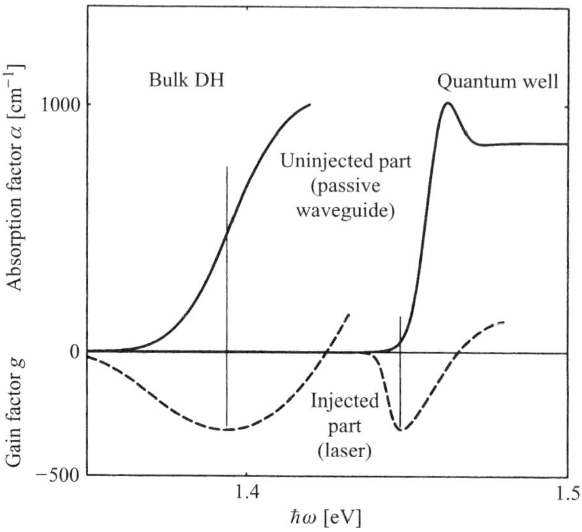

Figure 4.13 Comparison of the optical loss spectrum of passive waveguides in a quantum well structure and a bulk DH at a wavelength in the vicinity of the laser oscillation wavelength [12].

corresponding value for a bulk semiconductor. This means that a quantum well offers advantages in implementing distributed Bragg reflector lasers and monolithic integrated optical circuits, consisting of a laser-active section and a passive waveguide section in a waveguide of the same material.

It is known that, after a quantum well layer has been grown over a whole area of a semiconductor crystal substrate, the quantum well can be erased within a selected area. Erasing can be accomplished through intermixing of the well and barriers by impurity diffusion or ion implantation followed by thermal treatment, or deposition of the capping layer followed by thermal treatment. The processing is referred to as area-selective quantum disordering [13]. The disordered area can be used as a passive optical waveguide of low propagation loss at the wavelength of the gain maximum of the quantum well. Therefore, this processing offers an important technique for implementation of monolithic integrated optical circuits.

4.4 SPONTANEOUS EMISSION AND INJECTION CURRENT DENSITY

The rate of spontaneous emission of photons within the frequency range from ω to $\omega + d\omega$, per unit volume per unit time, can be calculated by summing the contributions by each subband given by Eq. (3.27), and then taking the convolution of the sum and the line-shape function describing the band-tail and relaxation effects. The result can be written as

$$\gamma_{\rm spt}(\hbar\omega)\,{\rm d}\omega = \frac{n_{\rm r}e^2\omega}{\pi m^2 c^3 \varepsilon_0}\,{\rm d}\omega \times \sum_n \int |M|^2 f_2(1-f_1)\frac{\rho_{2{\rm Dr}n}(\hbar\omega')}{L_z}L(\omega-\omega')\,{\rm d}\omega' \qquad (4.42)$$

The spontaneous emission spectrum can be calculated from the above expression. The total spontaneous emission rate $R_{\rm spt}$ per unit volume per unit time can be calculated by integrating $r_{\rm spt}(\hbar\omega)$ of Eq. (4.42) with respect to ω. The result is same as ω integration of $r_{\rm spt}(\hbar\omega)$ without taking the convolution with the line-shape function. The total spontaneous emission rate $R_{\rm spt}$ can be correlated with the injection current density J contributing to photon emission by

$$R_{\rm spt} = \int \gamma_{\rm spt}(\hbar\omega)\,{\rm d}\omega = \frac{J/q}{L_z} \qquad (4.43)$$

Therefore, the dependence of the gain on the injection current density can be calculated in the same manner as the calculation for the bulk semiconductor. For moderate injection current densities, this calculation procedure can be used to derive approximate results, since the spontaneous emission is dominated by the transitions of two-dimensional carriers described by Eqs (4.42) and (4.43). For higher injection current densities, however, the contribution of carriers in the quantum well and the SCH structure having energies higher than the quantum well barrier potential may not be neglected. Therefore, a complicated calculation taking this into account is required [4]. It is also necessary to consider the effects of the nonradiative recombination and carrier overflow in the same manner as in a bulk DH structure, in order to correlate accurately the calculated radiative recombination current density with the current density of an actual device.

As an example of the calculated dependence of the maximum gain on the injection current density (radiative recombination current density), a result for GRIN-SCH-SQW structures is shown in Fig. 4.14 [14]. The threshold current densities for Fabry–Perot lasers using these structures are also shown in the figure. The SQW thickness is denoted by L_z, and the structure approaches to ordinary DH structures for thickness $L_z \approx 400$ Å or larger. As also seen in this figure, the dependence of the maximum gain of quantum wells on the injection current density generally exhibits sublinear

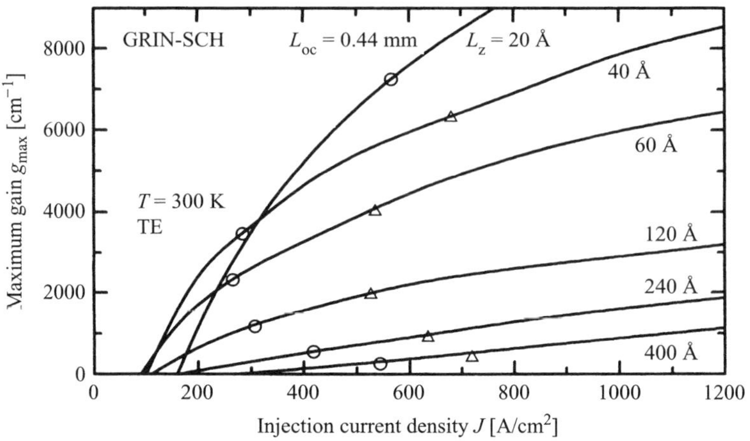

Figure 4.14 Dependence of maximum gain of GRIN-SCH-SQWs on the injection current density, and the threshold current densities of lasers using GRIN-SCH-SQW structures [14] for $\alpha = 40\ \text{cm}^{-1}$ (0) and $\alpha = 70\,\text{cm}^{-1}$.

characteristics. To describe such characteristic phenomenologically, expressions in the same form as Eq. (4.41) are used:

$$g_{\max} = g_0 \ln\left(\frac{J + J_s}{J_0 + J_s}\right) \tag{4.44a}$$

$$g_{\max} = g_0 \ln\left(\frac{J}{J_0}\right) \tag{4.44b}$$

A quantum well has a thickness smaller than that of an ordinary DH active layer and a small volume; therefore a large $g_{\max}$ is obtained with low injection current density J. Comparison between values of $g_{\max}$ for a same injection current density J/L_z per unit volume in thick DH structures and quantum well structures shows that $g_{\max}$ for quantum well structures is larger in a wide range of current densities, although the difference is not so significant as that of the J dependence of $g_{\max}$. In many cases, the same $g_{\max}$ can be accomplished in quantum well structures with J/L_z smaller than that required in DH structures. These advantages of quantum wells result from the effective contribution of the injected carriers to transitions at or near the subband edge relevant to $g_{\max}$, and the large value of the effective transition matrix element. It should be noted that the effective maximum gain for the optical wave in a laser is not $g_{\max}$ but $\Gamma g_{\max}$, where Γ is the optical confinement factor described in Chap. 5. As we see from Fig. 4.14, the Γg_{max} value required for laser oscillation can be obtained with an injection carrier density considerably lower than that required in ordinary DH structures, provided that the thickness of the quantum well and the SCH structure are appropriately designed so as to accomplish effective contribution of the injected carriers to the radiative recombination in the quantum well and to realize a large value of Γ.

4.5 STRAINED QUANTUM WELLS

It is usual for DHs and quantum wells to be constructed with mutually lattice-matched semiconductor alloy crystals. Quantum wells with a thickness of several tens of nanometers or less, however, can also be constructed with alloy compositions with slight lattice mismatch with the surrounding barriers. Such structures are called strained quantum wells, and they are classified into compressively strained quantum wells where the lattice constant of the well is larger than that of the barriers, and tensile-strained quantum wells where the lattice constant of the well is smaller. Representative example are $In_xGa_{1-x}As$ well with GaAs barriers, and $In_xGa_{1-x}As$ (or $In_{1-x}Ga_xAs_yp_{1-y}$) well with InP barriers; the former is a compressively strained quantum well, and the latter can be a tensile-strained or a compressively strained quantum well depending upon the composition. The strain gives rise to modification of the band structure of the quantum well, and appropriate design allows realization of characteristics useful for improvement in the performances of semiconductor lasers [15–19].

The modification of the band structure is shown schematically in Fig. 4.15. The major effect of strain is to decrease the effective masses of holes, associated with enhancement of curvature of the dispersion curves for the valence band. Although both compressive strain and tensile strain give rise to this tendency, the effect is more remarkable with the former. In semiconductors free from strain, the effective mass of electrons is nearly one order of magnitude smaller than that of holes, and the electron density of state is smaller. Therefore, the quasi-Fermi levels under electrical neutrality are at asymmetrical positions shifted upwards to the conduction band, as shown in Fig. 4.15(a). In strained quantum wells, on the other hand, the conduction and valence bands are better balanced, and the positions of the quasi-Fermi levels are closer to symmetrical positions, as shown in Figs 4.15(b) and 4.15(c). The improvement in balance leads to a reduction in carrier density required for a constant separation between quasi-Fermi levels for both bands, and hence a reduction in the carrier density required for population inversion. Therefore, the current density at transparency is

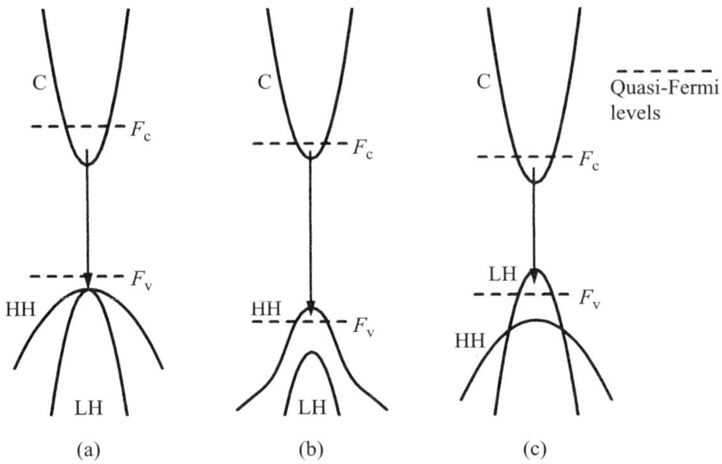

Figure 4.15 Band deformation and shift of quasi-Fermi levels by compressive and tensile strain: (a) without strain; (b) with compressive strain; (c) with tensilc strain.

reduced and the differential gain is enhanced. A significant enhancement of the differential gain by a factor of up to approximately 2 is feasible with appropriate design.

Whereas in bulk semiconductors free from strain the upper edges of the heavy hole and light hole bands are degenerate, in strained quantum wells the degeneracy is removed. As a result, the upper edge of the valence band is the edge of the heavy hole band for compressive strain and is the edge of the light hole band for tensile strain, as shown in Figs 4.15(b) and 4.15(c), respectively. Accordingly, the major transition is the electron–HH transition for compressive strain and is the electron–LH transition for tensile strain. The electron–HH transition contributes to amplification of the TE wave, and the electron–LH transition mainly to amplification of the TM wave. Therefore, it is possible to tailor the polarization characteristics of the amplification by giving an appropriate strain.

REFERENCES

1. Physical Society of Japan (ed.), *Physics and Applications of Superlattices* (in Japanese), Baifu-kan, Tokyo (1984).
2. H. Sakaki (ed.), *Superlattice Heterostructure Devices* (in Japanese), Kogyo Chosakai, Tokyo (1988).

3. H. Okamoto, *Optical Properties and Applications of Superlattice Structures* (in Japanese), Coronasha, Tokyo (1988).
4. C. Weisbuch and B. Vinter, *Quantum Semiconductor Structures*, Academic Press, New York (1991).
5. W. T. Tsang, *IEEE J. Quantum Electron.*, **QE-20**, 1119 (1984).
6. P. S. Zory, Jr (ed.), *Quantum Well Lasers*, Academic Press, New York (1993).
7. H. Okamura, S. Misewa, S. Yoshida, and S. Gonda, *Appl. Phys. Lett.*, **46**, 377 (1985).
8. S. R. Forrest, P. H. Schmidt, R. B. Wilson, and M. L. Kaplan, *Appl. Phys. Lett.*, **45**, 1199 (1984).
9. L. A. Coldren, and S. W. Corzine, *Diode Lasers and Photonic Integrated Circuits*, John Wiley, New York (1995), p. 502.
10. L. A. Coldren, and S. W. Corzine, *Diode Lasers and Photonic Integrated Circuits*, John Wiley, New York (1995), pp. 121, 518.
11. L. A. Coldren and S. W. Corzine, *Diode Lasers and Photonic Integrated Circuits*, John Wiley, New York (1995), pp. 164, 166.
12. S. Tarucha, Y. Horikoshi, and H. Okamoto, *Jpn. J. Appl. Phys.*, **22**, L482 (1983).
13. T. Venkatesan, S. A. Schwarz, D. M. Hwang, R. Bhat, M. Koza, H. W. Yoon, P. Mei, Y. Arakawa, and A. Yariv, *Appl. Phys. Lett.*, **49**, 701 (1986).
14. C. Weisbuch, *Proc. SPIE*, **869**, 155(1987).
15. K. J. Beerink, P. K. York, and J. J. Coleman, *Appl. Phys. Lett.*, **55**, 2582 (1989).
16. P. J. A. Thijs, L. F. Tiemeijer, P. I. Kuindersma, J. J. Binsma, and T. van Dongen, *IEEE J. Quantum Electron.*, **QE-27**, 1426 (1991).
17. G. P. Agrawal and N. K. Dutta, *Semiconductor Lasers*, second edition, Van Nostrand Reinhold, New York (1993).
18. W. W. Chow, S. W. Koch, and M. Sergeant III, *Semiconductor Laser Physics*, Springer, Berlin (1993).
19. L. A. Coldren and S. W. Corzine, *Diode Lasers and Photonic Integrated Circuits*, John Wiley, New York (1995).

5 Semiconductor Heterostructure Optical Waveguides

Optical waveguiding based on the confinement of an optical wave in a semiconductor heterostructure, as well as optical amplification based on stimulated emission, is an essential element for semiconductor lasers. This chapter presents optical waveguides consisting of semiconductor heterostructures for lasers. Analysis of passive optical characteristics [1–4] and optical characteristics under carrier injection are discussed. Analysis and characteristics of optical resonator for laser using such a waveguide are also presented.

5.1 OUTLINE OF OPTICAL WAVEGUIDES FOR SEMICONDUCTOR LASERS

Fundamental semiconductor lasers are implemented by sandwiching an active region for optical amplification between two regions of material with larger bandgap to construct a double hetero-structure (DH) for effective injection and confinement of high-density carriers. The thickness of the active layer of ordinary DH lasers is designed to be around 0.1 μm. Fortunately, in III–V compound alloy semiconductors, increasing the bandgap energy by changing the composition generally results in decreasing the refractive index for optical waves [5]. Therefore, an optical wave can be confined in the active layer and its vicinity. This means that a DH structure serves as an optical waveguide, which confines the wave in a narrow range with respect to the direction of the thickness, and guides it along the direction of the layer. In contrast with the free propagation of an optical wave in a homogeneous medium, where an optical wave cannot be concentrated into a region of submicron width over a distance of the order of millimeters, an optical waveguide can offer such confinement to enhance the interaction between optical wave and carriers. Thus practical semiconductor lasers can be implemented.

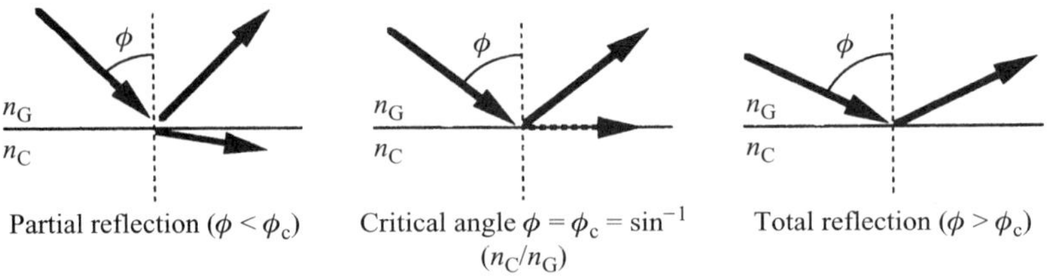

Figure 5.1 Reflection of a plane wave at a boundary between two media of different refractive indexes.

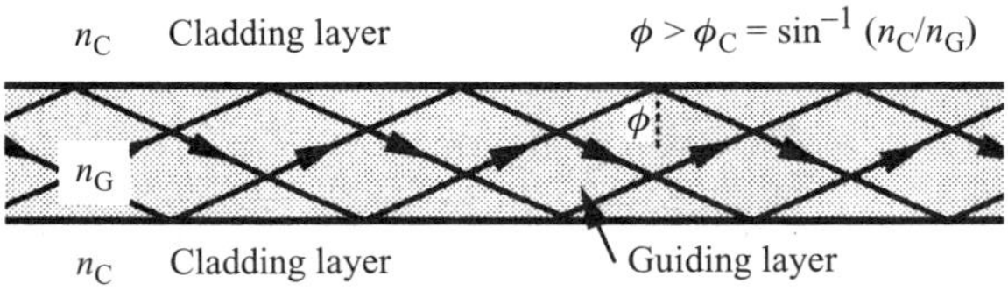

Figure 5.2 Ray optics model for the principle of an optical waveguide.

Consider an interface between two media having different refractive indexes n_G and n_C ($n_G > n_C$), and an optical plane wave incident on the interface at an angle ϕ from the medium of higher index n_G, as shown in Fig. 5.1. For small ϕ, the optical wave is transmitted into the medium of index n_C and propagates at a refraction angle $\sin^{-1}[(n_G/n_C)\sin\phi]$ determined by the Snell's law. If ϕ is larger than the critical angle ϕ_c given by

$$\phi_c = \sin^{-1}\left(\frac{n_C}{n_G}\right) \tag{5.1}$$

the optical wave is totally reflected back into the region of index n_G. Therefore, constructing a structure where a layer of index n_G is sandwiched between two layers of index n_C as shown in Fig. 5.2, optical waves with $\phi > \phi_c$ are confined in the layer of index n_G through successive total internal reflections and propagate in this layer with a zigzag path. Although the principle of optical waveguides can thus be illustrated by using the above-described ray optics model and total internal reflection, accurate analysis requires a wave optics treatment as discussed below.

A waveguides that confines the optical wave with respect to the direction perpendicular to the plane of substrate and guides it along the layer is called a planar waveguide, and the layer where the wave is confined is called the guiding layer. Although fundamental DH lasers use the active layer as a guiding layer, in many cases the guiding layer does not necessarily

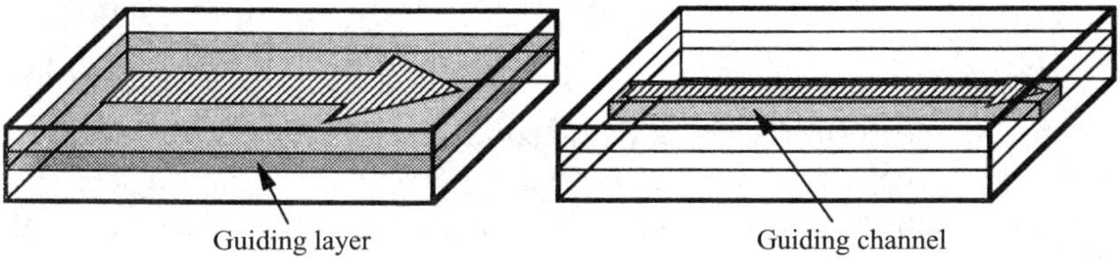

Figure 5.3 (a) Planar waveguide and (b) channel waveguide.

coincide with the active layer, since various multilayer waveguide structures are used for improvement in performances. For quantum well lasers, a separate confinement structure is required to attain efficient confinements of the carriers and the optical wave.

Semiconductor lasers using a planar waveguide, called broad area lasers, can easily be fabricated and provide a high output power. However, several factors including microdefects make it difficult to obtain stable laser oscillation with good uniformity along the lateral direction perpendicular to the optical axis. In many cases, they oscillate strongly in several narrow stripe regions. This phenomenon is called filamentation. Filamentation results in difficulty in focusing or collimating the output wave, limiting the applications. The wavelength spectrum of the output does not exhibit a single peak but a broad and complex distribution. This means that simple broad-area lasers cannot provide an output wave of high temporal and spatial coherence. To solve these problems, channel waveguides that confine the optical wave also in the lateral direction and guide it in a narrow region as shown in Fig. 5.3(b) are employed. Semiconductor lasers of this kind, called stripe lasers, can provide a stable and highly coherent output, and therefore they are at present the major semiconductor lasers.

5.2 FUNDAMENTAL EQUATIONS FOR THE OPTICAL WAVE

5.2.1 Maxwell Equations and Wave Equations

The Electric field $\boldsymbol{E}$, magnetic field $\boldsymbol{H}$, electric flux density $\boldsymbol{D}$, and magnetic flux density $\boldsymbol{B}$ of the optical wave, in general, together with the electric current density $\boldsymbol{J}$, satisfy the Maxwell equations

$$\nabla \times \boldsymbol{E} = -\frac{\partial \boldsymbol{B}}{\partial t} \tag{5.2a}$$

$$\nabla \times \boldsymbol{H} = \frac{\partial \boldsymbol{D}}{\partial t} + \boldsymbol{J} \tag{5.2b}$$

$$\nabla \cdot \boldsymbol{D} = 0 \tag{5.2c}$$

$$\nabla \cdot \boldsymbol{B} = 0 \tag{5.2d}$$

From these equations, the boundary conditions for optical waves are obtained. Let $\boldsymbol{e}_\mathrm{n}$ be a unit vector normal to a boundary, and 1 and 2 be subscripts for denoting values an both sides of the boundary; then the boundary conditions are given by

$$\boldsymbol{e}_\mathrm{n} \times (\boldsymbol{E}_1 - \boldsymbol{E}_2) = 0 \tag{5.3a}$$

$$\boldsymbol{e}_\mathrm{n} \times (\boldsymbol{H}_1 - \boldsymbol{H}_2) = 0 \tag{5.3b}$$

$$\boldsymbol{e}_\mathrm{n} \cdot (\boldsymbol{D}_1 - \boldsymbol{D}_2) = 0 \tag{5.3c}$$

$$\boldsymbol{e}_\mathrm{n} \cdot (\boldsymbol{B}_1 - \boldsymbol{B}_2) = 0 \tag{5.3d}$$

The above equations show that the tangential components of $\boldsymbol{E}$ and $\boldsymbol{H}$ and the normal components of $\boldsymbol{D}$ and $\boldsymbol{B}$ are continuous across the boundary.

Consider an optical wave of single angular frequency ω. Let $\exp(-\mathrm{i}\omega t)$ be the time dependence and omit it, and let ε and σ be the relative dielectric constant and conductivity, respectively, of the medium at ω. Then $\boldsymbol{D}$ and $\boldsymbol{E}$, $\boldsymbol{B}$ and $\boldsymbol{H}$, and $\boldsymbol{J}$ and $\boldsymbol{E}$ are correlated by

$$\boldsymbol{D} = \varepsilon_0 \varepsilon \boldsymbol{E} \tag{5.4}$$

$$\boldsymbol{B} = \mu_0 \boldsymbol{E} \tag{5.5}$$

$$\boldsymbol{J} = \sigma \boldsymbol{E} \tag{5.6}$$

Then Eq. (5.2) is rewritten as

$$\nabla \times \boldsymbol{E} = \mathrm{i}\omega\mu_0 \boldsymbol{H} \tag{5.7a}$$

$$\nabla \times \boldsymbol{H} = (-\mathrm{i}\omega\varepsilon_0\varepsilon + \sigma)\boldsymbol{E} \tag{5.7b}$$

$$\nabla \cdot \varepsilon \boldsymbol{E} = 0 \tag{5.7c}$$

$$\nabla \cdot \boldsymbol{H} = 0 \tag{5.7d}$$

Eliminating $\boldsymbol{H}$ from the above equations, we obtain a general wave equation for $\boldsymbol{E}$:

$$\nabla^2 \boldsymbol{E} + \nabla\left[\nabla\varepsilon \cdot \left(\frac{\boldsymbol{E}}{\varepsilon}\right)\right] + (k_0^2\varepsilon + \mathrm{i}\omega\nu_0\sigma)\boldsymbol{E} = 0 \tag{5.8}$$

where k_0 is the wave number defined by

$$k_0 = \omega(\varepsilon_0\mu_0)^{1/2} = \frac{\omega}{c} = \frac{2\pi}{\lambda} \tag{5.9}$$

Here c is the light velocity in vacuum, and λ the wavelength in vacuum. For a homogeneous medium we have $\nabla\varepsilon = 0$ and the wave equation is reduced to

$$\nabla^2 \boldsymbol{E} + (k_0^2\varepsilon + \mathrm{i}\omega\mu_0\sigma)\boldsymbol{E} = 0 \tag{5.10}$$

Even if the medium is not exactly homogeneous, we can use the above equation as an approximation, provided that $\nabla\varepsilon \approx 0$. As we see from Eqs. (5.7) and (5.10), we can use a complex relative dielectric constant $\tilde{\varepsilon}$ defined by

$$\tilde{\varepsilon} = \varepsilon + \frac{\mathrm{i}\sigma}{\omega\varepsilon_0} \tag{5.11}$$

substituting ε to include the σ term representing the losses in $\tilde{\varepsilon}$ as the imaginary part.

5.2.2 Description of a Carrier-injected Semiconductor

The above expressions can also be applied for semiconductor media by extending them to describe the stimulated emission and the refractive index change induced by carrier injection. The extension can be made by considering a polarization $\boldsymbol{P}_{\mathrm{act}}$ representing the excitation effects. Assuming that $\boldsymbol{P}_{\mathrm{act}}$ is correlated with the electric field $\boldsymbol{E}$ by

$$\boldsymbol{P}_{\mathrm{act}} = \varepsilon_0 \chi_{\mathrm{act}} \boldsymbol{E} \tag{5.12}$$

using a complex dielectric susceptibility χ_{act}, we can rewrite Eq. (5.4) as

$$\boldsymbol{D} = \varepsilon_0 \varepsilon \boldsymbol{E} + \boldsymbol{P}_{\mathrm{act}} = \varepsilon_0(\varepsilon + \chi_{\mathrm{act}})\boldsymbol{E} \tag{5.13}$$

in a form including $\boldsymbol{P}_{\mathrm{act}}$. Then the wave equation is rewritten as

$$\nabla^2 \boldsymbol{E} + k_0^2 \tilde{n}^2 \boldsymbol{E} = 0 \tag{5.14}$$

$$\tilde{n}^2 = \tilde{\varepsilon} = \varepsilon + \chi_{\mathrm{act}} + \frac{\mathrm{i}\sigma}{\omega\varepsilon_0} \tag{5.15}$$

where $\tilde{n}$ is a complex index of refraction. The index for cases without carrier injection is given by $n = \varepsilon^{1/2}$. Since in many cases, for semiconductors, $|\chi_{\mathrm{act}}| \ll \varepsilon$ and $\sigma/\omega\varepsilon_0 \ll \varepsilon$, we have

$$\mathrm{Re}\{\tilde{n}\} \approx n + \frac{\mathrm{Re}\{\chi_{\mathrm{act}}\}}{2n} \tag{5.16a}$$

$$\mathrm{Im}\{\tilde{n}\} \approx \frac{\mathrm{Im}\{\chi_{\mathrm{act}}\}}{2n} + \frac{\sigma}{2n\omega\varepsilon_0} \tag{5.16b}$$

Consider a plane wave propagating along the $+z$ direction in a homogeneous medium. Then substituting $\boldsymbol{E} = \boldsymbol{e}_x \exp(\mathrm{i}\tilde{\beta}z)$ with a complex propagation constant $\tilde{\beta}$ into Eq. (5.14) yields

$$\tilde{\beta} = k_0\tilde{n} = k_0(n + \Delta n) + \mathrm{i}\left(-\frac{g}{2} + \frac{\alpha}{2}\right) \tag{5.17}$$

$$\Delta n = \frac{1}{2n}\mathrm{Re}\{\chi_{\mathrm{act}}\} \tag{5.18}$$

$$g = -\frac{k_0}{n}\mathrm{Im}\{\chi_{\mathrm{act}}\} \tag{5.19}$$

$$\alpha = \frac{k_0\sigma}{n\omega\varepsilon_0} \tag{5.20}$$

Since this optical wave can be expressed as

$$\boldsymbol{E} = \boldsymbol{e}_x \exp[\mathrm{i}k_0(n + \Delta n)z] \exp\left(+\frac{gz}{2}\right) \exp\left(-\frac{\alpha z}{2}\right) \tag{5.21}$$

g is a power gain factor, Δn is the associated refractive index change, and α is a power attenuation factor.

5.3 OPTICAL WAVE IN A WAVEGUIDE

5.3.1 Optical Electromagnetic Modes in a Waveguide

An optical waveguide structure, in general, can be described by the distribution of the relative dielectric constant ε in the cross section. Taking the z axis of a coordinate system along the direction of optical wave propagation, the distribution for a waveguide that is uniform in the z direction can be written as

$$\varepsilon = \varepsilon(x, y) = [n(x, y)]^2 \tag{5.22}$$

The refractive index distribution $n(x, y)$ is assumed to have a maximum in the vicinity of the waveguide axis $(0, 0)$ and a smaller value for positions away from the axis. For the moment we assume a lossless waveguide ($\sigma = 0$). The electromagnetic fields of optical waves in this waveguiding structure can be written as

$$\boldsymbol{E}(x, y, z) = \boldsymbol{E}(x, y) \exp(\mathrm{i}\beta z) \tag{5.23a}$$

$$\boldsymbol{H}(x, y, z) = \boldsymbol{H}(x, y) \exp(\mathrm{i}\beta z) \tag{5.23b}$$

where β is a propagation constant. The above field must satisfy the Maxwell equations and boundary conditions. If $\nabla n^2(x, y) \approx 0$, $\boldsymbol{E}(x, y)$ satisfies the wave equation

$$\nabla_t^2 + [k_0^2 n^2(x, y) - \beta^2]\boldsymbol{E}(x, y) = 0 \tag{5.24}$$

where $\nabla_t = (\partial/\partial x, \partial y, 0)$.

Among many solutions of the above equation, solutions such that $\boldsymbol{E}(x, y)$ is finite at and near the waveguide axis and approaches zero asymptotically for positions far away from the axis are called guided modes, since they describe a confined optical wave. A finite number of guided modes, which are described by the discrete modal field distributions $\boldsymbol{E}(x, y)$ and $\boldsymbol{H}(x, y)$, and discrete propagation constants exist.

Solutions such that $\boldsymbol{E}(x, y)$ does not approach zero asymptotically for positions far away from the axis, on the other hand, are called radiation modes, since they describe optical waves propagating away from the axis. An infinite number of radiation modes exist, and the modal field distributions and the propagation constant are of continuous character.

It is convenient to express the propagation constant β of the mode by using the wave number k_0 of Eq. (5.9) in the form

$$\beta = Nk_0 \tag{5.25}$$

where N is called the effective index of refraction or the mode index.

5.3.2 Intermode Orthogonality and Power Flow

One of the most important general characteristics of optical waves in a lossless waveguide structure is intermode orthogonality. Let $\{\boldsymbol{E}_1, \boldsymbol{H}_1\}$ and $\{\boldsymbol{E}_2, \boldsymbol{H}_2\}$ be two sets of arbitrary electromagnetic fields; then from Eqs. (5.7a) and (5.7b) we obtain

$$\nabla(\boldsymbol{E}_1 \times \boldsymbol{H}_2^* + \boldsymbol{E}_2^* \times \boldsymbol{H}_1) = -2\sigma \boldsymbol{E}_1 \cdot \boldsymbol{E}_2^* = 0 \tag{5.26}$$

As for the two sets of field in the structure, consider two modes in the form of Eq. (5.23), and let $\{\boldsymbol{E}_m, \boldsymbol{H}_m\}$ and $\{\boldsymbol{E}_{m'}, \boldsymbol{H}_{m'}\}$ represent $\boldsymbol{E}(x, y)$ and $\boldsymbol{H}(x, y)$ for the two modes. Substituting them into Eq. (5.26) and integrating the resultant equation over the x–y plane yield

$$\begin{aligned}(\beta_m - \beta_{m'}) &\iint [\boldsymbol{E}_{tm} \times \boldsymbol{H}_{tm'}^* + \boldsymbol{E}_{tm'}^* \times \boldsymbol{H}_{tm'}]_z \,\mathrm{d}x\,\mathrm{d}y \\ &= \mathrm{i} \iint \nabla_t [\boldsymbol{E}_m \times \boldsymbol{H}_{m'}^* + \boldsymbol{E}_{m'}^* \times \boldsymbol{H}_m]_t \,\mathrm{d}x\,\mathrm{d}y\end{aligned} \tag{5.27}$$

where the subscript t, denotes components in the x–y plane, and the subscript z the z component.

The area integration on the right-hand side of the above equation can be rewritten as a closed-line integration along the rim of the infinitely large integration area. The value of the integration vanishes to zero, since $\boldsymbol{E}$, $\boldsymbol{H} \rightarrow 0$ at positions far away from the z axis for cases where either m or m' is a guided mode, and periodic boundary condition can be used to show its disappearance for cases where both m and m' are radiation modes. As a result, the area integration on the left-hand side of the above equation vanishes to zero when $\beta_m \neq \beta_{m'}$. This means that the transverse components of the electromagnetic fields of different modes are orthogonal to each other.

Let us consider next the optical power carried by the modal electromagnetic fields. Since the effective power flow density is given by the real part of the complex Poynting vector $\boldsymbol{S} = \frac{1}{2}(\boldsymbol{E} \times \boldsymbol{H}^*)$, the power flow transmitted by optical wave of mode m can be written as

$$\begin{aligned} P &= P_z \\ &= \mathrm{Re}\left\{ \iint \boldsymbol{S} \cdot e_z \, \mathrm{d}x \, \mathrm{d}y \right\} \\ &= \mathrm{Re}\left\{ \iint \tfrac{1}{2}(\boldsymbol{E}_t \times \boldsymbol{H}_t^*)_z \mathrm{d}x \, \mathrm{d}y \right\} \\ &= \tfrac{1}{4} \iint [\boldsymbol{E}_{tm} \times \boldsymbol{H}_{tm}^* + \boldsymbol{E}_{tm}^* \times \boldsymbol{H}_{tm}]_z \, \mathrm{d}x \, \mathrm{d}y \end{aligned} \tag{5.28}$$

The above quantity is such that $P > 0$ and $P < 0$ for $\beta_m > 0$ and $\beta_m < 0$, respectively.

It is convenient to use modal field expressions normalized in such a form that $P = 1\,\mathrm{W}$. The orthonormal relation can be written for guided modes as

$$\tfrac{1}{4} \iint [\boldsymbol{E}_{tm} \times \boldsymbol{H}_{tm'}^* + \boldsymbol{E}_{tm'}^* \times \boldsymbol{H}_{tm}]_z \, \mathrm{d}x \, \mathrm{d}y = \pm \delta_{mm'} \tag{5.29}$$

where the + sign should be taken on the right-hand side for $\beta_m = \beta_{m'} > 0$, and the − sign for $\beta_m = \beta_{m'} < 0$. For radiation modes that have a continuous spectrum, the Kronecker delta must be replaced by the Dirac delta function.

Optical modes have not only normality but also completeness; they form a complete normal system. This means that transverse components of an arbitrary optical electromagnetic fields can be expressed in an expansion form using the modal fields as

$$\boldsymbol{E}_t(x, y, z) = \sum_m a_m \boldsymbol{E}_{tm}(x, y) \exp(\mathrm{i}\beta_m z) \tag{5.30a}$$

$$\boldsymbol{H}_t(x, y, z) = \sum_m a_m \boldsymbol{H}_{tm}(x, y) \exp(\mathrm{i}\beta_m z) \tag{5.30b}$$

where $\sum_m$ denotes superposition with respect to all modes, i.e., summation over guided modes and integration over radiation modes. Thus $\boldsymbol{E}_t$ and $\boldsymbol{H}_t$ are described completely by the above expression. Since the z components $\boldsymbol{E}_z$ and $\boldsymbol{H}_z$ can be obtained from the Maxwell equations as

$$\nabla_t \times \boldsymbol{E}_t = \mathrm{i}\omega\mu_0 \boldsymbol{H}_z, \qquad \nabla_t \times \boldsymbol{H}_t = -\mathrm{i}\omega\varepsilon_0\varepsilon \boldsymbol{E}_z \tag{5.31}$$

all the components of arbitrary fields $\boldsymbol{E}$ and $\boldsymbol{H}$ can be described by a set of the expansion coefficients a_m.

5.4 PLANAR WAVEGUIDE

5.4.1 Wave Equation

The distribution of the relative dielectric constant of a planar waveguide of uniform layer structure can be written as

$$\varepsilon = \varepsilon(x) = [n(x)]^2 \tag{5.32}$$

Although the electromagnetic fields of optical modes propagating in the waveguide can be written in a form of Eq. (5.23), they can be simplified as function of x only, namely $\boldsymbol{E}(x)$ and $\boldsymbol{H}(x)$, because of uniformity along the y direction. Then decomposition of the Maxwell equations given by Eqs. (5.7a) and (5.7b) into components yields

$$-\mathrm{i}\beta E_y = \mathrm{i}\omega\mu_0 H_x \tag{5.33a}$$

$$-\frac{\mathrm{d}E_z}{\mathrm{d}x} + \mathrm{i}\beta E_x = \mathrm{i}\omega\mu_0 H_y \tag{5.33b}$$

$$\frac{\mathrm{d}E_y}{\mathrm{d}x} = \mathrm{i}\omega\mu_0 H_z \tag{5.33c}$$

$$-\mathrm{i}\beta H_y = -\mathrm{i}\omega\varepsilon_0 n^2 E_x \tag{5.33d}$$

$$-\frac{\mathrm{d}H_z}{\mathrm{d}x} + \mathrm{i}\beta H_x = -\mathrm{i}\omega\varepsilon_0 n^2 E_y \tag{5.33e}$$

$$\frac{\mathrm{d}H_y}{\mathrm{d}x} = -\mathrm{i}\omega\varepsilon_0 n^2 E_z \tag{5.33f}$$

From the above equations, we see that the optical modes can be classified into two categories. Putting $H_y = 0$ in Eq. (5.33), we have $E_x = E_z = 0$, and we see that H_x and H_z can be calculated from E_y as

$$H_x = -\frac{\beta}{\omega\mu_0} E_y, \qquad H_z = -\frac{\mathrm{i}}{\omega\mu_0}\frac{\mathrm{d}E_y}{\mathrm{d}x} \tag{5.34}$$

Modes of this category are called transverse electric (TE) modes, since the electric field vector is perpendicular to the propagation (z) axis. The electric field amplitude $E_y(x)$ satisfies the wave equation

$$\frac{\mathrm{d}^2 E_y}{\mathrm{d}x^2} + [k_0^2 n(x)^2 - \beta^2]E_y = 0 \tag{5.35}$$

On the other hand, putting $E_y = 0$, we have $H_x = H_z = 0$, and we see that E_x and E_z can be calculated from H_y as

$$E_x = \frac{\beta}{\omega\varepsilon_0 n^2} H_y, \qquad E_z = \frac{\mathrm{i}}{\omega\varepsilon_0 n^2}\frac{\mathrm{d}H_y}{\mathrm{d}x} \tag{5.36}$$

Modes of this category are called transverse magnetic (TM) modes, since the magnetic field vector is perpendicular to the propagation axis. The magnetic field amplitude $H_y(x)$ satisfies the wave equation

$$n^2 \frac{\mathrm{d}}{\mathrm{d}x}\left(n^{-2}\frac{Hy}{\mathrm{d}x}\right) + (k_0^2 n^2 - \beta^2)H_y = 0 \tag{5.37}$$

where $n(x)$ is simplified as n.

5.4.2 Step-Index Waveguides

As a fundamental planar waveguide, let us consider a structure consisting of a guiding layer of a high refractive index n_{G} and a thickness T sandwiched between two media of indexes n_{UC} and n_{LC}, as shown in Fig. 5.4(a). In ordinary DH lasers, the active layer serves as a guiding layer. Since the thickness of the low-index regions, called cladding layers, is practically designed to be sufficiently larger than the lateral expanse of the guided wave, the three-layer structure shown in Fig. 5.4 is appropriate as a model for the analysis. The structure has a step-type refractive index distribution $n(x)$ as shown in Fig. 5.4(b), and $n(x)$ can be written as

$$n(x) = \begin{cases} n_{\mathrm{UC}} & \left(\frac{T}{2} < x\right) \\ n_{\mathrm{G}} & \left(-\frac{T}{2} < x < \frac{T}{2}\right), \qquad n_{\mathrm{G}} > n_{\mathrm{UC}}, n_{\mathrm{LC}} \\ n_{\mathrm{LC}} & \left(x < -\frac{T}{2}\right) \end{cases} \tag{5.38}$$

When the refractive indexes are uniform in each region, the wave equations for TE and TM modes are in the same form. Put the propagation constant β

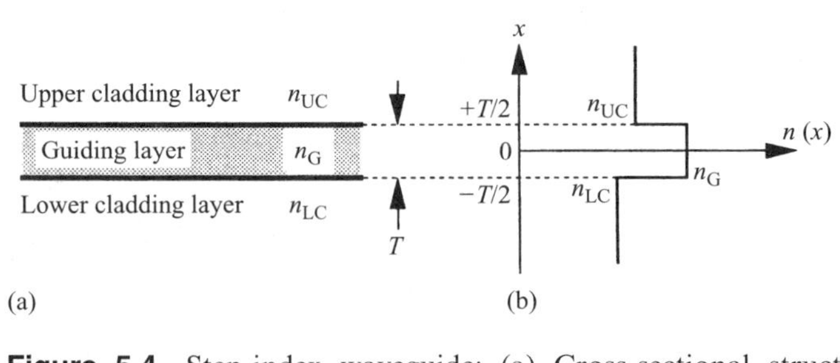

Figure 5.4 Step-index waveguide: (a) Cross-sectional structure; (b) refracture index distrubution.

as $\beta = Nk_0$, and put

$$\gamma_{UC} = k_0(N^2 - n_{UC}^2)^{1/2}, \qquad \gamma_{LC} = k_0(N^2 - n_{LC}^2)^{1/2}$$
$$k_G = k_0(n_G^2 - N^2)^{1/2} \tag{5.39}$$

Then the solutions of the wave equation are proportional to $\exp(\pm i\kappa x)$ in the guiding layer, and to $\exp(\pm\gamma x)$ in the cladding layers. For guided modes, κ and γ should be real, therefore the effective index N of refraction must be in the range

$$n_{UC}, \quad n_{LC} < N < n_G \tag{5.40}$$

Using an angle θ defined by

$$\frac{N}{n_G} = \frac{\beta}{k_0 n_G} = \cos\theta \tag{5.41}$$

spatial variation on the optical field amplitude in the guiding layer can be written as

$$\exp(\pm i\kappa x)\exp(i\beta z) = \exp\{ik_0 n_G[x\sin(\pm\theta) + z\cos(\pm\theta)]\} \tag{5.42}$$

The above expression describes two plane-wave components propagating at angles $\pm\theta$ with respect to the z axis. The plane-wave components are incident on to the boundary between the guiding and cladding layers at angle $\phi = \pi/2 - \theta$, and are totally reflected. The mathematical expressions for the guided modes are described in the following.

Transverse Electric Modes

Solutions $E_y(x)$ of Eq. (5.35), which approach zero asymptotically for $x \to \pm\infty$, can be written as

$$E_y(x) = \begin{cases} E_{UC}\exp\left[-\gamma_{UC}\left(x - \frac{T}{2}\right)\right] & \left(\frac{T}{2} < x\right) \\ E_G\cos\left[\kappa_G\left(x - \frac{T}{2}\right) + \Phi_{UC}\right] & \left(-\frac{T}{2} < x < \frac{T}{2}\right) \\ E_{LC}\exp\left[+\gamma_{LC}\left(x + \frac{T}{2}\right)\right] & \left(x < -\frac{T}{2}\right) \end{cases} \tag{5.43}$$

The other field components are given by Eq. (5.34) using this E_y. Here, from the boundary conditions, continuities of E_y and dE_y/dx are required.

Eliminating E_{UC}, E_G, and E_{LC} using the boundary conditions at $x = \pm T/2$ yields

$$\kappa_G T - \Phi_{UC} - \Phi_{LC} = m\pi \qquad (m = 0, 1, 2, \ldots) \tag{5.44}$$

where

$$\Phi_{UC} = \mathrm{Tan}^{-1}\left(\frac{\gamma_{UC}}{\kappa_G}\right), \qquad \Phi_{LC} = \mathrm{Tan}^{-1}\left(\frac{\gamma_{LC}}{\kappa_G}\right) \tag{5.45}$$

From the boundary condition we also obtain relations between E_{UC}, E_G, and E:

$$E_{UC} = E_G \cos\Phi_{UC}, \qquad E_{LC} = E_G(-1)^m \cos\Phi_{LC} \tag{5.46}$$

From Eq. (5.28) the power flow P per unit width along the y direction is

$$\begin{aligned} P &= \frac{2\beta}{\omega\mu_0}\int E_y^2(x)\,\mathrm{d}x \\ &= N\left(\frac{\varepsilon_0}{\mu_0}\right)^{1/2} E_G^2 T_{\mathrm{eff}} \end{aligned} \tag{5.47}$$

with

$$T_{\mathrm{eff}} = T + \frac{1}{\gamma_{UC}} + \frac{1}{\gamma_{LC}} \tag{5.48}$$

For symmetrical waveguides with $n_{UC} = n_{LC} = n_C$, Eq. (5.43) reduces to

$$E_y(x) = \begin{cases} E_G \cos\Phi_C \exp\left[-\gamma_C\left(x - \frac{T}{2}\right)\right] & \left(\frac{T}{2} < x\right) \\ E_G \cos\left(\kappa_G x - \frac{m\pi}{2}\right) & \left(-\frac{T}{2} < x < \frac{T}{2}\right) \\ E_G(-1)^m \cos\Phi_C \exp\left[+\gamma_C\left(x + \frac{T}{2}\right)\right] & \left(x < -\frac{T}{2}\right) \end{cases} \tag{5.43$'$}$$

with

$$K_G T - 2\Phi_C = m\pi, \qquad \Phi_C = \mathrm{Tan}^{-1}\left(\frac{\gamma_C}{\kappa_G}\right) \tag{5.44$'$}$$

and $E_y(x)$ are symmetrical functions for even m, and antisymmetrical functions for odd m. The former is called even modes and the latter odd modes.

Transverse Magnetic Modes

Solutions $H_y(x)$ of Eq. (5.37) which approach zero asymptotically for $x \to \pm\infty$, can be written as

$$H_y(x) = \begin{cases} H_{\mathrm{UC}} \exp\left[-\gamma_{\mathrm{UC}}\left(x - \frac{T}{2}\right)\right] & \left(\frac{T}{2} < x\right) \\ H_{\mathrm{G}} \cos\left[\kappa_{\mathrm{G}}\left(x - \frac{T}{2}\right) + \Phi_{\mathrm{UC}}\right] & \left(-\frac{T}{2} < x < \frac{T}{2}\right) \\ H_{\mathrm{LC}} \exp\left[+\gamma_{\mathrm{LC}}\left(x + \frac{T}{2}\right)\right] & \left(x < -\frac{T}{2}\right) \end{cases} \quad (5.49).$$

The other field components are given by Eq. (5.36). From the boundary conditions, continuities of H_y and $n^{-2}\mathrm{d}H_y/\mathrm{d}x$ are required. Eliminating H_{UC}, H_{G}, and H_{LC} yields

$$\kappa_{\mathrm{G}} T - \Phi_{\mathrm{UC}} - \Phi_{\mathrm{LC}} = m\pi \qquad (m = 0, 1, 2, \ldots) \tag{5.50}$$

$$\Phi_{\mathrm{UC}} = \mathrm{Tan}^{-1}\left[\left(\frac{n_{\mathrm{G}}}{n_{\mathrm{UC}}}\right)^2 \left(\frac{\gamma_{\mathrm{UC}}}{\kappa_{\mathrm{G}}^2}\right)\right] \tag{5.51a}$$

$$\Phi_{\mathrm{LC}} = \mathrm{Tan}^{-1}\left[\left(\frac{n_{\mathrm{G}}}{n_{\mathrm{LC}}}\right)^2 \left(\frac{\gamma_{\mathrm{LC}}}{\kappa_{\mathrm{G}}}\right)\right] \tag{5.51b}$$

$$H_{\mathrm{UC}} = H_{\mathrm{G}} \cos \Phi_{\mathrm{UC}}, \qquad H_{\mathrm{LC}} = H_{\mathrm{G}}(-1)^m \cos \Phi_{\mathrm{LC}} \tag{5.52}$$

The power flow P per unit width along the y direction is

$$\begin{aligned} P &= \frac{2\beta}{\omega u_0} \int \frac{E_y^2}{n^2} \mathrm{d}x \\ &= N\left(\frac{\mu_0}{\varepsilon_0}\right)^{1/2} H_{\mathrm{G}}^2 T_{\mathrm{eff}} \end{aligned} \tag{5.53}$$

with

$$T_{\mathrm{eff}} = T + \frac{1}{\gamma_{\mathrm{UC}} q_{\mathrm{UC}}} + \frac{1}{\gamma_{\mathrm{LC}} q_{\mathrm{LC}}}$$

$$q_{\mathrm{UC}} = \left(\frac{N}{n_{\mathrm{G}}}\right)^2 + \left(\frac{N}{n_{\mathrm{UC}}}\right)^2 - 1, \qquad q_{\mathrm{LC}} = \left(\frac{N}{n_{\mathrm{G}}}\right)^2 + \left(\frac{N}{n_{\mathrm{LC}}}\right)^2 - 1 \tag{5.54}$$

For symmetrical waveguides with $n_{UC} = n_{LC} = n_C$, Eq. (5.49) reduces to

$$H_y(x) = \begin{cases} H_G \cos\Phi \exp\left[-\gamma_C\left(x - \frac{T}{2}\right)\right] & \left(\frac{T}{2} < x\right) \\ H_G \cos\left(\kappa_G x - \frac{m\pi}{2}\right) & \left(-\frac{T}{2} < x < \frac{T}{2}\right) \\ H_G(-1)^m \cos\Phi_C \exp\left[+\gamma_C\left(x + \frac{T}{2}\right)\right] & \left(x < -\frac{T}{2}\right) \end{cases} \tag{5.49'}$$

with

$$\kappa_G T - 2\Phi_C = m\pi, \qquad \Phi_C = \mathrm{Tan}^{-1}\left[\left(\frac{n_G}{n_C}\right)^2 \frac{\gamma_C}{\kappa_G}\right] \tag{5.50'}$$

As we see above, the independent field component $E_y(x)$ for TE modes, and the independent field component $E_y(x)$, for TM modes are described by expressions of the same form, and they exhibit sinusoidal oscillatory distribution in the guiding layer, as shown in Fig. 5.5. In the guiding layer, two plane-wave components with propagation angles $\pm\theta$, corresponding to the zigzag rays in the ray optics model, propagate simultaneously.

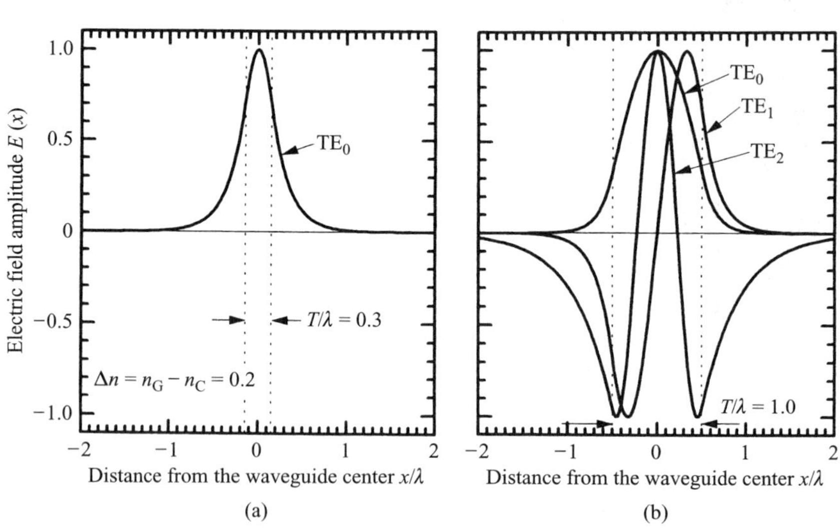

Figure 5.5 Examples of amplitude distributions of guided modes: (a) Single-mode; (b) multimode waveguide.

Interference of the two plane-wave components gives rise to formation of a standing wave with the oscillatory distribution. The field can be written as $\frac{1}{2}\exp(i\Phi_{UC})$ $(\exp[+i\kappa_G(x-T/2)]+\exp(-2i\Phi_{UC})$ $\exp[-i\kappa_G(x-T/2)])$. The first term represents a wave proceeding to the boundary with the upper cladding layer at angle $+\theta$, and the second term the wave proceeding at angle $-\theta$, after reflection. The complex amplitude of the latter at the boundary $x=T/2$ is $\exp(-2i\Phi_{UC})$ times the amplitude of the former. This means that the total internal reflection of the wave at the boundary with the upper cladding layer is associated with a phase shift of $-2\Phi_{UC}$. Reflection at the boundary with the lower cladding layer is associated with a phase shift of $-2\Phi_{LC}$. In the cladding layers the optical field amplitudes decay exponentially. The immersion of the field into the cladding layer, even in the total internal reflection, results from the wave nature of light. The wave immersed in the low-index region is called an evanescent wave. The depths to which the evanescent wave penetrates in the upper and lower cladding layers are $1/\gamma_{UC}$ and $1/\gamma_{UC}$, respectively.

It is known that the phase shifts $-2\Phi_{UC}$ and $-2\Phi_{LC}$ associated with the total internal reflection give rise to a shift in the appearance point of the reflected ray from the incident point along the z direction, as shown in Fig. 5.6. The magnitudes $2Z_{UC}$ and $2Z_{LC}$, of the shift can be obtained by calculating the derivatives of $-2\Phi_{UC}$ and $-2\Phi_{LC}$ with respect to β. The results are

$$2Z_{UC}=\frac{2}{\gamma_{UC}}\mathrm{Tan}\phi,\qquad 2Z_{LC}=\frac{2}{\gamma_{LC}}\mathrm{Tan}\phi \tag{5.55}$$

for TE modes, and

$$2Z_{UC}=\frac{2}{\gamma_{UC}q_{UC}}\mathrm{Tan}\phi,\qquad 2Z_{LC}=\frac{2}{\gamma_{LC}q_{LC}}\mathrm{Tan}\phi \tag{5.56}$$

for TM modes [6]. The ray shift is referred to as the Goos–Haenchen shift.

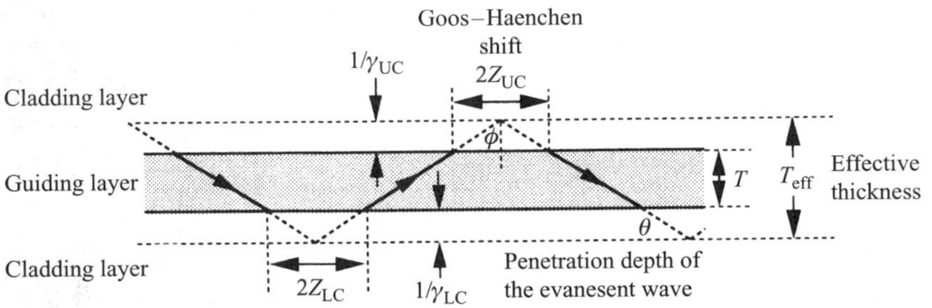

Figure 5.6 Rays in a waveguide, the Goos–Haenchen shift and the effective thickness of a waveguide.

The power flow expressions in Eqs (5.47) and (5.52) indicate that the guided wave power per unit width along the y direction is given by the product of the optical power density in free space and T_{eff}, which is the sum of the guiding layer thickness and the evanescent-wave penetration depths, and is called the effective thickness of the waveguide. The rays with successive total reflections proceed with bounces within a width of T_{eff}, as shown in Fig. 5.6.

The equations for determining the effective refractive index (Eqs (5.44) and (5.49)) are called characteristic equations of guided modes or dispersion relations. These equations indicate that the total phase shift for a lateral round trip of the optical wave, i.e., the sum of the phase shift $2\kappa_G T$ within the guiding layer and the phase shifts, $-2\Phi_{\mathrm{UC}}$ and $-2\Phi_{\mathrm{LC}}$ associated with the total reflection at the boundaries must equal the integer multiple of (m times) 2π. This means that self-consistent optical modes can exist only when the optical wave after the round trip coincides with the initial wave including the phase. From this requirement it can be seen that the values of the effective refractive index N of guided modes, are discrete. The integer m is the order of the mode.

As the characteristic equations show, N is determined depending upon the refractive indexes n_{G}, n_{UC}, and n_{LC} of each layer, the guiding layer thickness T, the optical wavelength λ, and the mode order m. Since the characteristic equations are transcendental equations, N cannot be determined analytically. However, the result of the calculation can be summarized in a convenient graphical form using normalized parameters, as described below.

Firstly we define a normalized guiding layer thickness V_T by

$$V_T = k_0 T (n_{\mathrm{G}}^2 - n_{\mathrm{LC}}^2)^{1/2} = \frac{\omega}{c} T (n_{\mathrm{G}}^2 - n_{\mathrm{LC}}^2)^{1/2} \tag{5.57}$$

The parameter V_T is also called the normalized frequency, since it is proportional to the optical frequency. We also define a parameter for describing the degree of asymmetry of the upper and lower cladding layers by

$$a_{\mathrm{TE}} = \frac{n_{\mathrm{LC}}^2 - n_{\mathrm{UC}}^2}{n_{\mathrm{G}}^2 - n_{\mathrm{LC}}^2} \tag{5.58}$$

The effective refractive index N is normalized into b by

$$b = \frac{N^2 - n_{\mathrm{LC}}^2}{n_{\mathrm{G}}^2 - n_{\mathrm{LC}}^2} \tag{5.59}$$

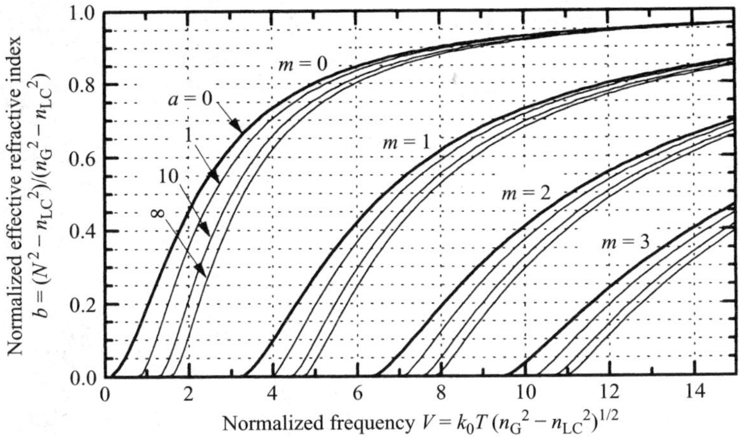

Figure 5.7 Relation between the normalized frequency and the normalized effective index of refraction.

Using these parameters, the characteristic equation for TE modes, namely Eq. (5.44), is rewritten as

$$V(1-b)^{1/2} - \mathrm{Tan}^{-1}\left(\frac{b}{1-b}\right)^{1/2} - \mathrm{Tan}^{-1}\left(\frac{a+b}{1-b}\right)^{1/2} = m\pi \qquad (5.60)$$

where V_T and a_{TE} are abbreviated as V and a. The calculated relations between V and b for the mth modes are plotted in Fig. 5.7. The graphs are called dispersion curves of guided modes. For a practical waveguide, one can calculate the values of V and a and read the value of b from the dispersion curves. Then the effective index can readily be obtained from

$$N = \left[n_{LC}^2 + b(n_G^2 - n_{LC}^2)\right]^{1/2} \approx n_{LC} + b(n_G - n_{LC}) \qquad (5.61)$$

As we see from the dispersion curves, the number of guided modes increases with increasing normalized guiding layer thickness V_T. The mode of lowest order $m=0$ is called the fundamental mode. A waveguide that supports only the fundamental mode is called a single-mode waveguide, and a waveguide that support several modes is called a multimode waveguide. The effective index of a higher-order mode is smaller than that for a lower-order mode in the same waveguide. The effective index N of each mode, increases numotonically with increasing V_T and approaches the refractive index n_G of the guiding layer asymptotically. Inversely, decreasing V_T results in a decrease in N. After the decrease in N down to either of the indexes n_{UC} or n_{LC} of the cladding layers, the mode disappears. This is called the cutoff.

The value of the normalized frequency corresponding to the cutoff V_{cm} of the mth mode, can be determined by putting $b \rightarrow 0$ to yield

$$V_{cm} = m\pi + \mathrm{Tan}^{-1}(a^{1/2}) \tag{5.62}$$

For symmetrical waveguides, we have $a=0$ and therefore $V_{c0}=0$, indicating that the fundamental mode exists in all cases. From Eq. (5.62), we see that the condition for the single-mode waveguide is

$$\mathrm{Tan}^{-1}(a^{1/2}) < V < \pi + \mathrm{Tan}^{-1}(a^{1/2}) \tag{5.63}$$

Although the above descriptions are for TE modes, a similar treatment can be carried out also for TM modes. Since the characteristic equation for TM modes is a little more complicated, accurate expressions cannot be given with V, a, and b only. However, by using a parameter defined by

$$a_{\mathrm{TM}} = \left(\frac{n_{\mathrm{G}}}{n_{\mathrm{UC}}}\right)^4 \frac{n_{\mathrm{LC}}^2 - n_{\mathrm{UC}}^2}{n_{\mathrm{G}}^2 - n_{\mathrm{LC}}^2} \tag{5.64}$$

instead of a_{TE} of Eq. (5.58), and using V_T of Eq. (5.57) and b of Eq. (5.59), the characteristic equation for TM modes can be expressed approximately by the same equation as for TE modes (Eq. (5.60)). Therefore, Fig. 5.7 can be used also to determine N approximately for TM modes. For symmetrical waveguides ($n_{\mathrm{LC}}=n_{\mathrm{UC}}$), the difference between N for TE and TM modes is very small, since $a_{\mathrm{TE}}=a_{\mathrm{TM}}=0$. For asymmetrical waveguides, we have $a_{\mathrm{TM}} > a_{\mathrm{TE}} > 0$, and therefore from Fig. 5.7 we see that N for a TM mode is somewhat smaller than N for a TE mode of the same order. The cutoff normalized frequency V_{cm} for the TM mode of each order is larger than that for the TE mode.

5.4.3 Graded-Index Waveguides

Waveguides such that the refractive index distribution $n(x)$ is a continuous function of x and varies gradually with position x are referred to as graded-index (or gradient-index) waveguides. A semiconductor waveguide of this kind can be formed by varying gradually the alloy composition during the epitaxy process. They are advantageous in accomplishing a low scattering loss.

Ray Trajectory

An example of a graded-index distribution $n(x)$ is shown in Fig. 5.8(a). A symmetrical waveguide with $n(x)=n(-x)$ is considered for simplicity.

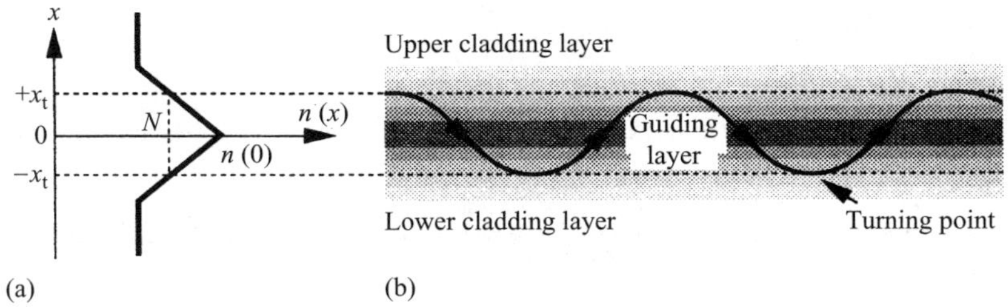

Figure 5.8 Graded-index waveguide: (a) refracture index distribution; (b) ray trace.

The effective index $N = \beta/k_0$ of guided modes is in the range

$$n(-\infty) = n(+\infty) < N < n(0) \tag{5.65}$$

and points $x = -x_t$ and $x = +x_t$ exist such that

$$N = n(-x_t) = n(+x_t) \tag{5.66}$$

At a position x within the region $-x_t < x < +x_t$ in the waveguide, the ray proceeds at an angle $\pm\theta(x)$ determined by

$$\frac{N}{n(x)} = \frac{\beta}{k_0 n(x)} = \cos[\theta(x)] \tag{5.67}$$

Therefore, the ray proceeds with a continuous and gradual change in direction, and the ray direction with respect to the x direction is inverted when the ray reaches $-x_t$ or $+x_t$ and the angle becomes $\theta = 0$. The ray trajectory is a smooth undulating curve as shown in Fig. 5.8(b). The points $x = \pm x_t$ are called turning points. The fact that the ray turns at these points, corresponding to the total internal reflection at the boundaries in a step-index waveguide, gives rise to the optical field confinement and guiding in the region where the refractive index is high.

Analysis by the Wentzel–Kramers–Brillanin Method

Although the guided mode fields and the dispersion characteristics of graded index waveguides in general cannot be derived analytically, the Wentzel–Kramers–Brillanin (WKB) method [7] can be employed for the approximate analysis to determine the general characteristics. When the index distribution is gradual and the spatial derivative is small, the wave equations for TE and TM modes (Eqs (5.35) and (5.37), respectively) reduce to the same form:

$$\frac{d^2G(x)}{dx^2} + [k_0^2 n(x)^2 - \beta^2]G(x) = 0 \tag{5.68}$$

Therefore, referring to solutions of the wave equation for uniform medium, and putting

$$\begin{aligned}
\kappa(x) &= [k_0^2 n(x)^2 - \beta^2]^{1/2} \\
\xi(x) &= \int_x^{x_t} \kappa(x)\,dx \qquad (0 < x < x_t) \\
\gamma(x) &= [\beta^2 - k_0^2 n(x)^2]^{1/2} \\
\eta(x) &= \int_{x_t}^{x} \gamma(x)\,dx \qquad (x_t < x)
\end{aligned} \tag{5.69}$$

we can assume that the solution $G(x)$ of Eq. (5.68) is in the form $\exp[\pm i\xi(x)]$ for $0 < x < x_t$, and $\exp[-\eta(x)]$ for $x_t < x$. In the WKB method, factors describing approximately the amplitude distribution are derived from the wave equation, and they are combined with the above factors to write $G(x)$ as

$$G(x) = \begin{cases}
A[\kappa(x)]^{-1/2}\cos\left(\xi(x) - \dfrac{\pi}{4}\right) & (0 < x < x_t) \\
\dfrac{A}{2}\left(\dfrac{2\pi\xi(x)}{3\kappa(x)}\right)^{1/2}\{J_{1/3}[\xi(x)] + J_{-1/3}[\xi(x)]\} & (x \lesssim x_t) \\
\dfrac{A}{2}\left(\dfrac{2\pi\eta(x)}{3\gamma(x)}\right)^{1/2}\{-I_{1/3}[\eta(x)] + I_{-1/3}[\eta(x)]\} & (x_t \lesssim x) \\
\dfrac{A}{2}[\gamma(x)]^{-1/2}\exp[-\eta(x)] & (x_t < x)
\end{cases} \tag{5.70}$$

where A is a constant, and J_m and I_m denote the Bessel function and the modified Bessel function respectively, of mth order. The first and fourth lines of the above expressions describe the solutions for the regions inside and outside the turning point, respectively (excluding the region close to the turning point). The second and third lines describe the solutions in the transient region around the turning point, and they are connected so as to satisfy the boundary condition at the turning point. The coefficients are chosen so that the asymptotic expressions of the second and third lines including J_m and I_m coincide with the first and fourth lines, respectively. Thus all lines are connected smoothly to describe the total field. We see from the first line that the phase shift associated with the total reflection at the turning point is $-\pi/2$. This corresponds to $-2\Phi_C$ for step-index waveguides. It should also be noted that $G(x)$ is a symmetrical or antisymmetrical function, and that the solution for $x<0$ is given by $G(-x)$ or $-G(-x)$. From the condition for this total field to be connected smoothly at $x=0$, we obtain, as a characteristic equation to determine the propagation constant, $\xi(0)-\pi/4=m\pi/2$, i.e.,

$$2\int_0^{x_t}[k_0^2 n(x)^2-\beta^2]^{1/2}\,\mathrm{d}x-\frac{\pi}{2}=m\pi \tag{5.71}$$

where m $(=0,1,2,\ldots)$ is the mode order. The first integral term and the second term $-\pi/2$ on the left-hand side correspond to $\kappa_G T$ and $-2\Phi_c$, respectively, in Eq. (5.44).

The field distribution $G(x)$ of graded-index waveguides is an oscillatory function inside the turning points and monotonically decaying function outside, as is the case with step-index waveguides. The variation, however, is affected by the index distribution and is compressed or expanded. A feature of graded-index waveguides is that the oscillation amplitude of $G(x)$ is enhanced in the vicinity of the turning points in comparison with that near the center of the waveguide. The position x_t of the turning points depends upon the mode order, and the distance of the turning points from the waveguide center is larger for modes of higher order. In the following, a few examples are discussed.

Parabolic Index Distribution

Consider a waveguide with a relative dielectric constant distribution in a form of parabolic function

$$\varepsilon(x)=n(x)^2=n_G^2-a^2x^2 \qquad (a>0) \tag{5.72}$$

The index distribution is also approximately parabolic. Then the solution of the wave equation Eq. (5.68) is given by Hermite–Gauss function:

$$G(x)=H_m[(k_0a)^{1/2}x]\exp\left(-\frac{k_0ax^2}{2}\right) \tag{5.73}$$

where H_m is Hermite polynomial of mth order, and $H_0(\xi)=1$, $H_1(\xi)=2\xi$, $H_2(\xi)=4\xi^2-2$, $H_3(\xi)=8\xi^3-12\xi,\ldots$. The function $G(x)$ is plotted in Fig. 5.9.

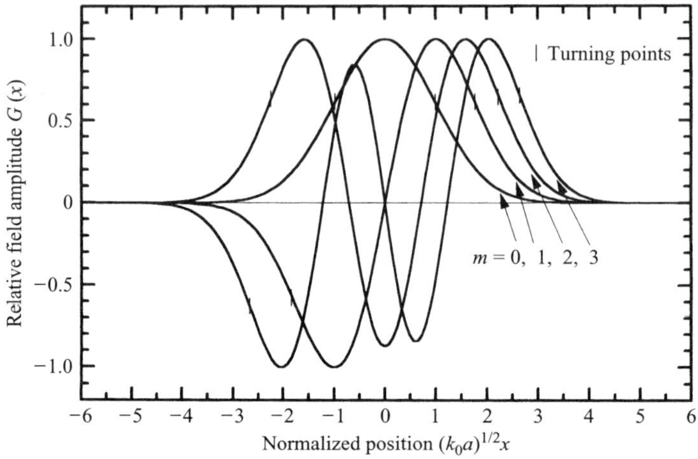

Figure 5.9 Field distribution of guided modes in a waveguide of parabolic index distribution.

The effective refractive index N_m is given by

$$N_m^2 = n_G^2 - (2m + 1)\left(\frac{a}{k_0}\right) \tag{5.74}$$

The turning point x_t is obtained from $n(x_t)^2 = N_m^2$ as

$$x_t = \left(\frac{2m + 1}{ak_0}\right)^{1/2} \tag{5.75}$$

It should be noted that the above expressions are valid only in a region where a^2x^2 is small. The refractive index of realistic waveguides approaches a finite value n_C for large $|x|$, and therefore, among the modes described by the above expressions, higher-order modes such that $N_m < n_c$ do not actually exist.

Symmetric Linear Index Distribution

Consider a waveguide of symmetrical linear index distribution:

$$\varepsilon(x) = n(x)^2 = n_G^2 - a|x| \qquad (a > 0) \tag{5.76}$$

Then, the solutions of the wave equation in Eq. (5.68) is given by

$$G(x) = \begin{cases} +Ai[k_0^{2/3}a^{1/3}(+x - x_t)] & (0 < x) \\ \pm Ai[k_0^{2/3}a^{1/3}(-x - x_t)] & (x < 0) \end{cases} \tag{5.77}$$

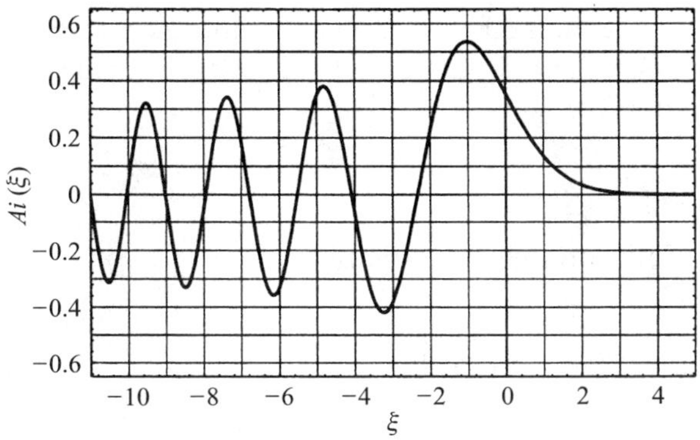

Figure 5.10 Airy function.

where $Ai(\xi)$ is the Airy function. The turning point x_t is correlated with effective index N by

$$x_t = \frac{n_G^2 - N^2}{a} \tag{5.78}$$

Airy functions are solutions of a differential equation $(d^2/d\xi^2 - \xi)Ai(\xi) = 0$. The graph is shown in Fig. 5.10 $Ai(\xi)$ is oscillatory for $\xi < 0$ and is monotonically decreasing for $\xi > 0$. The zero points of $Ai(\xi)$ and the zero points of $dAi(\xi)/d\xi$ are located alternately on the negative ξ axis. Let these points be $\cdots < -\xi_4 < -\xi_3 < -\xi_2 < -\xi_1 < -\xi_0 < 0$; then all ξ_m are positive, $-\xi_m$ for even m are zero points of $dAi(\xi)/d\xi$, and $-\xi_m$ for odd m are zero points of $Ai(\xi)$. For $G(x)$ and $dG(x)/dx$ to be continuous at $x = 0$, $-k_0^{2/3}a^{1/3}\,x_t = -\xi_m$ must hold, and, in the second line of the right-hand side of Eq. (5.77), the + sign should be taken for even-order modes (symmetrical modes) and the minus sign for odd-order modes (antisymmetrical modes). From these conditions, we obtain an expression for the turning point of the mth mode given by

$$x_t = \frac{\xi_m}{k_0^{2/3} a^{1/3}} \tag{5.79}$$

and an expression for the effective refractive index of the mth mode given by

$$N_m^2 = n_G^2 - \xi_m \left(\frac{a}{k_0}\right)^{2/3} \tag{5.80}$$

A WKB calculation shows that ξ_m can be approximated as

$$\xi_m \approx \left[\frac{3\pi}{4}\left(m + \frac{1}{2}\right)\right]^{2/3} \tag{5.81}$$

As in the previous subsection, it should be noted that the above expressions are valid only in a region where $a|x|$ is small, and higher-order modes do not actually exist.

Gaussian Index Distribution

Consider a waveguide of Gaussian index distribution:

$$\varepsilon(x) = n(x)^2 = n_0^2 + 2n_0\,\Delta n \exp\left[-\left(\frac{x}{\sigma}\right)^2\right] \tag{5.82}$$

Applying the WKB method, the field distribution $G(x)$ is given by Eq. (5.70) with the above $n(x)^2$, and from Eq. (5.70) the characteristic equation can be written as

$$2k_0 \int_0^{x_t} \left\{ n_0^2 + 2n_0\,\Delta n \exp\left[-\left(\frac{x}{\sigma}\right)^2\right] - N^2 \right\}^{1/2} \mathrm{d}x - \frac{\pi}{2} = m\pi \tag{5.83}$$

where

$$n_0 + \Delta n \exp\left[-\left(\frac{x_t}{\sigma}\right)^2\right] = N$$

Putting $N \to n_0$ and $x_t \to \infty$, we see that the cutoff condition is $2k_0 (2n_0\,\Delta n)^{1/2}\sigma(\pi/2)^{1/2} = (m+1/2)\pi$. Then, using a normalized guiding layer thickness V_σ defined by

$$V_\sigma = k_0(2\sigma)(2n_0\,\Delta n)^{1/2} \tag{5.84}$$

the cutoff condition is given by

$$V_\sigma = \left(m + \frac{1}{2}\right)(2\pi)^{1/2} \tag{5.85}$$

5.5 PERTURBATION THEORY AND THE OPTICAL CONFINEMENT FACTOR

The amplification gain and refractive index change induced by carrier injection in semiconductor waveguides, as well as propagation losses, can be analyzed on the basis of calculation of guided modes in waveguide described by a distribution of complex dielectric constant. However, analysis based upon the perturbation theory [8] is more useful for understanding the physical implications and also for practical calculation. This section presents the perturbation theory analysis of semiconductor DH waveguides.

5.5.1 Propagation Constant Change Due to Waveguide Deformation

An arbitrary waveguide can be described by a relative dielectric constant distribution

$$\varepsilon(x, y) = n^2(x, y) \tag{5.86}$$

Consider another waveguide described by a slightly modified distribution

$$\varepsilon(x, y) + \delta\varepsilon(x, y) = [n(x, y) + \delta n(x, y)]^2 \tag{5.87}$$

The modification in the dielectric constant distribution and the index distribution, denoted $\delta\varepsilon(x, y)$ and $\delta n(x, y)$, can be treated as small perturbations to $\varepsilon(x, y)$ and $n(x, y)$. The perturbations $\delta\varepsilon$ and δn do not necessarily have to be real; they can be complex values associated with imaginary parts describing gain or loss. Let $\{\boldsymbol{E}, \boldsymbol{H}\}$ and $\{\boldsymbol{E}', \boldsymbol{H}'\}$ be the electromagnetic fields of an optical wave propagating in the original waveguide and the perturbed waveguide, respectively. They satisfy the Maxwell equations:

$$\nabla \times \boldsymbol{E} = \mathrm{i}\omega\mu_0\boldsymbol{H}, \qquad \nabla \times \boldsymbol{H} = -\mathrm{i}\omega\varepsilon_0\varepsilon\boldsymbol{E} \tag{5.88a}$$

$$\nabla \times \boldsymbol{E}' = i\omega\mu_0\boldsymbol{H}', \qquad \nabla \times \boldsymbol{H}' = -\mathrm{i}\omega\varepsilon_0(\varepsilon + \delta\varepsilon)\boldsymbol{E}' \tag{5.88b}$$

From the above equations, using a formula for vector differential operation, we obtain

$$\nabla(\boldsymbol{E}' \times \boldsymbol{H}^* + \boldsymbol{E}^* \times \boldsymbol{H}') = \mathrm{i}\omega\varepsilon_0\boldsymbol{E}^* \cdot \delta\varepsilon\boldsymbol{E}' \tag{5.89}$$

Integration of the above equation in a thin region with infinitely wide cross-sectional area normal to the waveguide axis and infinitely small thickness along the axis yield

$$\iint \frac{\partial}{\partial z}[\boldsymbol{E}' \times \boldsymbol{H}^* + \boldsymbol{E}^* \times \boldsymbol{H}']_z \,\mathrm{d}x\,\mathrm{d}y = \mathrm{i}\omega\varepsilon_0 \iint \boldsymbol{E}^* \cdot \delta\varepsilon\, \boldsymbol{E}' \,\mathrm{d}x\,\mathrm{d}y \tag{5.90}$$

Now, to consider a guided mode, we put

$$\boldsymbol{E} = \boldsymbol{E}(x, y)\exp(\mathrm{i}\beta z), \qquad \boldsymbol{H} = \boldsymbol{H}(x, y)\exp(\mathrm{i}\beta z) \tag{5.91a}$$

$$\boldsymbol{E}' = \boldsymbol{E}'(x, y)\exp(\mathrm{i}\beta' z), \qquad \boldsymbol{H}' = \boldsymbol{H}'(x, y)\exp(\mathrm{i}\beta' z) \tag{5.91b}$$

Substituting the above expressions into Eq. (5.90), with the use of $E' \approx E$ and $H' \approx H$, we obtain a general expression for the change in the propagation constant caused by the perturbation $\delta\beta = \beta' - \beta$:

$$\delta\beta = \frac{(\omega\varepsilon_0/4)\iint \boldsymbol{E}^* \cdot \delta\varepsilon\, \boldsymbol{E}\,\mathrm{d}x\,\mathrm{d}y}{\frac{1}{4}\iint [\boldsymbol{E}\times\boldsymbol{H}^* + \boldsymbol{E}^*\times\boldsymbol{H}]_z\,\mathrm{d}x\,\mathrm{d}y} \tag{5.92}$$

The denominator in the above expression equals the guided wave power P of Eq. (5.28).

For TE modes in a planar waveguide, employing Eq. (5.34) the above expression is simplified as

$$\delta\beta = \frac{k_0^2}{2\beta}\frac{\int \delta\varepsilon(x)\,|E_y(x)|^2\,\mathrm{d}x}{\int |E_y(x)|^2\,\mathrm{d}x} \tag{5.93}$$

Using $\delta\varepsilon \approx 2n\,\delta n$, $\beta = k_0 N$ and $N \approx n$, Eq. (5.93) can be rewritten as

$$\delta\beta = k_0\,\delta N = k_0 \frac{\int \delta n(x)\,|E_y(x)|^2\,\mathrm{d}x}{\int |E_y(x)|^2\,\mathrm{d}x} \tag{5.94}$$

The above expressions show that the change δN in the guided mode index is given by the change δn in the refractive index averaged in the waveguide with optical intensity weighting. For TM modes, a similar treatment can be carried out, although the formulation is a little more complicated.

5.5.2 Carrier Injection Effects and Optical Losses

As was presented in Section 5.2.2, optical properties of semiconductors including the carrier injection effects and optical losses can be described by a complex relative dielectric constant $\tilde{\varepsilon}$ or a complex refractive index $\tilde{n}$:

$$\tilde{n}^2 = \tilde{\varepsilon} = \varepsilon + \chi_{\text{act}} + \frac{\mathrm{i}\sigma}{\omega\varepsilon_0} \tag{5.15}$$

Since the second and third terms describing the carrier injection effects and losses are in general smaller than the first term describing the dielectric properties, they can be treated as perturbations. First, the field distribution and the propagation constant are calculated for an unperturbed lossless waveguide. Then the change $\delta\tilde{\beta}$ in the propagation constant for a perturbation

$$\delta\tilde{\varepsilon} = \chi_{\text{act}} + \frac{\mathrm{i}\sigma}{\omega\varepsilon_0} \tag{5.95}$$

can be calculated using Eq. (5.92) to evaluate the change in the guided mode index, as well as the gain and attenuation. Using Eqs (5.16) and (5.17), the perturbation given by Eq. (5.95) can be converted into the perturbation in the complex refractive index:

$$\delta\tilde{n} = \Delta n - \frac{\mathrm{i}g}{2k_0} + \frac{\mathrm{i}\alpha}{2k_0} \tag{5.96}$$

where Δn, g, and α are the carrier-induced refractive index change, the gain factor, and the attenuation factor, respectively, and they are given by Eqs. (5.18), (5.19), and (5.20), respectively.

The second term on the right-hand side of Eq. (5.95) represents the optical absorption due to the free carriers in the waveguide structure. Substituting it into Eq. (5.92) to average over the waveguide cross section, we can evaluate the absorption loss α_{abs} for guided modes. In practical waveguides, the propagation loss includes not only this absorption loss but also a scattering loss due to imperfections in the waveguide structure. A phenomenological propagation loss factor representing both absorption and scattering losses given by

$$\alpha_{\mathrm{int}} = \alpha_{\mathrm{abs}} + \alpha_{\mathrm{scat}} \tag{5.97}$$

is usually used. The parameter α_{int} is referred to as the internal loss, since it describes the loss within the waveguide structure.

5.5.3 Confinement Factor

In a DH waveguide constructing is a semiconductor laser, the carrier injection effects are approximately uniform in the active layer and are represented by the first and second terms on the right-hand side of Eq. (5.96), while there is no substantial carrier effect outside the active layer. Let $\Delta n - \mathrm{i}g/2k_0$ be the value in the active layer; then the change in the propagation constant of a guided mode is given by the weighted average over the waveguide cross section and can be written as

$$\begin{aligned}\delta\tilde{\beta} &= k_0\Gamma\left(\Delta n - \frac{\mathrm{i}g}{2k_0}\right)\\ &= k_0\Gamma\,\Delta n - \frac{\mathrm{i}\Gamma g}{2}\end{aligned} \tag{5.98}$$

Since the guided mode field is distributed not only in the active layer but also in the cladding layers, the effective values of the gain and index change for the guided mode are reduced compared with those determined by the active layer. The parameter Γ in the above expression represents the reduction. For TE modes in a planar waveguide of thickness T, Eq. (5.94) can be used to obtain

$$\Gamma = \Gamma_T = \frac{\int_{-T/2}^{+T/2} |E_y(x)|^2 \, dx}{\int_{-\infty}^{+\infty} |E_y(x)|^2 \, dx} \tag{5.99}$$

Thus, Γ is given by the ratio of the power of the guided mode flowing in the active layer to the total power flow, and therefore Γ is called the confinement factor. Also, for TM modes, a confinement factor can be defined similarly, and Eq. (5.98) can be used as an approximate relation.

The complex propagation constant of a guided mode representing both the carrier injection effects and the propagation losses can be written as

$$\beta + \delta\tilde{\beta} + \frac{i\alpha_{\text{int}}}{2} = k_0(N + \Gamma\,\Delta n) - \frac{i\Gamma g}{2} + \frac{i\alpha_{\text{int}}}{2} \tag{5.100}$$

Γg in the above expression is an effective gain factor for the guided mode and is referred to as the mode gain. To distinguish the gain g from this mode gain, the gain g determined by the active layer only is referred to as the material gain. Thus, Γ offers a simple and convenient means for converting the values for the active layer to the effective values for the guided mode. For TE modes in a symmetrical step-index planar waveguide, from Eqs (5.43′) and (5.99) we obtain an expression for Γ_T:

$$\Gamma_T = \frac{1 + 2\gamma_C T / V_T^2}{1 + 2/\gamma_C T} \tag{5.101}$$

The value of Γ_T can be calculated by determining N using Fig. 5.7, determining γ_c and substituting it into the above expression. As a simpler method, it has been shown that Γ_T for the fundamental mode can be calculated with good accuracy by using [9]

$$\Gamma_T \approx \frac{V_T^2}{2 + V_T^2} \tag{5.102}$$

For higher-order modes, Γ_T decreases with increasing mode order. The above two expressions can also be used as an approximation for TM modes, although Γ_T for TM modes is slightly smaller than that for TE modes.

An important requirement in the design of semiconductor lasers is realization of large Γ to attain a large mode gain. A guideline for ordinary DH lasers is to choose the active (guiding) layer thickness T in such a manner that the product of A_g in Eq. (3.40) and Γ_T is maximized. Typical values are $T \approx 0.1$–$0.2\,\mu$m, $V_T \approx 1$, and $\Gamma_T \approx 1/3$. In quantum well lasers, on the other hand, use of a quantum well with a thickness of 10 nm or so as a guiding layer itself results in a very small Γ_T, since $V_T \ll 1$ and the confinement of the guided mode field is weak. To improve the situation, various structures, as shown in Fig. 4.8, are used. Figure 4.8(b) shows a separate-confinement heterostructure single quantum well (SCH-SQW). In the graded-index SCH single quantum well (GRIN-SCH-SQW) using a graded-index waveguide, as shown in Fig. 4.8(c), significant improvement can be expected. The values of Γ_T of these structures, however, are still smaller than those for ordinary DHs. The confinement factor $\Gamma_{T\,\mathrm{MQW}}$ of the multiple-quantum-well (MQW) structure, shown in Fig. 4.8(d), can be estimated as

$$\Gamma_{\mathrm{T\,MQW}} \approx \Gamma_{\mathrm{T\,av}} R_{\mathrm{W}} \tag{5.103}$$

where $\Gamma_{T\,\mathrm{AV}}$ is a confinement factor for an imaginary single-layer waveguide consisting of a guiding layer of a thickness corresponding to the total MQW thickness and a refractive index averaged over the MQWs with weighting by the thicknesses of the wells and barriers, and R_{W} is the ratio of the sum of the thicknesses of the wells to the total thickness of the MQWs. By optimizing the number of well, $\Gamma_{T\,\mathrm{MQW}}$ considerably larger than that for simple single-quantum-well structures can be obtained. Further enhancement of the confinement factor is possible with the modification as shown in Fig. 4.8(e).

5.6 CHANNEL WAVEGUIDES

5.6.1 The Marcatili Method

Lossless channel waveguides, in general, can be specified by the distribution of the refractive index over the cross section, and the guided mode can be examined by solving the wave equation in Eq. (5.24). As an example of fundamental channel waveguides, consider a channel waveguide of rectangular cross section of thickness T and width W, having a step-index profile, which is symmetrical with respect to the y direction, as shown in Fig. 5.11. The guided modes in this structure can be analyzed by the Marcatili [10] method.

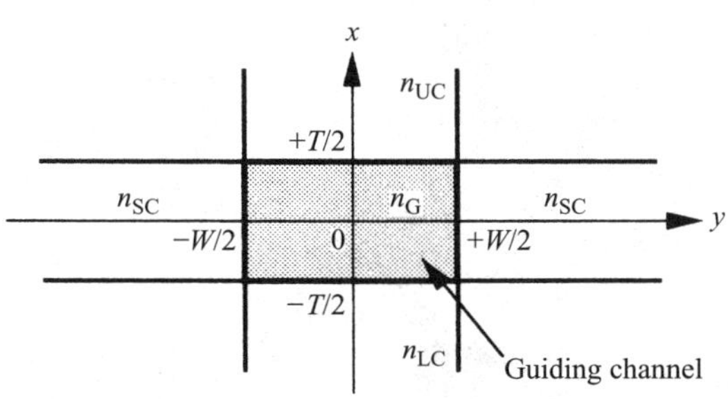

Figure 5.11 Cross section of a rectangular step-index channel waveguide.

Let us assume that the major electric field component is the y component $E_y(x, y)$, and the x component can be approximated as $E_y = 0$. Then, from the Maxwell equations, the other field components are given by

$$E_z = \frac{\mathrm{i}}{\beta}\frac{\partial E_y}{\partial y}, \qquad H_x = \frac{1}{\omega\mu_0\beta}\left(\frac{\partial^2 E_y}{\partial y^2} - \beta^2 E_y\right)$$
$$H_y = \frac{-1}{\omega\mu_0\beta}\frac{\partial^2 E_y}{\partial x\,\partial y}, \qquad H_z = \frac{1}{\mathrm{i}\omega\mu_0}\frac{\partial E_y}{\partial x} \tag{5.104}$$

When $E_y(x, y)$ satisfies the wave equation, all the field components are consistent with the Maxwell equations. By extending the expression for a planar waveguide, the major field amplitude $E_y(x, y)$ can be written as

$$E_y(x,y) = \begin{cases} E_\mathrm{G}\cos\left[\kappa_{\mathrm{G}x}\left(x-\frac{T}{2}\right)+\Phi_\mathrm{UC}\right]\cos\left(\kappa_{\mathrm{G}y}y+\frac{q\pi}{2}\right) \\ E_\mathrm{G}\cos\Phi_\mathrm{UC}\exp\left[-x-\gamma_\mathrm{UC}\left(\frac{T}{2}\right)\right]\cos\left(\kappa_{\mathrm{G}y}y+\frac{q\pi}{2}\right) \\ E_\mathrm{G}\cos\Phi_\mathrm{LC}\exp\left[+\gamma_\mathrm{LC}\left(x+\frac{T}{2}\right)\right]\cos\left(\kappa_{\mathrm{G}y}y+\frac{q\pi}{2}\right) \\ E_\mathrm{G}\cos\left[\kappa_{\mathrm{G}x}\left(x-\frac{T}{2}\right)+\Phi_\mathrm{UC}\right]\cos\left(+\frac{\kappa_{\mathrm{G}y}W}{2}+\frac{q\pi}{2}\right)\exp\left[-\gamma_\mathrm{sc}\left(y-\frac{W}{2}\right)\right] \\ E_\mathrm{G}\cos\left[\kappa_{\mathrm{G}x}\left(x-\frac{T}{2}\right)+\Phi_\mathrm{UC}\right]\cos\left(-\frac{\kappa_{\mathrm{G}y}W}{2}+\frac{q\pi}{2}\right)\exp\left[+\gamma_\mathrm{sc}\left(y+\frac{W}{2}\right)\right] \end{cases} \tag{5.105}$$

where the first line of Eq. (2.26) is for the core region and the following four lines are for the upper, lower, left and right cladding regions, i.e., G, UC, LC, and SC regions, respectively. For the four corner regions, no analytical solution satisfying the boundary condition with Eq. (5.105) is found. This problem can be omitted, since the field amplitude is small in these regions. The wave equations for the five regions require

$$\kappa_{\mathrm{G}x}^2 + \kappa_{\mathrm{G}y}^2 + \beta^2 = k_0^2 n_\mathrm{G}^2, \qquad \kappa_{\mathrm{G}x}^2 - \gamma_\mathrm{SC}^2 + \beta^2 = k_0^2 n_\mathrm{SC}^2$$
$$-\gamma_\mathrm{UC}^2 + \kappa_{\mathrm{G}y}^2 + \beta^2 = k_0^2 n_\mathrm{UC}^2, \qquad -\gamma_\mathrm{LC}^2 + \kappa_{\mathrm{G}y}^2 + \beta^2 = k_0^2 n_\mathrm{LC}^2 \tag{5.106}$$

From the boundary conditions at $x = \pm T/2$ and $y = \pm W/2$, we obtain

$$\kappa_{\mathrm{G}x}T - \Phi_\mathrm{UC} - \Phi_\mathrm{LC} = p\pi \qquad (p = 0, 1, 2, \ldots) \tag{5.107}$$

where

$$\Phi_{\mathrm{UC}} = \mathrm{Tan}^{-1}\left(\frac{\gamma_{\mathrm{UC}}}{\kappa_{\mathrm{G}x}}\right), \qquad \Phi_{\mathrm{LC}} = \mathrm{Tan}^{-1}\left(\frac{\gamma_{\mathrm{LC}}}{\kappa_{\mathrm{G}x}}\right)$$

and also

$$\kappa_{\mathrm{G}y} W - 2\Phi_{\mathrm{SC}} = q\pi \qquad (q = 0, 1, 2, \ldots) \tag{5.108}$$

where

$$\Phi_{\mathrm{SC}} = \mathrm{Tan}^{-1}\left[\left(\frac{n_{\mathrm{G}}}{n_{\mathrm{SC}}}\right)^2 \left(\frac{\gamma_{\mathrm{SC}}}{\kappa_{\mathrm{G}y}}\right)\right]$$

Equations (5.107) and (5.108) are characteristic equations for the direction of the guide thickness and the guide width, respectively, and the integers p and q are mode order numbers for the depth and width directions, respectively. The propagation constant β can be determined by solving the characteristic equations jointly with Eq. (5.106).

The guided modes obtained here are described by the product of guided modes for two planar waveguides perpendicular to the x and y directions, and they show that the optical wave is confined in and near the channel region and propagates along it. The major field components are E_y and H_x, similarly to TE guided modes in a planar waveguide of thickness T. The modes, however, are not exactly TE modes, because $E_z \neq 0$ as Eq. (5.104) shows and therefore the electric field is not perpendicular to the waveguide axis. In fact, in a channel waveguide, TE and TM modes cannot propagate independently, and all the guided modes are hybrid modes. The above-described modes with the major field E_y are designated as E^y_{pq} modes. The modes with the major field E_x are designated as E^x_{pq} modes. The expressions for the E^x_{pq} modes are obtained from those for the above E^y_{pq} modes, by replacing E_y by E_x, replacing the expressions for Φ_{UC} and Φ_{LC} by those for TM modes, and replacing the expression for Φ_{SC} by that for TE modes. The intensity distribution given by $|E_y(x, y)|^2$ is illustrated in Fig. 5.12, for lower-order modes. E^y and E^x modes are also called TE-like and TM-like modes. In semiconductor laser waveguides, the differences in the refractive index between the core and surrounding regions are small, the field amplitude is small except for the major components, and hence it is not important to recognize that the modes are hybrid modes. Therefore, the E^y and E^x modes are often called simply TE and TM modes.

5.6.2 The Effective Index Method

One of the methods for approximate calculation of the field distribution and the propagation constant of a channel waveguide is the effective

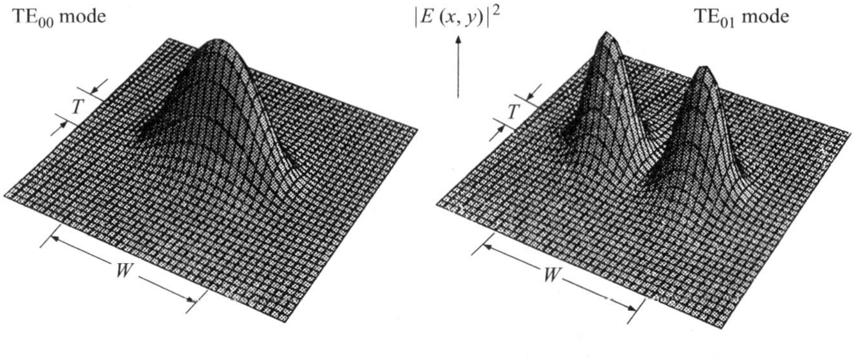

Figure 5.12 Examples of intensity distributions of guided modes in a channel waveguide.

index method. The field confinement is decomposed into those in the depth and width directions, and the structure is converted into two equivalent planar waveguides. Good accuracy is obtained for waveguides having a width larger than the thickness where the field varies slowly in the direction of the width. Since many semiconductor lasers are constructed with such a waveguide, this method offers a convenient and effective means for the analysis and design [10].

Consider a channel waveguide described by a refractive index distribution $n(x, y)$, as shown in Fig. 13. Then consider an imaginary planar guide that is defined by the a diffractive index distribution in the x direction, $n(x, y)$ with a constant y. Let $E(x; y)$ be the electric field amplitude of a TE guided mode in the imaginary planar guide, then satisfies the wave equation for a planar guide:

$$\left(\frac{\mathrm{d}^2}{\mathrm{d}x^2} + k_0^2[n^2(x, y) - N(y)^2]\right)E(x; y) = 0 \tag{5.109}$$

where $N(y)$ is the effective index of refraction for the planar guide. Here $N(y)$ can be considered a local effective index dependent upon the value of the constant y specifying the position in the y direction. $E(x; y)$ and $N(y)$ can readily be calculated by the methods described in Sections 5.4 and 5.5, for cases including those where $N(y)^2$ is complex.

We then use the above $E(x; y)$ and an unknown function $E(y)$ to write the guided mode field in the channel waveguide as

$$\boldsymbol{E}(x, y) \approx \boldsymbol{e}_y E(x; y)E(y) \tag{5.110}$$

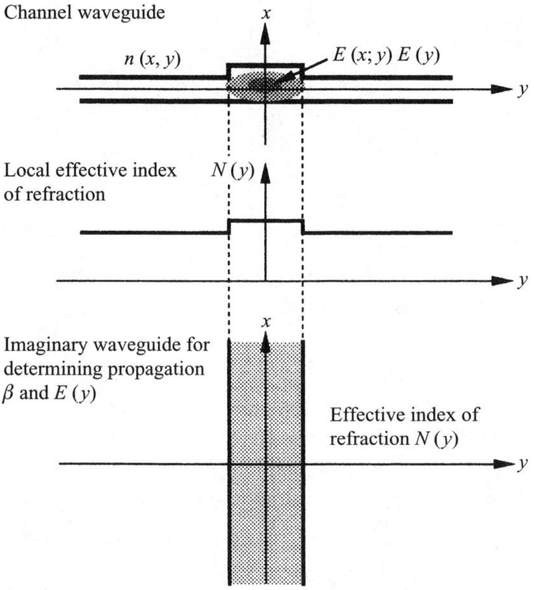

Figure 5.13 Schematic illustration of the effective index of refraction method.

where $\boldsymbol{e}_y$ is a unit vector in the y direction. From Eqs (5.109), (5.24) and (5.110), we obtain an equation for $E(y)$:

$$\left(\frac{d^2}{dy^2} + [k_0^2 N(y)^2 - \beta^2]\right) E(y) = 0 \tag{5.111}$$

We assumed that the variation in $E(x; y)$ with respect to y is slow, and used $\partial E(x; y)\partial y \approx 0$. The above equation is in exactly the same form as the wave equation for a planar waveguide, given by Eq. (5.35). This means that the effective index N of refraction and the field distribution $E(y)$ along the direction of the channel width are same as those for an imaginary planar waveguide specified by the index distribution $N(y)$, as shown at the bottom of Fig. 5.13. So far as the confinement in the direction of the width is concerned, the channel waveguide is equivalent to a medium that is uniform in the x direction and has an index distribution $N(y)$ along the y direction. Therefore, the field $E(y)$ and the propagation constant β can be calculated by again applying the method described in Section 5.4.

5.6.3 Gain-Guiding Waveguides

The simplest method to implement a stripe laser is to form a stripe electrode on the surface of a DH planar waveguide for current injection in a channel

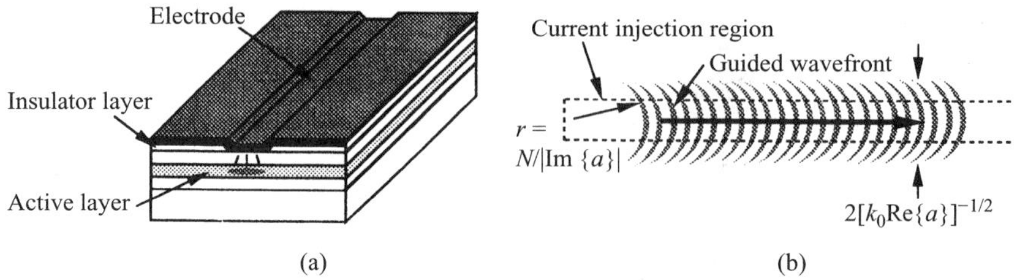

Figure 5.14 Gain waveguide: (a) structure of a gain-guiding laser; (b) intensity distribution and wavc-front of an optical wavc.

region, as shown in Fig. 5.14(a). The density of carriers injected from the stripe electrode through the cladding layer into the active layer forms a distribution with the peak at the center of the channel ($y = 0$) and monotonic decay slopes at both sides of it, as a result of the lateral diffusion. The corresponding complex effective index, in the region where $|y|$ is small, including the carrier-induced change δN can be written approximately as

$$N(y)^2 = N^2 + 2N\Gamma_T\left(\Delta n - \frac{ig}{2k_0}\right)\left[1 - \left(\frac{y}{W}\right)^2\right] \tag{5.112}$$

where y is the coordinate in the direction of the channel width, N is the effective index without carrier injection, Δn and g are the index change and the material gain, respectively, at the center of the channel ($y = 0$) in the active layer, and W is the width of the carrier distribution. Applying the effective index method, and making use of the solution given in Eqs (5.72) and (5.73) for a waveguide of parabolic index distribution, the distribution of the fundamental mode field in the direction of the guide width can be written as

$$G(y) = G_0 \exp\left(-\frac{k_0 \operatorname{Re}\{a\} y^2}{2}\right) \exp\left(-\frac{ik_0 \operatorname{Im}\{a\} y^2}{2}\right) \tag{5.113}$$

where

$$\begin{aligned} a^2 &= \frac{2N\Gamma_T(\Delta n - ig/2k_0)}{W^2} \\ &= \frac{-(N\Gamma_T/k_0)g(\alpha_c + i)}{W^2} \end{aligned} \tag{5.114}$$

Nothing that both Δn and g depend upon the carrier density, and that they are not independent of each other but are quantities proportional to the real and imaginary parts respectively, of the dielectric susceptibility χ_{act}, we define a parameter α_c [12] by

$$\alpha_c = \frac{\mathrm{Re}\{\chi_{act}\}}{\mathrm{Im}\{\chi_{act}\}} \tag{5.115}$$

In Eq. (5.114), we used a relation $\Delta n = -(\alpha_c/2k_0)g$ obtained from Eqs (5.18) and (5.19). Usually, α_c is a positive value of the order of unity. The complex parameter a must satisfy $\mathrm{Re}\{a\} > 0$, and from Eq. (5.114) we have $\mathrm{Im}\{a\} < 0$. Therefore, we see from Eq. (5.113) that the optical wave is confined also in the direction of the stripe width with a 1/e full width of $2[k_0 \mathrm{Re}\{a\}]^{-1/2}$, and the wavefront is convex towards the propagation ($+z$) direction with a curvature radius $r = N/|\mathrm{Im}\{a\}|$, as shown in Fig. 5.14(b). This lateral guiding phenomena, resulting from the strong amplification on the channel axis that keeps the field amplitude large in the vicinity of the axis, is called gain guiding. The diffraction effect that allows the divergence of the wave propagation accounts for the curvature of the wavefront. The parameter α_c defined by Eq. (5.115) is called the antiguiding factor, since it represents the relative index decrease that prevents the guiding associated with the gain.

The gain-guiding waveguides can easily be fabricated, and therefore they played an important role in the process of development of semiconductor lasers [5,13,14]. However, they suffer from several drawbacks, such as weak lateral confinement, poor controllability of this confinement, and difficulty in obtaining single-mode oscillation; therefore at present their importance is low.

5.6.4 Index-Guiding Waveguides

To overcome the drawback of gain-guiding waveguides, it is necessary to construct a channel waveguide having a index structure for lateral field confinement. Various types of index-guiding waveguide are used for semiconductor lasers [5,13–17]. They can be classified into ridge waveguides, rib waveguides, and buried heterostructure waveguides. Examples of the cross sections of these waveguiding structures are illustrated in Fig. 5.15.

Ridge waveguides consists of a ridge structure formed by etching to remove the upper cladding layer on both sides of a channel region. The current is injected within this narrow ridge region. Rib waveguides have a structure where the active layer or cladding layer is thinned on both sides of a channel region and the resultant structure is buried by overgrowing the upper cladding layer. The current is injected within the channel region through the upper cladding layer. These waveguides can be analyzed and

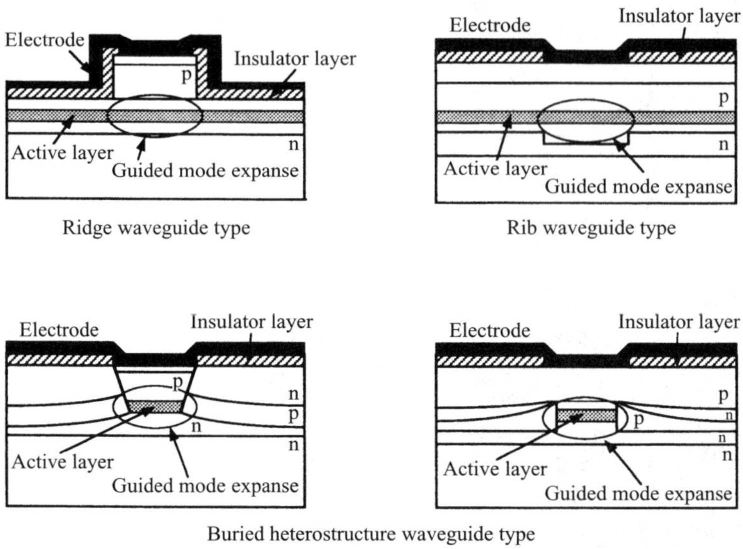

Figure 5.15 Examples of cross-sectional structures of index-guiding semiconductor lasers.

designed with good accuracy by employing the effective index method, since the lateral field confinement is relatively weak. First the theory for planar waveguides is used for design of the multiplayer waveguiding structure. For step-index waveguides, expressions using the normalized guide thickness V_T and Fig. 5.7 can be used to determine the effective indexes N_c and N_s for the regions within and outside the channel, respectively. Using the values of N_c and N_s, a normalized guide width is defined by

$$V_W = k_0 W (N_C^2 - N_S^2)^{1/2} \tag{5.116}$$

For waveguides that are symmetrical with respect to the direction of the guide width, the asymmetry parameter is $a = 0$. Then the effective index N for the channel waveguide can readily be calculated by using Fig. 5.7 again and

$$b = \frac{N^2 - N_S^2}{N_C^2 - N_S^2} \tag{5.117}$$

To obtain good laser performances, it is necessary to design a single-mode waveguide by choosing the thickness and the width so that both V_T and V_W satisfy the single-mode condition in Eq. (5.63). The confinement factor Γ is given approximately by

$$\Gamma = \Gamma_T \Gamma_W \tag{5.118}$$

where Γ_T and Γ_W are confinement factors for the directions of the guide thickness and width, respectively, and they can be evaluated by the method described in Section 5.5.3.

Buried heterostructure waveguides have a structure where the guiding layer is preserved in a channel region and is completely removed on both sides of the channel; the resultant structure is buried by an overgrown cladding layer. They offer advantages such as strong lateral field confinement and high stability of the lateral mode. The approximate analysis and design can be achieved by employing the Marcatili method or the effective index method. In the effective index method, the effective index for the regions on both sides of the channel cannot be determined by the procedure described in Section 5.6.2, since there is no waveguiding layer present in the side regions. The difficulty, however, can be solved by the following method. Let n_C and n_S be the refractive indexes in the guiding channel and the side claddings, respectively, and let N_C be the effective index for the channel region. Here we can use the field distribution $E(x)$ in the thickness direction for the region within the channel, instead of $E(x; y)$ in Eq. (5.110). Then, for the region on both sides of the channel, the equation that results when $N(y)^2$ in Eq. (5.111) is replaced by $n_S^2 - (n_C^2 - N_C^2)$ holds, and the effective index N_S for the side regions is given by

$$N_S = [n_S^2 - (n_C^2 - N_C^2)]^{1/2} \tag{5.119}$$

By applying the effective index method with N_S and N_C determined in this way, the buried heterostructure waveguides can be analyzed and designed in a similar manner to ridge and rib waveguides.

5.7 REFLECTION AT WAVEGUIDE FACETS

A facet perpendicular to the optical axis of a semiconductor DH waveguide, formed by cleaving the waveguide, functions as a mirror that partially reflects a guided mode power. Such facet mirrors are used to construct a resonator for semiconductor lasers. The reflectivity of the facet mirror is an important parameter for analyzing the laser performances.

5.7.1 Reflection of a Plane Wave

As the simplest model for facet reflection, consider an interface between a semiconductor of refractive index n and the air, and a plane wave incident on to the interface from the semiconductor region at an incidence angle θ.

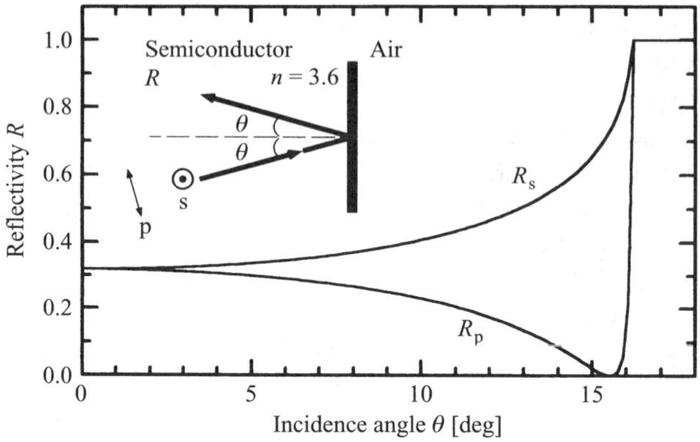

Figure 5.16 Reflectivity for a plane wave reflecting at a boundary between semiconductor and air.

The power reflectivity is given by the well-known Fresnel formula

$$R_p = \frac{\mathrm{Tan}^2(\theta - \theta')}{\mathrm{Tan}^2(\theta + \theta')} \tag{5.120a}$$

$$R_s = \frac{\sin^2(\theta - \theta')}{\sin^2(\theta + \theta')} \tag{5.120b}$$

for the p polarization with the electric field parallel to the plane of incidence, and the s polarization with the electric field perpendicular to the plane of incidence, respectively [18]. From the Snell law, the angle of refraction in the air is given by $\theta' = \sin^{-1}(n \sin \theta)$. The dependence of the reflectivities R_p and R_s for $n = 3.6$ on the incidence angle θ are shown in Fig. 16. For $\theta = 0$, the reflectivities are the same for both polarizations and are approximately 32%. Except for $\theta = 0$ and $\theta > 16°$ where total internal reflection takes place for both polarizations, the relation $R_p < R_s$ holds.

A guided mode in the waveguide can be described by two plane-wave components propagating at angles $\pm\theta$, and these plane waves for TE modes correspond to s waves, and those for TM modes to p waves. This consideration suggests that the reflectivity for a TE mode is higher than that for a TM mode of the same order. Decreasing the guide thickness leads to an increase in θ and would result in an increase in the TE reflectivity and a decrease in the TM reflectivity. These predictions are correct for cases where the guide thickness is sufficiently large. However, they do not necessarily apply generally, since the guided modes have evanescent tails in the cladding regions.

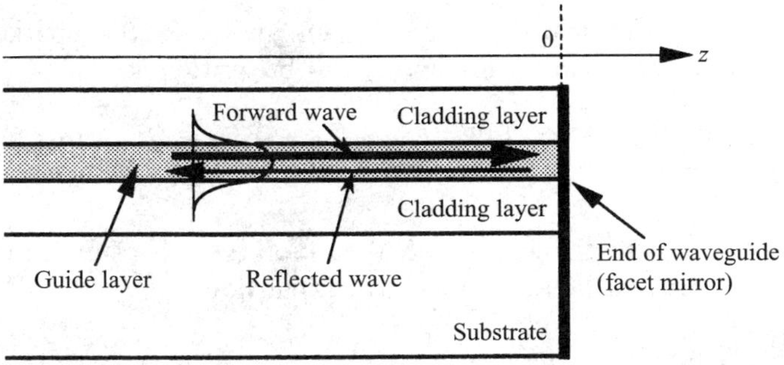

Figure 5.17 Reflection of a guided mode at a waveguide facet.

5.7.2 Reflectivity for Guided Modes

Consider a guided wave of TE_m mode impinging on a waveguide facet at $z=0$ after forward propagation in the waveguide in the $+z$ direction, as shown in Fig. 5.17. Backward-propagating waves produced by the reflection at the facet include not only a guide wave of the same mode as the forward wave but also another mode. Based upon Eq. (5.30), the total field in the waveguide ($z<0$) may be written in a form of mode expansion including the forward wave and the backward wave as [19]

$$E_y(x,z) = \Psi_m(x)\exp(i\beta_m z) + \sum_l A_{ml}\Psi_l(x)\exp(-i\beta_l z) \tag{5.121}$$

Although radiation modes should be included in the exact expansion, here we assume that conversion to radiation modes is negligible and the total field can be described approximately by an expansion with guided modes only. The major magnetic field component associated with Eq. (5.121) can be written as

$$H_x(x,z) = -\frac{1}{\omega\mu_0}\left[\Psi m(x)\beta_m\exp(i\beta_m z) - \sum_l A_{ml}\Psi_l(x)\beta_l\exp(-i\beta_l z)\right] \tag{5.122}$$

The function $\Psi_m(x)$ is the field profile for the TE_m mode given by Eq. (5.43) and can be written as

$$\Psi_m(x) = \int \Psi_m(k)\exp(ikx)\,dk \tag{5.123}$$

where $\Psi_m(k)$ is the Fourier transform of $\Psi_m(x)$, and k is the spatial frequency. The orthonormal relation for $\Psi_m(x)$ can be written as

$$\int \Psi^*_{m'}(x)\Psi_m(x)\,\mathrm{d}x = 2\pi \int \Psi^*_{m'}(k)\Psi_m(k)\,\mathrm{d}k = \delta_{mm'} \tag{5.124}$$

In the free space outside the waveguide ($0 < z$), on the other hand, the electric field can be expressed by the plane-wave expansion

$$E_y(x, z) = \int R(k) \exp[\mathrm{i}kx + \mathrm{i}\beta(k)z]\,\mathrm{d}k \tag{5.125}$$

where

$$\beta(k) = (k_0^2 - k^2)^{1/2}$$

and an associated expression for the major magnetic field:

$$H_x(x, z) = \frac{-1}{\omega\mu_0} \int R(k)\beta(k) \exp[\mathrm{i}kx + \mathrm{i}\beta(k)z]\,\mathrm{d}k \tag{5.126}$$

Since E_y and H_x should be continuous at the boundary $z = 0$, we obtain from Eqs (5.121)–(5.126)

$$\Psi_m(k) + \sum_l A_{ml}\Psi_l(k) = R(k) \tag{5.127a}$$

$$\beta_m \Psi_m(k) - \sum_l A_{ml}\beta_l \Psi_i(k) = R(k)\beta(k) \tag{5.127b}$$

Eliminating $R(k)$ from the above equations yields

$$\beta_m \Psi_m(k) - \sum_l A_{ml}\beta_l \Psi_l(k) = \beta(k)\left[\Psi_m(k) + \sum_l A_{ml}\Psi_l(k)\right] \tag{5.128}$$

and, integrating with respect to k the above equation multiplied by $2\pi\ \Psi^*_{m'}(k)$, and using the orthonormal relation, we obtain

$$\sum_l A_{ml}(\beta_l \delta_{m'l} + \beta_{m'l}) = \delta_{m'm}\beta_m - \beta_{m'm} \tag{5.129}$$

where

$$\beta_{m'm} = 2\pi \int \Psi^*_{m'}(k)\beta(k)\Psi_m(k)\,\mathrm{d}k$$

The solution of the above equation can be written as

$$[A] = ([\beta] - [B]^{\mathrm{t}})([\beta] + [B]^{\mathrm{t}})^{-1} \tag{5.130}$$

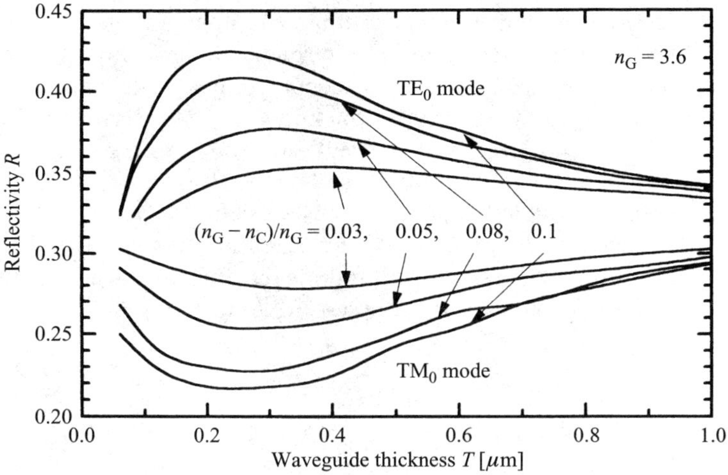

Figure 5.18 Example of the calculated facet reflectivity of guided wave [19].

where $[A]$ is a matrix with $A_{mm'}$ as the elements, $[B]$ a matrix with $B_{m'm}$ as the elements, and $[\beta]$ a diagonal matrix with β_m as the diagonal elements. Since the amplitude of the backward TE$_{m'}$ mode produced from the forward TE$_m$ with an amplitude of unity is $A_{mm'}$, the power reflectivity for the relevant modes is given by $|A_{mm'}|^2$. A similar calculation can be made for TM modes, by using $\Psi_m(x)$ for TM modes and a diagonal matrix $[\beta]$ with β_m/N_m^2 (N_m is the effective index of the TM$_m$ mode) as the diagonal elements.

Figure 5.18 shows the reflectivities for fundamental TE and TM modes calculated by the above procedure. As expected, decreasing the guide thickness results in an increase in the TE reflectivity and a decrease in the TM reflectivity in the region of large guide thicknesses. In the region of small guide thicknesses, however, the reflectivities exhibit opposite dependences and, in the limiting case for small thicknesses, the reflectivities approach each other again. This is because, an decreasing the guide thickness, the evanescent penetration increases and for the limiting case the guided mode profile is dominated by the evanescent wave. Even so, the TE reflectivity remains larger. The difference between the TE and TM reflectivities is the factor dominating the oscillation polarization of DH semiconductor lasers of facet mirror type. The results presented here for planar waveguides apply also for channel waveguides as an approximation, since channel guides for semiconductor lasers have a width considerably larger than the thickness.

In the above discussion, use of an as-cleaved facet as a mirror was assumed. It is also possible to tailor the reflectivity by coating the facet with a transparent dielectric thin film. It is known that, for a plane wave incident normally to an interface between a medium of refractive index n_s and the air, the reflectivity can be reduced to zero by depositing a thin film of a refractive index $n_f = n_s^{1/2}$ and an optical thickness of $\lambda/4$. Although for waveguide facets the reflectivity cannot be reduced exactly to zero, similar design allows reduction in the reflectivity to 10^{-4} or less. It is also possible to realize a high reflectivity of 90% or larger, using a multilayer structure consisting of high-index and low-index layers deposited alternately. The facet coating technique offers an important means for improvement in semiconductor laser performances and implementation of traveling-wave-type semiconductor optical amplifiers.

5.8 WAVEGUIDE FABRY–PEROT RESONATOR

In a structure that results when a waveguide is cleaved to cut it into a finite length and to form facets at the both ends, the guided mode propagates back and forth with successive reflections at the facets. The optical power is accumulated substantially, when a resonance condition is satisfied. The structure is called a waveguide Fabry–Perot (FP) resonator, since it is a waveguide implementation of the FP resonator consisting of two mirrors aligned parallel to each other [18]. Providing this resonator structure with an optical gain by current injection results in a FP-type semiconductor laser. This section presents the passive characteristics of a waveguide FP resonator.

5.8.1 Transmission Characteristics

Consider a waveguide FP resonator of length L and an optical wave incident into it through the left facet, as shown in Fig. 5.19. Here we omit the carrier

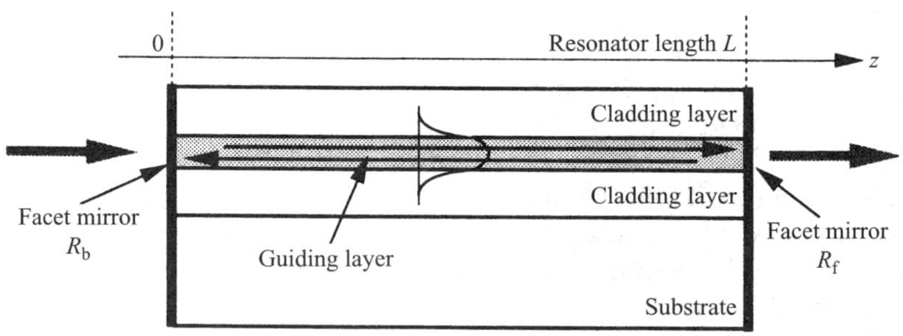

Figure 5.19 Waveguide Fabry–Perot resonator.

injection effects to examine the passive characteristics. Then the complex propagation constant $\tilde{\beta}$, generally given by Eq. (5.100), can be written as

$$\tilde{\beta} = \beta + \frac{\mathrm{i}\alpha_{\mathrm{int}}}{2}, \qquad \alpha_{\mathrm{int}} > 0, \qquad \beta = Nk_0 = \frac{N\omega}{c} \tag{5.131}$$

The optical wave in the waveguide resonator can be written as

$$\boldsymbol{E}(x, y, z) = \boldsymbol{E}(x, y)[A^+ \exp(+\mathrm{i}\tilde{\beta}z) + A^- \exp(-\mathrm{i}\tilde{\beta}z)] \tag{5.132}$$

where the first and second terms in the square brackets represent the forward and backward waves, respectively. Let R_b and R_f be the power reflectivities of the left and right facet mirrors, respectively. Then, neglecting the mode conversion, the power transmissivities are $1 - R_\mathrm{b}$ and $1 - R_\mathrm{f}$. Let A_{in} and A_{tr} be the amplitudes of the wave incident from the left-hand side of the resonator and the wave transmitted into the right-hand side, respectively; then from the boundary conditions for both facets we obtain the relations

$$\begin{aligned} &A^+ = (1 - R_\mathrm{b})^{1/2} A_{\mathrm{in}} + R_\mathrm{b}^{1/2} A^-, \qquad A^- \exp(-\mathrm{i}\tilde{\beta}L) = R_\mathrm{f}^{1/2} A^+ \exp(+\mathrm{i}\tilde{\beta}L) \\ &A_{\mathrm{tr}} = (1 - R_\mathrm{f})^{1/2} A^+ \exp(+\mathrm{i}\tilde{\beta}L) \end{aligned} \tag{5.133}$$

The incident wave is assumed to have a field distribution proportional to $\boldsymbol{E}(x, y)$ on the left facet, and the incident and transmitted waves are normalized so that the powers are given by $|A_{\mathrm{in}}|^2$ and $|A_{\mathrm{tr}}|^2$, respectively.

From the above equations, the power transmissivity T of the resonator is calculated as

$$\begin{aligned} T &= \frac{|A_{\mathrm{tr}}|^2}{|A_{\mathrm{in}}|^2} \\ &= \frac{(1 - R_\mathrm{f})(1 - R_\mathrm{b}) \exp(-\alpha L)}{1 + R_\mathrm{f} R_\mathrm{b} \exp(-2\alpha L) - 2(R_\mathrm{f} R_\mathrm{b})^{1/2} \exp(-\alpha L) \cos(2\beta L)} \end{aligned} \tag{5.134}$$

where α_{int} is abbreviated as α. The transmissivity T takes the maximum periodically at $2\beta L = 2m\pi$ (m is an integer). The maximum condition is that all the waves produced by the successive mirror reflections and the round trips in the waveguide are superimposed in phase. When this condition is satisfied, an increase in the amplitudes of the wave in the waveguide and the transmitted wave, i.e., resonance, occurs. The dependence of T on the optical frequency ω is shown in Fig. 5.20. Since from Eq. (5.131) $\delta(2\beta L) = 2L(\partial\beta/\partial\omega)\delta\omega$, the separation $\Delta\omega$ in angular frequency between adjacent peaks is given by

$$\Delta\omega = \frac{2\pi}{2L\,\partial\beta/\partial\omega} = 2\pi\frac{v_\mathrm{g}}{2L} = 2\pi\frac{c}{2LN_\mathrm{g}} \tag{5.135}$$

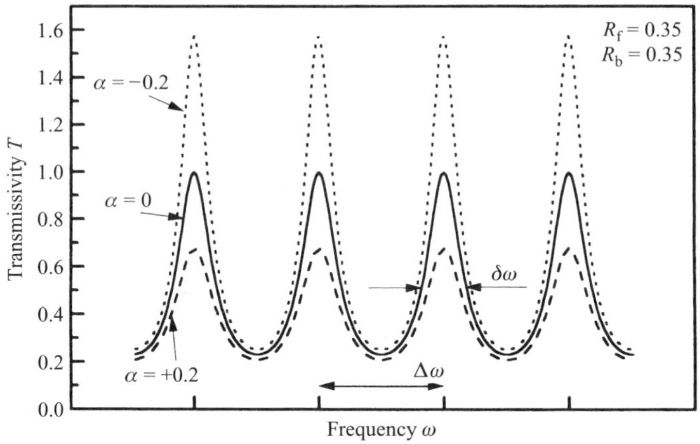

Figure 5.20 Spectral transmissivity of a waveguide Fabry-Perot resonator.

$$v_g = \frac{1}{\partial\beta/\partial\omega} = \frac{c}{N_g}, \qquad N_g = N + \omega\frac{\partial N}{\partial\omega} \tag{5.136}$$

where v_g is the group velocity, at which the optical power is transmitted in the waveguide, and N_g is the group index of refraction. The full width $\delta\omega$ at half–maximum can be calculated from Eq. (5.134); the ratio of $\delta\omega$ to $\Delta\omega$ is given approximately by

$$F = \frac{\Delta\omega}{\delta\omega} = \frac{\pi R^{1/2}}{1-R}, \qquad R = (R_f R_b)^{1/2}\exp(-\alpha L) \tag{5.137}$$

The ratio F is referred to as the finesse. The above expression for F, which applies for cases where R is not substantially small compared with unity, shows that resonators consisting of a low-loss waveguide and mirrors of high reflectivities R have a large F and exhibit a sharp resonance characteristic.

5.8.2 Optical Power Accumulation

Consider a case where there is no external incidence of an optical wave to the resonator, but a source that emits an optical wave of frequency ω propagating in the $+z$ direction is positioned at $z = \xi$ in the waveguide. Then the wave in the waveguide can be written as

$$\boldsymbol{E}(x, y, z) = \begin{cases} \boldsymbol{E}(x, y)[A^+\exp(+i\tilde{\beta}z) + A^-\exp(-i\tilde{\beta}z)] & (0 < z < \xi) \\ \boldsymbol{E}(x, y)[B^+\exp(+i\tilde{\beta}z) + B^-\exp(-i\tilde{\beta}z)] & (\xi < z < L) \end{cases} \tag{5.138}$$

The boundary conditions at the facets ($z=0$, L) and $z=\xi$ can be written as

$$A^+ = R_b^{1/2} A^-, \qquad B^- \exp(-i\tilde{\beta}L) = R_f^{1/2} B^+ \exp(+i\tilde{\beta}L)$$
$$A^+ \exp(+i\tilde{\beta}\xi) + a\,d\xi = B^+ \exp(+i\tilde{\beta}\xi), \qquad A^- = B^- \tag{5.139}$$

where $a\,d\,\xi$ represents the amplitude of the source. By determining A^+, A^-, B^+, and B^- from above equations, the optical power P_f transmitted to the right-hand side of the waveguide is calculated as

$$P_f = \frac{(1 - R_f)\exp[-\alpha(L - \xi)]|a\,d\xi|^2}{1 + R_f R_b \exp(-2\alpha L) - 2(R_f R_b)^{1/2} \exp(-\alpha L)\cos(2\beta L)} \tag{5.140}$$

The power transmitted P_b to the left-hand side of the waveguide is given by a similar expression with the same denominator as the above expression. The ω dependences of P_f and P_b are similar to that of the transmissivity T shown in Fig. 5.20; they have maxima periodically at $2\beta L = 2m\pi$ (m is an integer). This means that the resonance takes place and the optical power is accumulated substantially in the resonator, when the optical waves are superimposed in phase.

5.8.3 Longitudinal Resonance Modes and Photon Lifetime

An optical wave of steady state cannot exist in a passive FP resonator, provided that there is no external incidence, nor internal generation of an optical wave. This is evident from the fact that on putting $A_{in}=0$ in Eq. (5.133) we have $A^+=A^-$. However, a field satisfying the boundary conditions can be found, if we extend the optical angular frequency ω to a complex frequency $\omega - i\gamma$ and assume an optical wave having a time dependence in the form of $\exp[-i(\omega - i\gamma)t] = \exp(-i\omega t)\exp(-\gamma t)$. Let us write the complex propagation constant dependent on the frequency ω as $\tilde{\beta}(\omega)$, then the value for a frequency $\omega - i\gamma$ is given by

$$\tilde{\beta}(\omega - i\gamma) = \tilde{\beta}(\omega) + \frac{\partial \tilde{\beta}}{\partial \omega}(-i\gamma) = \tilde{\beta} - \frac{i\gamma}{v_g} \tag{5.141}$$

where v_g is the group velocity of the guided mode. The validity of the above simple treatment can be shown by using the time-dependent wave equations. Then the optical wave in the waveguide can be written as

$$\boldsymbol{E}(x, y, z; t) = \boldsymbol{E}(x, y)E(z)\exp(-i\omega t)\exp(-\gamma t) \tag{5.142}$$

and $E(z)$ can be expressed by using the complex propagation constant $\tilde{\beta}(\omega - i\gamma)$ as

$$E(z) = A^+ \exp[+i\tilde{\beta}(\omega - i\gamma)z] + A^- \exp[-i\tilde{\beta}(\omega - i\gamma)z] \tag{5.143}$$

Using the power reflectivities R_b and R_f of the left- and right-hand facet mirrors, respectively, the boundary conditions can be written as

$$A^+ = R_b^{1/2} A^-, \qquad A^- = R_f^{1/2} A^+ \exp[+2i\tilde{\beta}(\omega - i\gamma)L] \tag{5.144}$$

For the above equations to have nontrivial solutions except for $A^+ = A^- = 0$,

$$(R_f R_b)^{1/2} \exp[2i\tilde{\beta}(\omega - i\gamma)L] = 1 \tag{145}$$

is required, and the above relation is decomposed into the real and imaginary parts as follows:

$$2\,\mathrm{Re}\{\tilde{\beta}(\omega - i\gamma)\}L = 2m\pi \qquad (m \text{ is an integer}) \tag{5.146a}$$

$$(R_f R_b)^{1/2} \exp[-2\mathrm{Im}\{\tilde{\beta}(\omega - i\gamma)\}L] = 1 \tag{5.146b}$$

Equation (5.145) indicates that in the resonator there can exist only such waves that are superimposed in phase after a round trip. Each of the optical waves satisfying this condition is called a longitudinal mode. The integer m is the order of the longitudinal mode. From Eq. (5.146), we see that the longitudinal mode function $E(z)$ can be written as

$$E(z) = A^+ \exp\left[+\left(\frac{\alpha_{\mathrm{mir}}}{2} + \frac{im\pi}{L}\right)z\right] + A^- \exp\left[-\left(\frac{\alpha_{\mathrm{mir}}}{2} + \frac{im\pi}{L}\right)z\right]$$

$$\alpha_{\mathrm{mir}} = \frac{1}{2L}\ln\left(\frac{1}{R_f R_b}\right) \tag{5.147}$$

To distinguish from the longitudinal mode presented here, the guided mode of the waveguide described by $\boldsymbol{E}(x, y)$ is referred to as the lateral mode. From Eq. (5.146a) we see that, for cases where $\tilde{\beta}$ is given by Eq. (5.131), the angular frequencies ω_m for the mth longitudinal mode are aligned with a separation

$$\Delta\omega = \frac{2\pi}{2L\,\partial\beta/\partial\omega} = 2\pi\frac{v_g}{2L} = 2\pi\frac{c}{2LN_g} \tag{5.148}$$

which is the same as the separation of the transmission peak frequencies. The separation is called the longitudinal mode separation. If the wavelength dependence of the effective index is neglected, the longitudinal modes are aligned with a constant separation.

When the lateral mode $\boldsymbol{E}(x, y)$ is normalized, the power flows P^+ and P^-, for the forward and backward waves, respectively, at $t=0$ are given by the absolute squares of the first and second terms of the right-hand side of Eq. (5.147), i.e.,

$$P^+(z) = |A^+|^2 \exp(+\alpha_{\text{mir}} z) \tag{5.149a}$$

$$P^-(z) = |A^-|^2 \exp(-\alpha_{\text{mir}} z) \tag{5.149b}$$

The energies per unit length of the forward and backward waves stored in the waveguide are given by P^+/v_g and P^-/v_g, respectively, with the group velocity v_g; the total optical energy stored in the resonator is given by the sum of P^+/v_g and P^-/v_g integrated over the waveguide section $0 < z < L$. By normalizing the field so that the total stored energy equals to 1 J, A^+ and A^- are determined as

$$\begin{aligned}\frac{|A^+|^2}{R_b} &= |A^-|^2 \\ &= \frac{v_g \alpha_{\text{mir}} R_f^{1/2}}{[1-(R_f R_b)^{1/2}](R_f^{1/2} + R_b^{1/2})}\end{aligned} \tag{5.150}$$

The normalized resonance mode is given by $\boldsymbol{E}(x, y)E(z)$ with the lateral mode $\boldsymbol{E}(x, y)$ and the longitudinal mode $E(z)$ given by Eq. (4.147) with the above A^+ and A^- substituted.

Equation (5.142) indicates that the power in the resonator decays with time in a form of $\exp(-2\gamma t)$. From Eqs (5.141) and (5.146b), the decay factor γ for cases where $\tilde{\beta}$ is given by Eq. (5.131) can be written as

$$\begin{aligned}2\gamma &= v_g\left[\alpha_{\text{int}} + \frac{1}{2L}\ln\left(\frac{1}{R_f R_b}\right)\right] \\ &= v_g(\alpha_{\text{int}} + \alpha_{\text{mir}})\end{aligned} \tag{5.151}$$

$$\alpha_{\text{mir}} = \frac{1}{2L}\ln\left(\frac{1}{R_f R_b}\right)$$

If there is no optical power source, the optical wave can exist only in such a form that the power stored in the resonator in the past is consumed and attenuates with time. Equation (5.151) shows that the decay factor is given by the sum of the internal loss α_{int} of the waveguide and the mirror loss α_{mir}. The latter represents the attenuation due to partial power outgoing through the facet mirrors and is given by the round-trip power multiplication factor $R_f R_b$ converted into a decay factor by taking the logarithm and dividing by the round-trip length $2L$ for averaging. In the right-hand side of the above expression, the group velocity is included as a factor to convert the decay per unit length to that per unit time. Rewriting 2γ given by Eq. (5.151) as $1/\tau_{ph}$, we have

$$\frac{1}{\tau_{ph}} = v_g\left[\alpha_{int} + \frac{1}{2L}\ln\left(\frac{1}{R_f R_b}\right)\right] \tag{5.152}$$

Then τ_{ph}, the time for which the power in the resonator decays to $1/e$ times the initial value, is referred to as the photon lifetime and is an important parameter for the rate equation analysis of semiconductor lasers.

Figure 5.21 illustrates the distributions of the forward wave power P^+ and backward wave power P^-, given by Eq. (5.149), and $P^+ + P^-$. As shown in this figure, P^+ and P^- observed at an instant exhibit distributions that increase with position shift in the direction of propagation, to satisfy the boundary conditions at the partially reflecting facet mirrors. However, the optical waves actually propagate with time at the group velocity v_g (proceed a distance $v_g t$), and the entire optical power attenuates in the form of $\exp(-2\gamma t)$. This means that the propagating optical power attenuates to be $\exp(+\alpha_{mir} v_g t)\exp(-2\gamma t) = \exp(-\alpha_{int} v_g t)$ times the initial value. Therefore, the results obtained here are consistent with the physical insights. The intensity distribution $|E(z)|^2$ of a longitudinal mode is that of a standing wave formed by the interference of the forward and backward waves, as

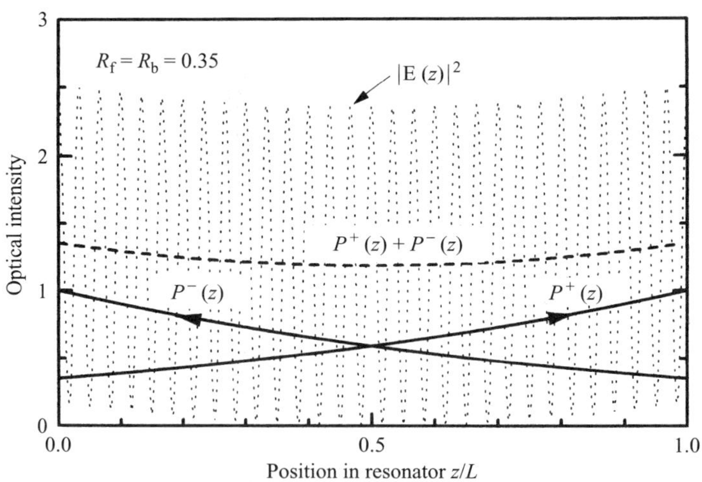

Figure 5.21 Optical power flow distribution P^+ and P_- and intensity distribution $|E(z)|^2$ of a resonance mode in a waveguide Fabry–Perot resonator.

shown in Fig. 5.21. The number of standing-wave periods coincides with the longitudinal mode order m.

5.9 FAR-FIELD PATTERNS

The optical wave transmitted through the waveguide is partly reflected at the facet; simultaneously, it is partly transmitted through the facet and radiates into the free space. The transmitted wave, observed just outside the facet, is in a form proportional to the lateral mode $\boldsymbol{E}(x, y)$ of the waveguide and the intensity distribution is called the near-field pattern (NFP). The output wave propagates in free space, with evolution of the wavefront and the intensity distribution. The intensity distribution observed at a plane sufficiently away from the facet is called the far-field pattern (FFP). Analysis of the wavefront and the intensity distribution in the far field provides important insights useful for the design of focusing, collimating and coupling into an optical fiber of semiconductor laser light for practical applications.

We put the near-field distribution, given by a lateral mode $\boldsymbol{E}(x, y)$, as

$$\boldsymbol{E}(x, y) = \boldsymbol{e}\Psi(x, y) \tag{5.153}$$

where $\boldsymbol{e}$ is a unit vector for describing the polarization. Note that the spatial expanse of $\Psi(x, y)$ is comparable with the effective thickness T_{eff} (or width) of the waveguide. We take a plane of observation (X–Y plane) perpendicular to the z axis at a distance D from the facet, as shown in Fig. 5.22. According to the scalar diffraction theory [18], the optical field amplitude $\Psi(X, Y)$ in the Fraunhofer region satisfying $D \gg 2T_{\text{eff}}^2/\lambda$ is given by

$$\begin{aligned}\Psi(X, Y) = {} & (-\mathrm{i}2\pi k_0 \cos\delta)\frac{\exp(\mathrm{i}k_0 r)}{r} \\ & \times \left(\frac{1}{2\pi}\right)^2 \iint \Psi(x, y)\exp[-\mathrm{i}k_0(x\sin\theta_x + y\sin\theta_y)]\,\mathrm{d}x\,\mathrm{d}y\end{aligned} \tag{5.154}$$

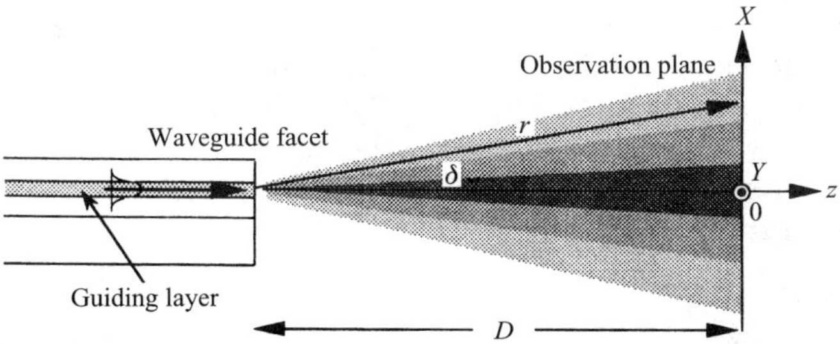

Figure 5.22 Radiation of guided wave through a facet.

$$\cos\delta = \frac{D}{r}, \qquad \sin\theta_x = \frac{X}{r}, \qquad \sin\theta_y = \frac{Y}{r}$$

where r denotes the distance from the center of the facet to the observation point (X, Y). On the right-hand side of the above expression, $\exp(ik_0 r)/r$ describes a wavefront of a spherical wave diverging from the center of the facet. The second line indicates that the field amplitude distribution in the far field is proportional to the two-dimensional Fourier transform of that in the near field. In many cases, this distribution can be written in a form of a product of a function of X and a function of Y. The relative intensity profile in the X direction is given by $|\cos\delta\, \Psi(k_0 \sin\theta_x)|^2$ using the Fourier transform of the mode function $\Psi(x)$ given by

$$\Psi(k) = \frac{1}{2\pi}\int \Psi(x)\exp(-ikx)\,\mathrm{d}x$$

with substitution of $k = k_0 \sin\theta_x$ into it. For a symmetrical step-index waveguide, for example, the Fourier transform of the mode function is calculated as

$$\Psi(k) = \frac{2^{1/2}}{\pi T_{\mathrm{eff}}^{1/2}}\cos\left(\frac{\kappa T}{2}\right)k_0^2(n_{\mathrm{G}}^2 - n_{\mathrm{c}}^2)\frac{\gamma\cos(kT/2) - k\sin(kT/2)}{(\gamma^2 + k^2)(\kappa^2 - k^2)}$$

$$\text{(TE mode)} \tag{5.155a}$$

$$\Psi(k) = \frac{2^{1/2}}{\pi T_{\mathrm{eff}}^{1/2}}\cos\left(\frac{\kappa T}{2}\right) \times \left(\frac{\gamma(n_{\mathrm{G}}/n_{\mathrm{c}})^2\cos(kT/2) - k\sin(kT/2)}{\kappa^2 - k^2} + \frac{\gamma\cos(kT/2) - k\sin(kT/2)}{\gamma^2 + k^2}\right)$$

$$\text{(TM mode)} \tag{5.155b}$$

The intensity distributions $|\cos\delta\, \Psi(k_0 \sin\theta_x)|^2$ for the fundamental modes are single-peak functions, as shown in Fig. 5.23. The distributions of higher-order modes have zeros of the same number as the mode order. A simple Gaussian approximation $\Psi(x) = \exp[-(x/w)^2]$ for the fundamental mode gives the Fourier transform:

$$\Psi(k_0 \sin\theta) = \frac{w}{2\pi^{1/2}}\exp\left[-\left(\frac{wk_0 \sin\theta}{2}\right)^2\right] \tag{5.156}$$

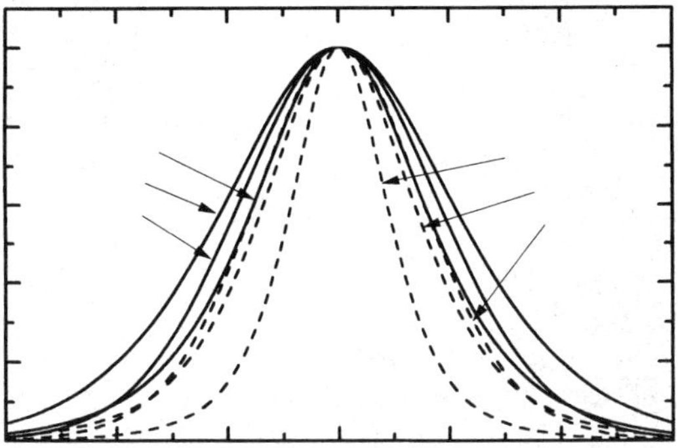

Figure 5.23 Example of calculated far-field intensity distributions of a guide mode (fundamental mode).

Therefore, using $\cos\delta \approx 1$, the far field pattern is also given approximately by a Gauss function, and the $1/e^2$ half-width θ_w, of the angular divergence in the far field can be calculated simply by

$$\theta_w = \sin^{-1}\left(\frac{\lambda}{w\pi}\right) \tag{5.157}$$

As the above equation shows, the smaller the near-field mode size, the larger is the far-field angular divergence. Since semiconductor lasers usually have an active channel width larger than the active layer thickness, they exhibit an elliptic far-field pattern with a divergence angle $\theta_\perp$ in the direction perpendicular to the active layer larger than the divergence angle $\theta_{//}$ that in the parallel direction, as shown in Fig. 6.1 in the next chapter.

As Eqs (5.155) and (5.156) are real functions, the wavefront of the output wave in the far field is identical with that of a spherical wave diverging from the center of the facet. This is true for many lasers using an index-guiding waveguide. Therefore, the output wave can readily be focused or collimated by a single lens. The guided mode wavefront in lasers using a gain-guiding waveguide, however, is curved in a convex shape extended toward the direction of propagation with a radius r of curvature, as shown in Fig. 5.14(b). Therefore, the output wavefronts observed in the plane of waveguide are those of a spherical wave diverging from a virtual diverging point shifted from the facet towards inside the waveguide by r/N (N is the effective index). On the other hand, the diverging point for the output wavefronts observed in the plane perpendicular to the waveguide plane is at the center of the facet. The separation between the two diverging points gives rise to astigmatism in the output wave focused by an ordinary lens. Another drawback of gain-guiding waveguides is that the refractive index change related to the antiguiding factor α_c makes it difficult to obtain a stable single-peak far-field pattern.

REFERENCES

1. D. Marcuse, *Theory of Dielectric Optical Wavegudies*, Academic Press, New York (1974).
2. T. Tamir (ed.), *Integrated Optics*, Springer, Berlin (1975).
3. S. Kawakami, *Optical Waveguides* (in Japanese), Asakura, Tokyo (1980).
4. H. Nishihara, M. Haruna, and T. Suhara, *Integrated Optical Circuits*, McGraw-Hill, New York (1989); Japanese version, Ohmsha, Tokyo (1993).
5. H. C. Casey, Jr, and M. B. Panish, *Heterostructure Lasers*, Academic Press, New York (1978).
6. H. Kogelnik, T. P. Sosnofski, and H. P. Weder, *IEEE J. Quantum Electron*, **QE-9**, 795 (1973).
7. L. I. Schiff, *Quantum Mechanics*, McGraw-Hill, New York (1968).
8. P. M. Morse and H. Feshbach, *Methods of Theoretical Physics*, McGraw-Hill, New York (1953).
9. D. Botez, *IEEE J. Quantum Electron*, **QE-17**, 178 (1981).
10. E. J. Marcatili, *Bell Syst. Tech. J.*, **48**, 2071 (1969).
11. W. Streifer and E. Kapon, *Appl. Opt.*, **18**, 3724 (1979).
12. D. D. Cook and F. R. Nash, *J. Appl. Phys.*, **46**, 1660 (1975).
13. R. Ito and M. Nakamura, *Semiconductor Lasers—Fundamentals and Applications* (in Japanese), Baifuban, Tokyo (1989).
14. G. P. Agrawal and N. K. Dutta, *Semiconductor Lasers*, second edition, Van Nostrand Reinhold, New York (1993).
15. Y. Suematsu (ed.) *Semiconductor Lasers and Optical Integrated Circuits* (in Japanese), Ohmsha, Tokyo (1984).
16. Japan Society of Applied Physics (ed.), *Fundamentals of Semiconductor Lasers* (in Japanese), Ohmsha, Tokyo (1987).
17. L. A. Coldren and S. W. Corzine, *Diode Lasers and Photonic Integrated Circuits*, John Wiley, New York (1995).
18. M. Born and E. Wolf, *Principle of Optics*, Pergamon, Oxford (1970).
19. T. Ikegami, *IEEE J. Quantum Electron.*, **QE-8**, 470 (1972).

6
Characteristics of Semiconductor Lasers

This chapter presents fundamental theory and characteristics of semiconductor injection lasers from the device point of view. First, the structure of the fundamental Fabry–Perot (FP)-type semiconductor lasers and the outline of the oscillation are described. Then rate equations are introduced, and the phenomenological analysis of the semiconductor laser performances is presented. The laser performances clarified through the rate equation analysis are discussed in detail. Distributed feedback lasers are presented in the next chapter.

6.1 SEMICONDUCTOR LASER STRUCTURE AND OUTLINE OF OSCILLATION

6.1.1 Fabry–Perot-type Semiconductor Laser Structure

Figure 6.1 illustrates the structure of a FP-type semiconductor laser, which is a representative semiconductor laser. We consider a GaAs laser as an example. Usually, a n-type GaAs crystal is used as a substrate and, on the substrate, an $Al_xGa_{1-x}As$ lower cladding layer (n type), a GaAs active layer (p type or n type), an $Al_xGa_{1-x}As$ upper cladding layer (p type), and a GaAs contact layer (n^+ type) are sequentially grown by epitaxy. The thickness of the active layer for ordinary DH lasers is 0.1–0.2 μm. The active layer serves as an optical waveguide, as described in Chap. 4. Except for broad-area lasers, a channel waveguide of 2–5 μm width as shown in Fig. 5.15 is formed by etching after the growth of the active layer. Then, a structure for injecting the current within a narrow channel region is formed by etching the upper cladding layer, deposition of an insulator layer and the patterning it. Then the substrate is thinned by grinding to a thickness of approximately 100 μm. Ohmic electrodes are formed by deposition of AuGe/Au and Cr/Au

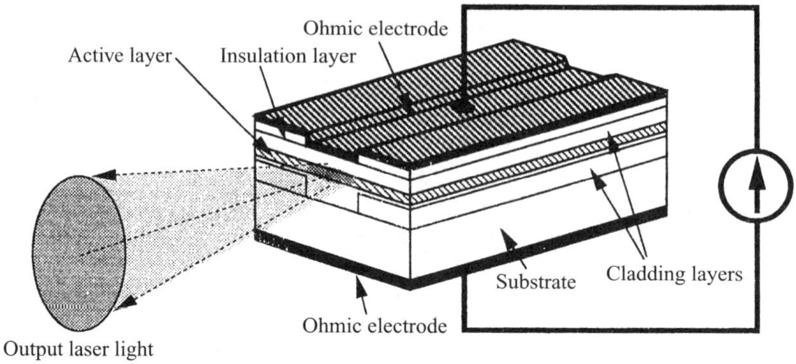

Figure 6.1 Schematic illustration of a double-heterostructure (DH) FP semiconductor laser.

layers on the back surface of the substrate and the top contact layer, followed by thermal treatment. The crystal surface is cleaved to form facets perpendicular to the channel axis, and then laser chips of a channel length of several hundred microns are obtained. However, as cleaved facets are often used as the resonator mirrors, the facet may be coated with dielectric films to obtain the facet reflectivity optimized for accomplishing a high output power. Semiconductor lasers for practical application are mounted on a heat sink and packaged with wire bonding.

Quantum well lasers are fabricated by growing a separate-confinement heterostructure (SCH) consisting of a single quantum well (SQW) or multiple quantum wells (MQWs), as shown in Fig. 4.8, instead of a simple DH structure. The following process is similar to that described above.

6.1.2 Oscillation Conditions

As discussed in Chaps 3 and 4, amplification of optical waves takes place in a semiconductor laser, when it is excited by carrier injection into the active region with a current across the p–n junction. The laser structure also serves as an optical waveguide FP resonator. If the amplification gain becomes sufficiently high, as a result of successive amplification of the optical wave traveling back and forth with feedback by the reflection at the facet mirrors, optical energy is accumulated and laser oscillation occurs.

Here we consider conditions for the laser oscillation. Laser oscillation, in general, is no more than maintaining a constant power of a coherent optical wave propagating back and forth in the resonator. Therefore, the oscillation condition is the condition required for the complex amplitude of the wave after a round trip between the mirrors to equal the initial complex amplitude.

Let g be the gain factor of the active region (material gain), and Γ be the confinement factor for the guided wave; then the effective gain factor for a guided mode (mode gain) is given by Γg. Let L be the length of the active region (waveguide length), α_{int} be the factor representing the propagation loss for the guided wave due to absorption and scattering, and R_{f} and R_{b} be the power reflectivities of the front and back facet mirrors, respectively. Then the condition for the absolute amplitude of the wave after a round trip to equal the initial amplitude is given by $R_{\text{f}}R_{\text{b}}\exp[2(\Gamma g-\alpha_{\text{int}})L]=1$, i.e.,

$$\Gamma g = \alpha_{\text{int}} + \frac{1}{2L}\ln\left(\frac{1}{R_{\text{f}}R_{\text{b}}}\right) \tag{6.1}$$

Let ω be the angular frequency of the optical wave and $\beta(\omega)$ be the propagation constant of the guided mode. For the moment we omit the carrier-induced refractive index change. Then the condition for the phase of the wave after a round trip to equal the initial phase is given by

$$2\beta(\omega)L = 2\pi m \quad (m \text{ is an integer}) \tag{6.2}$$

The frequencies ω_m satisfying the above condition, are the longitudinal mode frequencies explained in Section 5.8, and the longitudinal mode separation is given by

$$\Delta\omega = \frac{2\pi}{(2L\,\partial\beta/\partial\omega)} = 2\pi\frac{c}{2LN_{\text{g}}} \tag{6.3}$$

where N_{g} is the effective group index.

The first and second terms on the right-hand side of the amplitude condition for the laser oscillation, in Eq. (6.1) represent the propagation loss and the loss due to the leakage of the optical power through the facet mirrors, respectively. The equation indicates that an effective gain to compensate these losses is required for the oscillation. The phase condition (Eq (6.2)), being a positive feedback condition or a resonance condition, indicates that the oscillation frequency is limited to the discrete longitudinal mode frequencies with a separation given by Eq. (6.3). The laser oscillation takes place only for optical waves satisfying both conditions. Therefore, when the injection current is increased, and the mode gain $\Gamma g(\omega_m)$ at the longitudinal mode frequency ω_m closest to the peak of the gain spectrum $g(\omega)$ reaches the value given by the right-hand side of Eq. (6.1), the oscillation starts with this mode. Note that $\Gamma g(\omega_m)$ approximately equals the maximum mode gain Γg_{max}. The oscillation condition is illustrated in Fig. 6.2.

The value of the injection current at which the laser oscillation starts is called the threshold current. The value of the gain that enables the oscillation to occur, given by the right-hand side of Eq. (6.1), is referred to

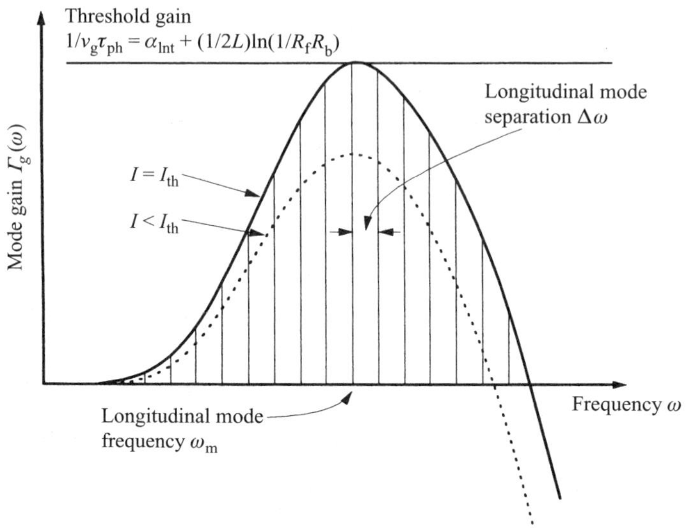

Figure 6.2 Oscillation condition of a FP semiconductor laser.

as the threshold gain. The dependence of the maximum gain g_{max} before the start of oscillation on the injection current density is given phenomenologically by Eq. (3.45) or (3.44). Therefore, the threshold current can be found by determining the injection current density at which g_{max} equals the threshold gain.

The gain spectrum, injection-current-density-dependent maximum gain, and the internal loss of the waveguide are important data for the performance analysis and the laser design presented in this chapter. Methods for determining these experimentally are presented in Appendix 4.

6.1.3 Injection Current and Optical Output Power

Once the oscillation starts when the injection current I reaches the threshold current I_{th}, a further increase in I is not associated with an increase on the gain g; the gain g is kept constant to maintain the condition given by Eq. (6.1). Since the gain g is a function of the carrier density, this means that the carrier density is also kept constant, under the assumption that the energy distribution of the carriers maintains the quasithermal equilibrium even after the start of oscillation. Similarly, the number of spontaneous emissions is also kept constant. Therefore, the injected carriers corresponding to the injection current exceeding the threshold, $I - I_{th}$, are depleted by the stimulated emission recombination. Let P_{out} be the number of photons per unit time representing the optical output power of the laser; then P_{out} can be written as

$$P_{\text{out}} = \begin{cases} F_{\text{spt}}\eta_{\text{spt}}\dfrac{I}{q} & (I < I_{\text{th}}) \\ F_{\text{spt}}\eta_{\text{spt}}\dfrac{I_{\text{th}}}{q} + F_{\text{stm}}\eta_{\text{stm}}\dfrac{I - I_{\text{th}}}{q} & (I_{\text{th}} < I) \end{cases} \tag{6.4}$$

The first line and the first term in the second line in the above expressions represent the incoherent spontaneous emission component, and the second term in the second line represents the coherent stimulated emission component. The factors F_{spt} and F_{stm} represent the ratios of the power coupled out of the laser to the power emitted in the laser. The spontaneous emission diverges in all direction from the active region, while stimulated emission occurs for resonance mode(s) of the waveguide FP resonator and propagates along the waveguide axis. Therefore, $F_{\text{spt}} \ll F_{\text{stm}}$ holds for observation of the output on the waveguide axis. On the other hand, η_{spt} and η_{stm} denote the internal quantum efficiencies. Note that $\eta_{\text{spt}} = \eta_{\text{stm}} = 1$ if the nonradiative recombination and the carrier leakage are negligible, but $\eta_{\text{spt}} < 1$, $\eta_{\text{stm}} < 1$, if they are not negligible. As Eq. (6.4) indicates, the output power versus injection current characteristic ($I - P$ characteristic) is indicated by a line with a kink at I_{th}, as illustrated in Fig. 6.3. The slope of the line in the oscillation region

$$\eta_{\text{D}} = \frac{\mathrm{d}P_{\text{out}}}{\mathrm{d}(I/q)} = F_{\text{stm}}\eta_{\text{stm}} \tag{6.5}$$

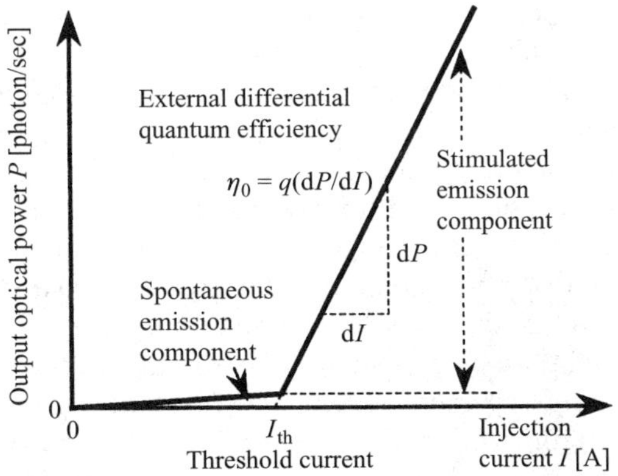

Figure 6.3 Injection current dependence of the output optical power of a semiconductor laser.

is referred to as the external differential quantum efficiency. The external differential efficiency η_D and the threshold current I_{th} are important parameters that characterize the laser oscillation. For a given laser, these parameters can easily be measured. The optical power generated by the stimulated emission is divided into the loss in the laser, and outputs transmitted through the facet mirror at the both ends of the laser. As Eq. (6.1) shows, the ratio of the loss to the total output is $\alpha_{int}/[(1/2L)\ \ln(1/R_f R_b)]$ and, when $R_f = R_b = R$, the total output is halved for both facets. Then the output coupling coefficient F_{stm} is given by

$$F_{stm} = \frac{1}{2} \frac{(1/L)\ln(1/R)}{\alpha_{int} + (1/L)\ln(1/R)} \tag{6.6}$$

Expressions for general cases will be derived in Section 6.3.

6.2 RATE EQUATIONS

6.2.1 Outline of the Rate Equation Analysis

The density matrix analysis discussed in Section 3.2 enabled a simple calculation of the amplification gain in a semiconductor to be made and clarified nonlinear phenomena such as hole burning and the associated gain saturation. Density matrix analysis, however, would require a very complicated formulation, if considerations on the guided mode and the resonance mode discussed in Chap. 5 are incorporated. Analysis of the behavior of many modes and dynamic characteristics would be even more complicated. To facilitate the analysis of practical laser operation, a simplified method should be developed on the basis of the results of density matrix analysis and optical mode analysis.

The electrons in the valence and conduction bands each form energy distributions depending on the state of laser operation. The electron states in a band, however, tend to approach the equilibrium distribution through relaxation with a time constant of the order of 0.1 ps. Therefore, for discussion of the laser dynamics in the time range much longer than the relaxation time, the electrons in the valence band as a whole can be treated as if they are in a single energy level, and those in the conduction band as if they were in another level. Thus, we approximate the electron energy distribution in the semiconductor by two energy levels, one representing the conduction band and the other representing the valence band, as shown in Fig. 6.4. The optical wave, on the other hand, is described by separate mathematical expressions for the intensity and the phase. Then, the fundamental equations that govern the operation of a semiconductor laser

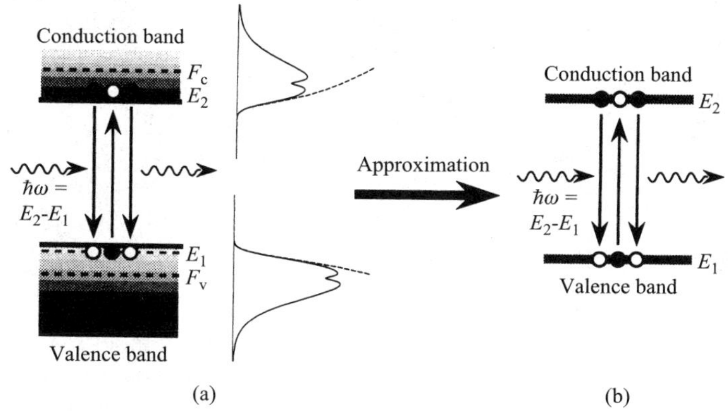

Figure 6.4 Two-level approximation for carrier state in the active layer of a semiconductor laser: (a) band structure; (b) two-level approximation.

are rewritten as the rate equations that describe the time variations in the carrier density and the optical intensity. In the following, we derive the rate equations and discuss the physical implications.

6.2.2 Rate Equation for the Carrier Density

We first describe the minority-carrier density in a semiconductor by using the density matrix ρ discussed in Section 3.6. Let N_{vc} be the total number of electrons in the conduction and valence bands of a semiconductor of unit volume. The number of electrons in the conduction band (denoted by subscript n) is given by $N_c = N_{vc}\sum_n \rho_{nn}$, the number of electrons in the valence band (denoted by subscript m) by $N_v = N_{vc}\sum_m \rho_{mm}$, and the total number $N_c + N_v = N_{vc}$ is constant. Consider a p-type semiconductor, and let N be the density of the minority carriers (electrons in the conduction band). Then, from $N = N_c = N_{vc}\sum_n \rho_{nn}$ and $N = N_{vc} - N_v = N_{vc} - N_{vc}\sum_m \rho_{mm}$, we obtain

$$N = \frac{1}{2N_{vc}}\left(\sum_n \rho_{nn} - \sum_m \rho_{mm} + 1\right) \tag{6.7}$$

The zeroth-order term $\rho_{nn}^{(0)}$ in the expansion expression for the diagonal elements ρ_{nn} of the density matrix satisfies Eq. (3.66). From Eqs (3.66b), (3.56), (3.68), and (3.74), we see that the second-order term $\rho_{nn}^{(2)}$ satisfies

$$\begin{aligned}\frac{d}{dt}\rho_{nn}^{(2)} = &-\frac{\pi}{2\hbar^2}\sum_m\left(\rho_{nn}^{(0)} - \rho_{mm}^{(0)}\right)|\boldsymbol{P}_{nm}|^2|\boldsymbol{E}|^2 L(\omega - \omega_{nm}) \\ &- \rho_{nn}^{(2)}\left(\frac{1}{\tau_c} + \frac{1}{\tau_s}\right)\end{aligned} \tag{6.8}$$

Omitting the terms of orders higher than fourth order, we can express ρ_{nn} by $\rho_{nn}^{(0)}$ and $\rho_{nn}^{(2)}$, and from Eqs (3.66a) and (6.8) we see that the diagonal element ρ_{nn} for the conduction band electrons satisfies the equation

$$\begin{aligned}\frac{\mathrm{d}}{\mathrm{d}t}\rho_{nn} = &-\frac{\pi}{2\hbar^2}\sum_m \left(\rho_{nn}^{(0)} - \rho_{mm}^{(0)}\right)|\boldsymbol{P}_{nm}|^2|\boldsymbol{E}|^2 L(\omega - \omega_{nm}) \\ &-\frac{\rho_{nn} - \rho_{nn}^{\mathrm{QE}}}{\tau_{\mathrm{c}}} - \frac{\rho_{nn} - \rho_{nn}^{\mathrm{TE}}}{\tau_{\mathrm{s}}} + \Lambda_n \end{aligned} \tag{6.9}$$

Similarly, the diagonal element ρ_{mm} for the valence-band electrons satisfies the equation

$$\begin{aligned}\frac{\mathrm{d}}{\mathrm{d}t}\rho_{mm} = &-\frac{\pi}{2\hbar^2}\sum_n \left(\rho_{mm}^{(0)} - \rho_{nn}^{(0)}\right)|\boldsymbol{P}_{nm}|^2|\boldsymbol{E}|^2 L(\omega - \omega_{nm}) \\ &-\frac{\rho_{mm} - \rho_{m}^{\mathrm{QE}}}{\tau_{\mathrm{v}}} - \frac{\rho_{mm} - \rho_{mm}^{\mathrm{TE}}}{\tau_{\mathrm{s}}} + \Lambda_m \end{aligned} \tag{6.10}$$

Calculating the time variation in N using the above two equations, we obtain

$$\begin{aligned}\frac{\mathrm{d}}{\mathrm{d}t}N = &-\frac{n_{\mathrm{r}}c\varepsilon_0}{2\hbar\omega} g|\boldsymbol{E}|^2 \\ &-\frac{1}{2N_{\mathrm{vc}}}\sum_n \frac{\rho_{nn} - \rho_{nn}^{\mathrm{QE}}}{\tau_{\mathrm{c}}} + \frac{1}{2N_{\mathrm{vc}}}\sum_m \frac{\rho_{mm} - \rho_{mm}^{\mathrm{QE}}}{\tau_{\mathrm{v}}} \\ &-\frac{N - N^{\mathrm{TE}}}{\tau_{\mathrm{s}}} + N_{\mathrm{vc}}\sum_n \Lambda_n \end{aligned} \tag{6.11}$$

Equation (3.73) was used in the first line of the above equation, and the linear gain $g^{(1)}(\hbar\omega)$ was written simply as g. This stimulated emission recombination term can be rewritten as $-(n_{\mathrm{r}}n_{\mathrm{g}}\varepsilon_0/2\hbar\omega)|\boldsymbol{E}|^2G$, by using the relation $g = G/v_{\mathrm{g}} = G(n_{\mathrm{g}}/c)$ between the gain g per unit length and the gain G per unit time. Here, $(n_{\mathrm{r}}n_{\mathrm{g}}\varepsilon_0/2\hbar\omega)|\boldsymbol{E}|^2$ equals the photon number per unit volume, since it is the energy density $(n_{\mathrm{r}}n_{\mathrm{g}}\varepsilon_0/2)|\boldsymbol{E}|^2$ of the optical field, divided by the photon energy $\hbar\omega$. For general cases where the carrier density is not spatially uniform, the second line on the right-hand side of Eq. (6.11) represents the carrier diffusion effect. The spontaneous emission term $(N - N^{\mathrm{TE}})/\tau_{\mathrm{s}}$ in the third line can be approximated as N/τ_{s}, since the carrier density N^{TE} in thermal equilibrium is much smaller than the carrier density N under carrier injection. As the carrier injection term $N_{\mathrm{vc}}\sum_{\mathrm{n}}\Lambda_n$, is equal to the number of carriers injected in unit time and in unit volume, it can be replaced by J/dq using the injection current density J, the elementary charge q $(= -e)$, and the active layer thickness d. Through the above changes of the expressions, we obtain a rate equation for the carrier density:

$$\frac{\mathrm{d}}{\mathrm{d}t}N = -\frac{n_\mathrm{r}n_\mathrm{g}\varepsilon_0}{2\hbar\omega}|\boldsymbol{E}|^2G - \frac{N}{\tau_\mathrm{s}} + \frac{J}{dq} \tag{6.12}$$

where the diffusion term is omitted. For an n-type semiconductor, exactly the same rate equation holds, although here the minority carriers with density N are the holes in the valence band.

6.2.3 Photon Density

Next we simplify Eq. (6.12) under the approximation and the assumption that the injection current density J is uniform. Whereas the optical intensity $|\boldsymbol{E}|^2$ is a function of position in the semiconductor laser and it forms a distribution of a standing wave with a submicron period, the spatial distribution of the carrier density N is affected by the diffusion effect and tends to be smoothed out. Noting this, we replace the stimulated emission recombination term proportional to $|\boldsymbol{E}|^2$ in Eq. (6.12) by the value averaged in the active region. We also replace the carrier density N by the value averaged in the active region, and consider it a position-independent quantity. Then, the gain G is also considered position independent. These approximations and assumptions do not conflict with each other.

The optical field $\boldsymbol{E}$ in the laser must satisfy the boundary conditions in the resonator, and therefore it can be expressed in terms of the resonance mode discussed in Section 5.8.3. Assuming oscillation in a single optical mode, the field can be written as

$$\boldsymbol{E}(\boldsymbol{r}, t) = \boldsymbol{E}(\boldsymbol{r})\,\exp(-\mathrm{i}\omega t)\,E(t), \qquad \boldsymbol{E}(\boldsymbol{r}) = \boldsymbol{E}(x, y)E(z) \tag{6.13}$$

where $\boldsymbol{E}(\boldsymbol{r})$ is a normalized resonance mode function. Let V_c and V_a be the volumes of the resonator and the active region, respectively. We define a photon density S by $S =$ (number of photons in the resonator/volume of the resonator). Noting that $\boldsymbol{E}(\boldsymbol{r})$ is normalized, we can write

$$\begin{aligned} S(t) &\equiv \int_{V_\mathrm{c}} \frac{n_\mathrm{r}n_\mathrm{g}\varepsilon_0}{2}\frac{|\boldsymbol{E}(\boldsymbol{r}, t)|^2\,\mathrm{d}\boldsymbol{r}}{\hbar\omega V_\mathrm{a}} \\ &= \frac{|E(t)|^2}{\hbar\omega V_\mathrm{a}} \end{aligned} \tag{6.14}$$

The stimulated emission recombination term in Eq. (6.12) averaged in the active region can be written as

$$\begin{aligned} \left\langle\left(\frac{n_\mathrm{r}n_\mathrm{g}\varepsilon_0}{2\hbar\omega}\right)|\boldsymbol{E}|^2\right\rangle_\mathrm{a} &= \int_{V_\mathrm{a}} \frac{n_\mathrm{r}n_\mathrm{g}\varepsilon_0}{2}\frac{|\boldsymbol{E}(\boldsymbol{r})|^2\,\mathrm{d}\boldsymbol{r}}{\hbar\omega V_\mathrm{a}} \\ &= \Gamma S \end{aligned} \tag{6.15}$$

$$\Gamma = \frac{\int_{V_a} n_r n_g |\boldsymbol{E}(\boldsymbol{r})|^2 \, d\boldsymbol{r}}{\int_{V_c} n_r n_g |\boldsymbol{E}(\boldsymbol{r})|^2 \, d\boldsymbol{r}} \approx \frac{\int_{V_a} |\boldsymbol{E}(\boldsymbol{r})|^2 \, d\boldsymbol{r}}{\int_{V_c} |\boldsymbol{E}(\boldsymbol{r})|^2 \, d\boldsymbol{r}} \tag{6.16}$$

where Γ is the factor of field confinement in the active region discussed in Section 5.5. Using Eqs (6.14)–(6.16), the rate equation for the carrier density given in Eq. (6.12) is rewritten as the simpler form

$$\frac{d}{dt} N = -\Gamma G S - \frac{N}{\tau_s} + \frac{J}{dq} \tag{6.17}$$

The first, second and third terms on the right-hand side of the above equation represent the change in the minority-carrier density due to the stimulated emission recombination, the spontaneous emission recombination, and the carrier injection, respectively.

6.2.4 Rate Equation for the Photon Density

As we saw in Section 5.8.3, the amplitude of a resonance mode field varies with time in the form $\exp(-\gamma t)$, and γ is correlated with the complex propagation constant $\tilde{\beta}$ of the guided mode in the resonator by Eqs (5.141) and (5.145). For a passive resonator where $\tilde{\beta}$ is given by Eq. (5.131), it has been shown that γ is given by Eq. (5.151). For a laser resonator with a carrier-injected active region, $\tilde{\beta}$ is given by Eq. (100) consisting of the carrier-induced mode gain Γg and the mode index change $\Gamma k_0 \, \Delta n$, i.e.,

$$\tilde{\beta} = \beta + \Gamma\left(k_0 \, \Delta n - \frac{ig}{2}\right) + \frac{i\alpha_{int}}{2} \tag{6.18}$$

and then, from Eqs (5.141) and (5.146b), and $g = G/v_g$, γ is given by

$$2\gamma = -\Gamma G + v_g(\alpha_{int} + \alpha_{mir}) = -\Gamma G + \frac{1}{\tau_{ph}} \tag{6.19}$$

The parameter τ_{ph} is the photon lifetime defined by

$$\frac{1}{\tau_{ph}} = v_g(\alpha_{int} + \alpha_{mir}) \tag{6.20a}$$

$$\alpha_{mir} = \frac{1}{2L} \ln\left(\frac{1}{R_f R_b}\right) \tag{6.20b}$$

where R_f and R_b are the reflectivities of the facet mirrors, and L is the resonator length. Since, from Eqs (6.13) and (6.14), the photon density S is proportional to the absolute square of the field amplitude, the time dependence of S is $\exp(-2\gamma t)$. Therefore, the time derivative of S equals $-2\gamma S = (\Gamma G - 1/\tau_{ph})S$. This result based on semiclassical theory, however, does not include the effect of spontaneous emission. The second term on the right-hand side of Eq. (6.17) indicates that N/τ_s photons are spontaneously emitted per unit time per unit volume of the active region. Let C_s be the ratio of the number of photons belonging to the same mode as the laser oscillation mode to the total number of spontaneous emission photons, then the spontaneous emission gives rise to an additional increase by $C_s N/\tau_s$ photons per unit time in the photon density S, defined as the number of photons in the resonator per unit volume of the active region additional increase. Therefore the rate equation for the photon density S is given by

$$\frac{d}{dt}S = \Gamma G S - \frac{S}{\tau_{ph}} + \frac{C_s N}{\tau_s} \tag{6.21}$$

6.2.5 Rate Equation for the Optical Phase

Although Eq. (6.21) is appropriate for describing the temporal variation in the optical intensity, it is not capable of describing the phase of the optical field. In order to carry out the analysis on the laser characteristics related to the optical phase, another equation is required for describing the time variation in the phase. Noting that the optical field $\boldsymbol{E}$ in the laser can be written as Eq. (6.13), we consider $E(t)$ a complex function, let $\phi(t)$ be the phase, and put $E(t) = |E(t)| \exp[-i\phi(t)]$. Then, the instantaneous frequency of $\boldsymbol{E}$ is given by

$$\omega(t) = \omega + \delta\omega(t), \quad \delta\omega(t) = \frac{d\phi(t)}{dt} \tag{6.22}$$

The laser oscillation frequency ω, on the other hand, is determined from the complex propagation constant given Eq. (6.18) and the longitudinal mode resonance condition given by Eq. (5.146a), by

$$2L\left(\beta(\omega) + \frac{\Gamma\omega}{c}\Delta n\right) = 2m\pi \tag{6.23}$$

Generally, Δn varies with time and, accordingly, ω determined by the above equation varies with time. Let ω_{th} and Δn_{th} be the values of ω and Δn, respectively, at the threshold, and put $\omega(t)=\omega_{\text{th}}+\delta\omega(t)$ and $\Delta n(t)=\Delta n_{\text{th}}+\delta n(t)$; then from the above equations we obtain

$$\beta(\omega_{\text{th}})+\frac{\Gamma\omega_{\text{th}}}{c}\Delta n_{\text{th}}=\frac{2m\pi}{2L} \tag{6.24a}$$

$$\delta\omega=-\frac{\Gamma\omega_{\text{th}}}{c}\frac{\delta n}{\partial\beta/\partial\omega}=-\frac{\Gamma\omega_{\text{th}}}{N_{\text{g}}}\delta n \tag{6.24b}$$

where N_{g} is the effective group index of refraction for the guided mode, and use has been made of a relation $1/(\partial\beta/\partial\omega)=v_{\text{g}}=c/N_{\text{g}}$. We identify ω in Eqs (6.13) and (6.22) with w_{th} determined by Eq. (5.24a). We also identify $\delta\omega(t)$ in Eq. (6.22) with the frequency change $\delta\omega(t)$ given by Eq. (6.24b), and rewrite it using the phase $\phi(t)$. Then, from Eqs (6.22) and (6.24b), we obtain

$$\frac{\text{d}}{\text{d}t}\phi=-\frac{\Gamma\omega}{N_{\text{g}}}\delta n \tag{6.25}$$

The refractive index change δn is correlated with the gain change $\delta G=G-G_{\text{th}}$ through the Kramers–Kronig relation (see Appendix 3), and the former is proportional to the latter. Using the antiguiding factor α_{c} defined by Eq. (5.115), δn can be written as

$$\delta n=-\frac{\alpha_{\text{c}}c}{2\omega}\frac{\delta G}{v_{\text{g}}}=-\frac{\alpha_{\text{c}}N_{\text{g}}}{2\omega}\delta G=-\frac{\alpha_{\text{c}}N_{\text{g}}}{2\omega}\left(G-\frac{1}{\Gamma\tau_{\text{ph}}}\right) \tag{6.26}$$

and therefore Eq. (6.25) can be rewritten as

$$\frac{\text{d}}{\text{d}t}\phi=\frac{\alpha_{\text{c}}}{2}\left(\Gamma G-\frac{1}{\tau_{\text{ph}}}\right) \tag{6.27}$$

The above equation is the rate equation for the optical phase.

The rate equations, given by Eqs (6.17), (6.21), and (6.27), were derived using the boundary conditions under the assumption that $\tilde{\beta}$ and ω are independent of time. Therefore, they do not hold exactly for cases where the laser is modulated with an ultrafast signal. However, they give sufficient accuracy for analyzing the laser dynamics in time range longer than the photon lifetime τ_{ph} (several picosecands or less for ordinary semiconductor lasers).

6.2.6 Diagram for and Extension of Rate Equations

A combination of the rate equations for the carrier density N and the photon density S given by

$$\frac{\text{d}}{\text{d}t}N=-\Gamma GS-\frac{N}{\tau_{\text{s}}}+\frac{J}{dq} \tag{6.28a}$$

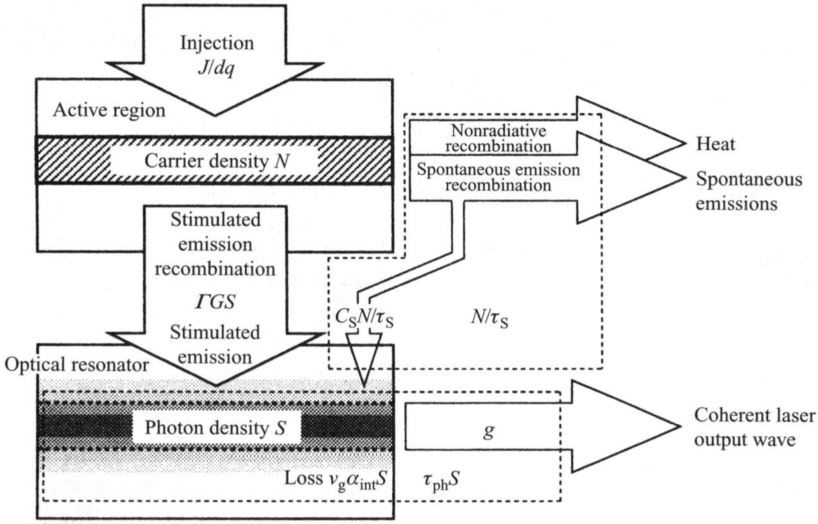

Figure 6.5 Diagram of the rate equations for a semiconductor laser.

$$\frac{\mathrm{d}}{\mathrm{d}t}S = \Gamma GS - \frac{S}{\tau_{\mathrm{ph}}} + \frac{C_{\mathrm{s}}N}{\tau_{\mathrm{s}}} \tag{6.28b}$$

simply describes the mutual correlation and the time variances of N and S and provides a very powerful means for analysis of laser characteristics. The rate equations can be depicted as a diagram as shown in Fig. 6.5.

The gain factor G in the rate equations is a function of the carrier density N. If the gain saturation effect is taken into account, the gain G depends not only on N but also on the photon density S. Although thus far laser oscillation in a single optical mode has been assumed, in general a laser may oscillate in several modes. In order to make an analysis of the multimode oscillation, we express the optical field $\boldsymbol{E}(\boldsymbol{r}, t)$ in the form of summation of the mode components as Eq. (6.13) and define the photon densities S_m for each mode m. Thus, taking into account the saturation effect and the possibility of multimode oscillation, we obtain a set of extended rate equations given by

$$\frac{\mathrm{d}}{\mathrm{d}t}N = -\sum_m \Gamma_m\left(G_m - \sum_{m'} \xi_{mm'}S_{m'}\right)S_m - \frac{N}{\tau_{\mathrm{s}}} + \frac{J}{dq} \tag{6.29a}$$

$$\frac{\mathrm{d}}{\mathrm{d}t}S_m = \Gamma_m\left(G_m - \sum_{m'} \xi_{mm'}S_{m'}\right)S_m - \frac{S_m}{\tau_{\mathrm{ph}}} + \frac{C_{\mathrm{s}m}N}{\tau_{\mathrm{s}}} \tag{6.29b}$$

The coefficients $\xi_{mm'}$ in the above equations represent the gain saturation in terms of the photon density (note that the same saturation is represented by ζ in Eq. (3.81)), and they satisfy $\xi_{mm} < \xi_{mm'} (m \neq m')$.

6.3 STEADY-STATE OSCILLATION CHARACTERISTICS

This section presents the rate equation analysis to clarify in detail the steady-state oscillation characteristics, which were outlined in Section 6.1.

6.3.1 Single-mode Model and Approximate Solution of the Rate Equations

As the simplest model, let us consider first steady-state operation of a semiconductor laser under the assumption that the oscillation takes place with a single mode. We omit gain saturation. Noting that the maximum gain $G_{\max} = v_g g_{\max}$ can be approximately given by a linear function of the carrier sensity N as Eq. (3.40) or (4.41a), we put

$$G = G_{\max} = A(N - N_0) \tag{6.30}$$

and put $\mathrm{d}/\mathrm{d}t = 0$ in the rate equations given by Eq. (6.28) to obtain

$$0 = -\Gamma G S - \frac{N}{\tau_s} + \frac{J}{dq} \tag{6.31a}$$

$$0 = \Gamma G S - \frac{S}{\tau_{ph}} + \frac{C_s N}{\tau_s} \tag{6.31b}$$

As we see soon, the optical output power is proportional to S. The above equations can easily be solved and, in particular, the approximate solution for cases where C_s is small can be expressed by a simple formula. Omitting the C_s term in Eq. (6.31b), we see that $S = 0$ or $\Gamma G = 1/\tau_{ph}$. The former result $S = 0$, represents the subthreshold situation before the oscillation starts. In this situation, putting the terms representing the stimulated emission and the stimulated emission recombination in Eq. (6.31) yields

$$N = \frac{J}{dq}\tau_s, \qquad S = C_s \tau_{ph} \frac{J}{dq} \qquad (J < J_{th}) \tag{6.32}$$

which indicates that, whereas the carrier density N increases with increasing injection current density J, the output light is only the spontaneous emission component. On the other hand, the latter result $\Gamma G = 1/\tau_{ph}$ is equivalent to Eq. (6.1) representing the oscillation condition. From this equation and

Eqs (6.30) and (6.31a), the threshold carrier density N_{th} and the threshold current density J_{th} are given by

$$N_{th} = N_0 + \frac{1}{\Gamma A \tau_{ph}} \tag{6.33a}$$

$$J_{th} = \frac{dq}{\tau_s} N_{th} \tag{6.33b}$$

After the start of the oscillation ($J_{th} < J$), from $\Gamma G = 1/\tau_{ph}$ being maintained and Eqs (6.30) and (6.31), the carrier density N and the photon density S are given by

$$N = N_{th} \tag{6.34a}$$

$$S = \tau_{ph}\left(C_s \frac{J_{th}}{dq} + \frac{J - J_{th}}{dq} \right) \tag{6.34b}$$

While N stays constant, S increases in proportion to $J - J_{th}$. Obviously, this result is consistent with the discussion presented in Section 6.1.

6.3.2 Threshold Current and Output Power

The expression for the threshold current density in Eq. (6.33) shows that attainment of a low threshold requires an active layer of small carrier density N_0 at transparency and a large differential gain coefficient A. From this viewpoint, quantum well structures are advantageous compared with bulk DHs. The internal loss α_{int} of the waveguide must be minimized. Guided mode analysis shows that the confinement factor Γ is close to unity for large active layer thicknesses d, and, with decreasing d, Γ decreases approximately in proportion to d^2, requiring a large material gain g for a small d as shown by Eq. (6.1). On the other hand, as Eq. (6.33b) shows, increasing d results in an increase, in proportion to d, in the current density J required to obtain a constant g. Accordingly, there exists a minimum of J_{th} at a value of the thickness d. The active layer thickness of DH lasers is usually designed at the value that minimizes J_{th}. For quantum well lasers, the structures shown in Fig. 4.8 are adopted to attain a low threshold.

Since the optical field in the resonator can be expressed as Eq. (6.13) using the normalized mode function $\boldsymbol{E}(\boldsymbol{r})$, the stored optical energy is given by $|E(t)|^2 = \hbar\omega V_a S$ with the photon density S defined by Eq. (6.14). On the other hand, the value of the power $P^+(L)$, flowing in the $+z$ direction of the normalized mode field $\boldsymbol{E}(\boldsymbol{r})$ with unit energy at $z = L$, is given by Eq. (5.149a), and $1 - R_f$ times the value is transmitted through the front facet mirror of reflectivity R_f. Therefore, using Eqs (5.149a), (5.150), and (6.20b), we obtain a relation between the forward optical output power P_f and the photon density S:

$$P_f = \hbar\omega V_a S P^+(L)(1 - R_f) = \frac{\hbar\omega V_a S v_g \alpha_{mir}}{1 + (R_f/R_b)^{1/2}(1 - R_b)/(1 - R_f)} \tag{6.35}$$

Similarly, the backward output power is given by the above expression with R_f and R_b exchanged. Comparison between the above expression, with S given by Eq. (6.34) substituted, and Eq. (6.4) yields an expression for the output coupling coefficient:

$$F_{stm} = \frac{v_g \tau_{ph} \alpha_{mir}}{1 + (R_f/R_b)^{1/2}(1 - R_b)/(1 - R_f)} \tag{6.36}$$

For $R_f = R_b = R$, the above expression coincides with Eq. (6.6).

Basic FP lasers use facet mirrors with $R \approx 0.32$ that can easily be formed by simple cleaving. However, the laser performances can be improved by optimizing the mirror reflectivities R_f and R_b, since the threshold current and the output power depend upon the reflectivities, as the above expressions show. For ordinary applications, a high backward output power is not required, and the forward power P_f can be enhanced by increasing the back mirror reflectivity R_b to a value close to unity. Taking the front mirror reflectivity R_f as a parameter, Eqs (6.20) and (6.36) show that enhancing R_f results in a long photon lifetime τ_{ph}, indicating enhancement of the optical power accumulation in the resonator and, in turn, reduction in the output coupling coefficient F_{stm}. Accordingly, enhancing R_f results in reduction in the threshold current density but it is associated with deterioration in the external differential quantum efficiency, as shown in Fig. 6.6. In many cases, the maximum available output power is limited by the catastrophic optical damage (COD) that occurs at the facet. The optimization procedure for high-power lasers is to determine R_f in such a manner that the injection current density required to obtain an output power prescribed from the COD level of the material is minimized.

6.3.3 Effects of Spontaneous Emission

Spontaneous emission was omitted in the approximate analysis which gave the result in Eq. (6.34). The effects of spontaneous emission can be examined by using Eq. (6.31). A modification of Eq. (3.31b) yields

$$S = \frac{C_s N/\tau_s}{1/\tau_{ph} - \Gamma G} \tag{6.37}$$

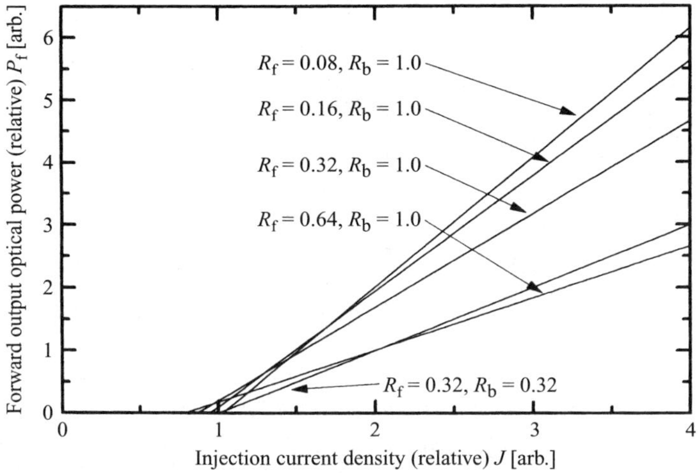

Figure 6.6 Output optical power versus injection current density depending on the facet mirror reflectivity.

which indicates that the resonance mode component $C_s N/\tau_s$ of the spontaneous emission as a seed is amplified to cause oscillation, and that the photon density S increases rapidly as the gain approaches the threshold gain given by $\Gamma G = 1/\tau_{ph}$. Substituting Eq. (6.37) into Eq. (6.31a), with the use of Eq. (6.30), we can easily deduce the solutions for N and S. The result can be written as

$$N = \frac{N_{th}\left\{J/J_{th} + C_s'' - \left[\left(J/J_{th} + C_s''\right)^2 - 4(J/J_{th})C_s'\right]^{1/2}\right\}}{2C_s'}$$

$$S = \frac{N_{th}\tau_{ph}}{2\tau_s}\left\{\frac{J}{J_{th}} - C_s'' + \left[\left(\frac{J}{J_{th}} + C_s''\right)^2 - 4\frac{J}{J_{th}}C_s'\right]^{1/2}\right\} \tag{6.38}$$

with

$$C_s' = 1 - C_s, \qquad C_s'' = 1 - C_s\frac{N_0}{N_{th}}$$

The dependences of the carrier density and the photon density on the injection current density are plotted in Fig. 6.7 using normalized variables. The result shows that, with a large contribution of the spontaneous emission represented by a large C_s, the laser output power starts to increase at values of N smaller than the threshold value N_{th} given by Eq. (6.33a), and as a

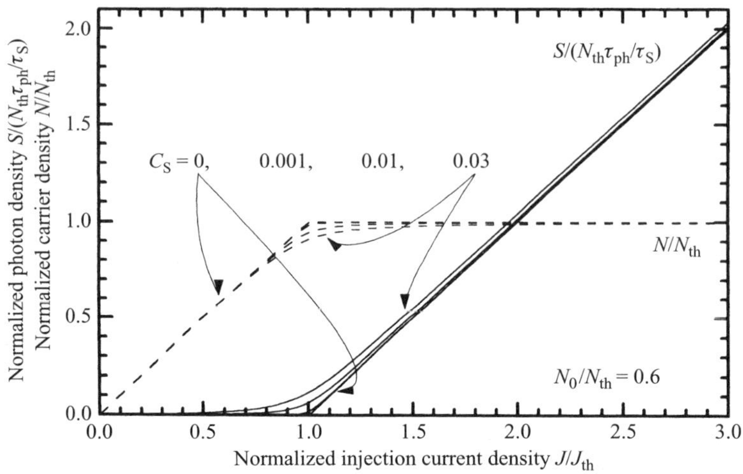

Figure 6.7 Dependences of the carrier density and photon density on the injection current density.

result the kink in the graph for the output power versus the injection current density at the threshold is rounded. A semilogarithmic plot of the injection current density dependence of the output power is shown in Fig. 6.8. The graph shows that the output power changes by several orders of magnitude depending on C_s in the vicinity of the threshold.

The value of the spontaneous emission coupling coefficient C_s can be estimated by an approximate expression derived for index-guiding lasers in Appendix 5:

$$C_s = \frac{\Gamma \lambda^4}{4\pi^2 n_r n_g n_e V_a \Delta\lambda_{sp}} \tag{6.39}$$

where Γ is the confinement factor, $\Delta\lambda_{sp}$ the spectrum width of the spontaneous emission, and V_a the volume of the active region [1,2]. The value of C_s is of the order of $10^{-5}-10^{-4}$ for ordinary index-guiding lasers and is several times this value for gain-guiding lasers. However, there exist cases where the experimental result does not agree well with the theoretical prediction on the injection current density dependence of the output power using C_s estimated by the above expression. Since the injection current density dependence of the output power can easily be measured for a given laser, the value of C_s for the laser can be determined by comparing the experimental result with the theoretical curves in Fig. 6.8. The spontaneous emission not only affects the behavior of the optical output power and the

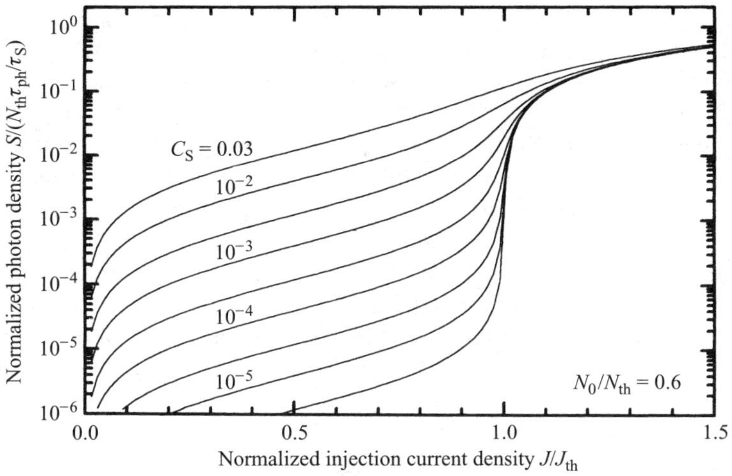

Figure 6.8 Photon density versus injection current density depending on the spontaneous emission coupling coefficient.

carrier density in the vicinity of the threshold but also may dominate the oscillation mode characteristics, as will be discussed in Section 6.3.4.

6.3.4 Lateral Mode Characteristics

The lateral and longitudinal modes of a resonator with which a FP-type semiconductor laser was constructed were discussed in detail in Section 5.8. If the above-described laser oscillation conditions are satisfied with several modes, oscillation consisting of several modes would take place. The mode competition that determines the actual behavior of the laser in this situation which allows multimode oscillation can be analyzed by using the oscillation condition given by Eq. (6.1) and multimode rate equations given by Eq. (6.29).

The lateral modes can be classified into transverse electric (TE) and transverse magnetic (TM) modes. For ordinary bulk DH lasers, the material gains g are the same for TE and TM modes, and there is no significant difference in the confinement factors Γ and also in the internal losses α_{int}. The facet reflectivity R for TE modes is considerably larger than that for TM modes, as shown in Section 5.7, and, therefore, the threshold gain determined by the right-hand side of Eq. (6.1) is smaller for TE modes. Therefore, FP lasers oscillate with TE mode(s). For quantum well lasers, the gain is also larger for TE modes, and the TE oscillation takes place with a TM mode suppression ratio higher than that of bulk DH lasers.

We next consider cases where the waveguide with which a semiconductor laser is constructed is of index guiding type and supports several lateral modes of different orders. With low injection current levels, only the fundamental lateral mode with the smallest internal loss α_{int} and the largest confinement factor Γ oscillates. However, with increase in the injection current, the output power increases and the increase in the stimulated emission reduces the carrier density on the center axis of the active channel, giving rise to saturation of the increase in the fundamental mode gain relative to the higher-order mode gain. Thus, the higher-order modes start to oscillate. The reduction in the carrier density is referred to as spatial hole burning. The change in the lateral mode, which is associated with a complex change in the far-field pattern, is not desirable from an application point of view. The lateral multimode oscillation makes it difficult to attain single-longitudinal-mode oscillation. Therefore, it is required to design and implement a waveguide that supports a single lateral mode or one close to that for obtaining stable single-longitudinal-mode oscillation. In a narrow active channel width, however, the carrier diffusion effect dominates the carrier distribution and suppresses spatial hole burning, allowing the practical fundamental mode output power to be achieved below the injection level before the start of the oscillation in higher-order modes. Thus a single-mode waveguide is not a strictly necessary condition.

In gain-guiding lasers, the lateral mode itself changes with the change in the carrier injection level. Lateral mode confinement may show a complicated change with the injection level through the gain guiding and the antiguiding due to the carrier-induced refractive index change. Moreover, spatial hole burning promotes the transition to multiple-lateral-mode oscillation. Therefore single-longitudinal-mode oscillation can hardly ever be obtained. Another drawback is the problem in the far-field wavefront pointed out in Section 5.9. Although gain-guiding lasers are easy to fabricate, they involve the drawbacks described above, and nowadays the technique for fabrication of index-guiding lasers has been well established. Thus gain-guiding lasers are less important. In the following two sections, the longitudinal mode characteristics of single-lateral-mode lasers will be discussed in detail.

6.3.5 Longitudinal Mode Characteristics

As a simple model, we first consider steady-state oscillation without the spontaneous emission coupled to the oscillation modes ($C_s = 0$). In the region of small injection current densities J, where no mode started to oscillate, the stimulated emission recombination term $\sum_m \Gamma_m G_m S_m$ in Eq. (6.29a) and the saturation effect term $\sum_{m'} \xi_{mm'} S_{m'}$ in Eqs (6.29a) and (6.29b) are zero, and Eqs (6.29a) and (6.29b) reduce to $0 = -N/\tau_s + J/dq$ and $0 = (\Gamma_m G_m - 1/\tau_{ph})S_m$, respectively. Then, with increase in J, the carrier density N increases as $N = \tau_s J/dq$; the gain G_m for each mode also increases, and oscillation starts with the mode that satisfies first the oscillation condition $\Gamma_m G_m = 1/\tau_{ph}$. After the start of oscillation in a mode, even with further increase in J the oscillation conditions for the other modes are not satisfied, since the increment is consumed by transformation to the optical output of this mode and N is kept at the value to maintain the oscillation condition for this mode. This means that, without coupling of the spontaneous emission, the laser oscillates in a single mode.

To examine the oscillation mode characteristics of a realistic laser, the effect of spontaneous emission must be considered [3]. If the coupling of the spontaneous emission ($C_s > 0$) exists, the process is started by the spontaneous emission and then amplified; therefore the laser output is obtained even before $\Gamma_m G_m = 1/\tau_{ph}$ is exactly satisfied, as shown by Eq. (6.37). Consider a case where several longitudinal modes belonging to a single lateral mode are simultaneously oscillating in a steady state. Then, by substituting 0 into the left-hand side of Eq. (6.29), we obtain equations to determine the carrier density N and the photon densities S_m of each longitudinal mode:

$$\frac{J}{dq} = \frac{N}{\tau_s} + \sum_m \Gamma_m \left(G_m - \sum_{m'} \xi_{mm'} S_{m'} \right) S_m \tag{6.40a}$$

$$S_m = \frac{\tau_{ph} C_s N/\tau_s}{1 - \tau_{ph} \Gamma_m \left(G_m - \sum_{m'} \xi_{mm'} S_{m'} \right)} \tag{6.40b}$$

By using the approximate parabolic expression for the gain spectrum $G(\omega)$ shown in Fig. 6.2, the gains G_m for each mode can be written as

$$G_m = G(\omega_m) = G_{max} \left[1 - \left(\frac{\omega_m - \omega_0}{\Delta\omega_G} \right)^2 \right] \tag{6.41a}$$

$$G_{max} = G(\omega_0) = A(N - N_0) \tag{6.41b}$$

where ω_m denotes the longitudinal mode frequencies, and $\Delta\omega_G$ the half-width of the gain band. Assuming that the maximum gain frequency ω_0 coincides with one of the longitudinal mode frequencies, we number the longitudinal mode order m with reference to this central longitudinal mode to obtain $\omega_m - \omega_0 = m\,\Delta\omega$. Solving Eq. (6.40) for all modes m in the gain bandwidth jointly, we obtain photon densities for each mode. The optical output power P_m of each mode is proportional to S_m as shown by Eq. (6.35).

For ordinary FP semiconductor lasers, the longitudinal mode separation $\Delta\omega$ is narrower by more than one order of magnitude than the gain bandwidth $2\,\Delta\omega_G$, and the mode dependences of G_m and τ_{ph} are small. Therefore in the vicinity of the threshold the laser oscillates in several longitudinal modes. Omitting the saturation effect term $\sum_{m'}\xi_{mm'}S_{m'}$ in Eq. (6.40) for simplicity, from Eqs (6.40b) and (6.41a) we obtain

$$S_m = \frac{\tau_{ph} C_s N/\tau_s}{1 - \tau_{ph}\Gamma G_{max} + (m\Delta\omega/\Delta\omega_G)^2} \tag{6.42}$$

where use has been made of the relation $\tau_{ph}\Gamma G_{max} \approx 1$ obtained from the oscillation condition. Putting $m = 0$ in the above equation, and using $N \approx N_{th}$ at the threshold or in the vicinity of the threshold, we obtain $S_0 = (\tau_{ph}C_sN_{th}/\tau_s)(1 - \tau_{ph}\Gamma G_{max})$. Therefore the intensities of each mode relative to the central mode intensity are calculated as

$$\frac{S_m}{S_0} = \left[1 + \frac{S_0\tau_s}{C_sN_{th}\tau_{ph}}\left(\frac{m\,\Delta\omega}{\Delta\omega_G}\right)^2\right]^{-1} \tag{6.43}$$

The above expression indicates that, the larger the spontaneous emission coupling coefficient C_s, the stronger is the tendency of multiple-longitudinal-mode oscillation. Consider next a case where the injection current is increased over the threshold. The output power of the major oscillation mode ($m = 0$) with the maximum gain increases approximately in proportion to the increase in the injection current, as the denominator of the right-hand side of Eq. (6.42) approaches zero, while the powers of other modes saturate. From Eq. (6.43), the full width at half-maximum of the envelope of the longitudinal mode spectrum is given by

$$2\Delta\omega_{env} = 2\Delta\omega_G\left(\frac{C_sN_{th}\tau_{ph}}{S_0\tau_s}\right)^{1/2} \tag{6.44}$$

Note that the oscillation spectrum width $2\Delta\omega_{env}$ is narrower than the gain bandwidth $2\Delta\omega_G$, and that $2\Delta\omega_{env}$ narrows further with increasing output power. Thus, even if the laser oscillates in multiple modes at injection currents slightly higher than the threshold, the increase in the injection current results in a pronounced enhancement of the major oscillation mode power and saturation in the relative powers of other modes. Thus the laser oscillation approaches single-mode oscillation. The evolution of the oscillation spectrum of single-lateral-mode FP semiconductor lasers is shown in Fig. 6.9. The quasisingle-mode oscillation is obtained in index-guiding lasers, where C_s is small, 10^{-5}–10^{-4} [4]. From Eq. (6.43), the side

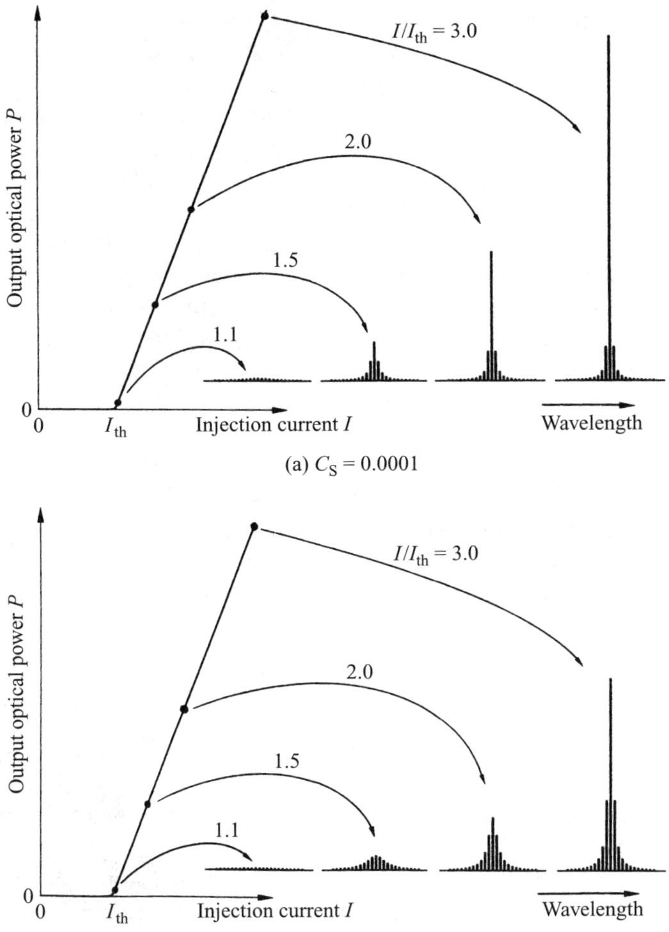

Figure 6.9 Calculated evolution of the longitudinal mode spectrum of a semiconductor laser (an index-guiding facet mirror laser, with a single lateral mode).

mode suppression ratio is given by

$$R_{sms} = \frac{S_0}{S_1} = 1 + \frac{S_0 \tau_s}{C_s N_{th} \tau_{ph}} \left(\frac{\Delta\omega}{\Delta\omega_G} \right)^2 \tag{6.45}$$

For cases where $R_{sms} \gg 1$, the results of the analysis using the single-mode model apply with good accuracy. In quantum well lasers with gain bandwidth $\Delta\omega_G$ narrower than that of bulk DH lasers, a larger side mode suppression ratio is obtained. Gain-guiding lasers having C_s larger than 10^{-4}, on the other hand, usually oscillate in multiple modes even at high injection level [5].

The above-described mode characteristics and the expressions in Eqs (6.42)–(6.45) are the results of an analysis with the gain saturation omitted. Since the gain saturation coefficients $\xi_{mm'}$ in Eq. (6.40) satisfy $\xi_{mm} < \xi_{mm'} (m \neq m')$, the more strongly a mode oscillates, the more strongly the gains for other modes saturate, suppressing the powers of the modes. Thus the gain saturation enhances the tendency discussed above. When the injection current is further increased to obtain even higher output powers, however, the strong hole burning caused in the vicinity of the major oscillation mode gives rise to gain saturation and therefore a relative increase in the gains for other modes, and the tendency of the multiple-longitudinal-mode oscillation may become stronger again.

6.3.6 Intermode Competition

We next discuss the temporal competition between modes. For simplicity, we assume that two longitudinal modes p and q can oscillate. Then the rate equations for the photon densities can be written as

$$\frac{\mathrm{d}}{\mathrm{d}t} S_p = \left[\Gamma_p (G_p - \xi_{pp} S_p - \xi_{pq} S_q) - \frac{1}{\tau_{\mathrm{ph}}} \right] S_p \tag{6.46a}$$

$$\frac{\mathrm{d}}{\mathrm{d}t} S_q = \left[\Gamma_q (G_q - \xi_{qq} S_q - \xi_{qp} S_p) - \frac{1}{\tau_{\mathrm{ph}}} \right] S_q \tag{6.46b}$$

where the spontaneous emission terms are omitted. Let X_p and X_q be equal to $\Gamma_p(G_p - \xi_{pp}S_p - \xi_{pq}S_q) - 1/\tau_{\mathrm{ph}}$ and $\Gamma_q(G_q - \xi_{qq}S_q - \xi_{pq}S_p) - 1/\tau_{\mathrm{ph}}$, respectively. Oscillation of the mode p is temporally attenuated (annihilated) if $X_p < 0$, and is increased (created) if $X_p > 0$. The situation is same for the mode q. Here, we draw lines indicating $X_p = 0$ and $X_q = 0$ on a plane with S_p and S_q as the ordinate and the abscissa, as shown in Fig. 6.10. Then, $X < 0$ in the upper right region of these lines, and $X > 0$ in the lower left region of them. In the plane, annihilation and creation of the oscillations in modes p and q can be sectioned with these lines as the borders between them, and therefore a steady-state point can be determined graphically [6]. As shown in Figs 6.10(a)–6.10(d), the steady state is oscillation of mode p only, oscillation of mode q only, simultaneous oscillation of modes p and q, or bistable oscillation of mode p or q (steady-state oscillation of mode p or q is selected by the initial condition). The steady-state oscillation mode that the laser reaches can be determined by knowing which of Figs 6.10(a)–6.10(d) in the figure represents the relation of the two lines for the given parameters. The graphical method is useful for analyzing the change in the oscillation

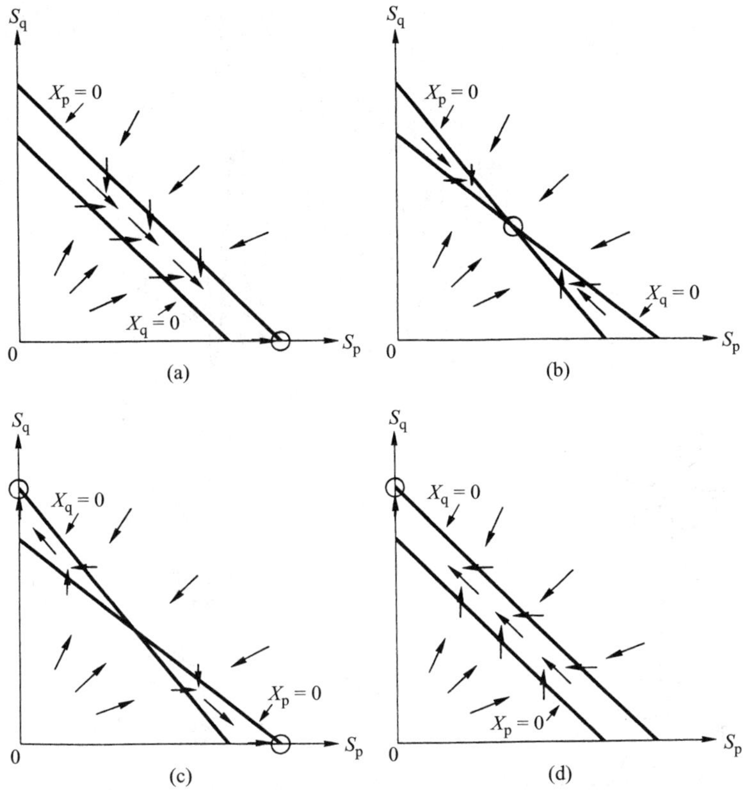

Figure 6.10 Graphical analysis of longitudinal mode competition of a semiconductor laser: (a) single-mode oscillation of mode p; (b) two-mode oscillation of modes p and q; (c) single-mode oscillation of mode p or q (bistable oscillation); (d) single-mode oscillation of mode q. S_p and S_q represent the photon densities of the longitudinal modes p and q, respectively. (S_p, S_q) moves towards the direction indicated by arrows and settles at the point indicated by the open circle.

mode associated with the changes in ambient temperature and injection current. The transition of the oscillation mode with hysteresis under parameter changes can also be explained using this figure [6].

6.4 MODULATION CHARACTERISTICS

The output power of a semiconductor laser can be modulated directly by driving the laser with a signal wave superimposed on the injection current density, as shown in Fig. 6.11. The direct modulation is simple, efficient, and

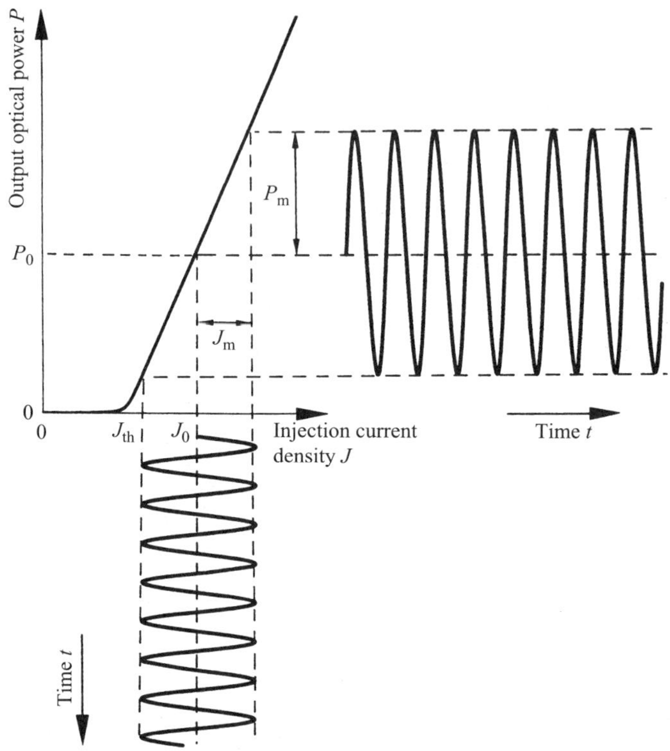

Figure 6.11 Direct modulation of a semiconductor laser.

applicable to high-speed modulation. It is an important feature of semiconductor lasers that allows a variety of applications. A semiconductor laser under direct modulation driving exhibits a complicated behavior unlike steady-state oscillation, because of the interaction between temporal changes in the carriers and photons. This chapter presents a theoretical analysis of the direct modulation to clarify the dynamic characteristics of semiconductor lasers.

6.4.1 Rate Equation for Small-Signal Modulation

The rate equations for analyzing the temporal variations in the carrier density $N(t)$ and the photon density $S(t)$ for a given time-dependent injection current density $J(t)$, under the assumption of single-mode oscillation, are given by Eq. (6.28). They are nonlinear differential equations and cannot be solved analytically. Therefore, we first consider cases where the modulation amplitudes are not large and separate J, N, and S into constant bias components representing the operating point and time-dependent modulation components:

$$J(t) = J_0 + \delta J(t) \qquad (J_0 \gg |\delta J(t)|) \tag{6.47a}$$

$$N(t) = N_0 + \delta N(t) \qquad (N_0 \gg |\delta N(t)|) \tag{6.47b}$$

$$S(t) = S_0 + \delta S(t) \qquad (S_0 \gg |\delta S(t)|) \tag{6.47c}$$

Assuming that the modulation components are much smaller than the bias components, we can perform an approximate but analytical small-signal analysis [7]. The bias components J_0, N_0, and S_0 are values for injection without modulation signal ($\delta J(t) = 0$), and they satisfy the steady-state rate equations:

$$0 = -\Gamma G_0 S_0 - \frac{N_0}{\tau_s} + \frac{J_0}{dq} \tag{6.48a}$$

$$0 = +\Gamma G_0 S_0 - \frac{S_0}{\tau_{ph}} + \frac{C_s N_0}{\tau_s} \tag{6.48b}$$

The gain G, given as a function of the carrier density N and the photon density S, can be approximated by the constant and linear terms in the Taylor expansion around the bias point as

$$G = G(N, S) = G_0 + G_N\,\delta N + G_S\,\delta S \tag{6.49}$$

$$G_N = \left[\frac{\partial G}{\partial N}\right]_0 (= A > 0), \qquad G_S = \left[\frac{\partial G}{\partial S}\right]_0 (< 0)$$

where $[\ ]_0$ denotes the values at the bias point. The coefficients G_N and G_S represent the differential gain and the gain saturation, respectively, and they usually satisfy $G_N \gg |G_S|$. Substituting Eqs (6.47) and (6.49) into Eq. (6.28), using Eq. (6.48), and omitting the small quantities of the second and higher orders, we obtain differential equations for the modulation components:

$$\begin{aligned} \frac{d}{dt}\delta N &= -\Gamma(G_0 + G_S S_0)\,\delta S - \left(\Gamma G_N S_0 + \frac{1}{\tau_s}\right)\delta N + \frac{\delta J}{dq} \\ \frac{d}{dt}\delta S &= -\left(\frac{C_s N_0}{\tau_s S_0} - \Gamma G_S S_0\right)\delta S + \left(\Gamma G_N S_0 + \frac{C_s}{\tau_s}\right)\delta N \end{aligned} \tag{6.50}$$

Eliminating δN from the above equations yields an equation for the modulation in the photon density:

$$\frac{d^2}{dt^2}\delta S + 2\Gamma_R \frac{d}{dt}\delta S + \Omega_R^2\,\delta S = \Gamma G_N S_0 \frac{\delta J}{dq} \tag{6.51}$$

where Γ_R and Ω_R are given by

$$2\Gamma_R = \frac{1}{\tau_s} + \Gamma(G_N - G_S)S_0 + \frac{C_s N_0}{\tau_s S_0} \tag{6.52}$$

$$\Omega_R^2 = \Gamma\left(\frac{G_N}{\tau_{ph}} - \frac{G_S}{\tau_s}\right)S_0 + \frac{C_s}{\tau_s}\left(\frac{1}{\tau_{ph}} + \Gamma G_S S_0 + \frac{N_0}{\tau_s S_0}\right) \tag{6.53}$$

Equation (6.51) is a differential equation for a system with a resonance frequency Ω_R and a damping factor Γ_R driven by a motive force represented by the right-hand side. The spontaneous emission terms with the coupling coefficient C_s, on the right-hand sides of Eqs (6.52) and (6.53), are small compared with other terms and can be omitted. The values of Γ_R and Ω_R are positive and, for realistic lasers, $\Gamma_R \ll \Omega_R$ except in the vicinity of the threshold.

6.4.2 Step Response and Relaxation Oscillation

As a modulation signal $\delta J(t)$, consider a step increment in the injection current density by J_m at $t = 0$:

$$\delta J(t) = J_m u(t), \qquad u(t) = \begin{cases} 0 & (t < 0) \\ 1 & (0 < t) \end{cases} \tag{6.54}$$

The small-signal transient response can be examined by solving Eq. (6.51) with initial conditions $\delta N(0) = 0$ and $\delta S(0) = 0$. The result is given by

$$\delta N(t) = \frac{1}{\Omega_{R'}} \frac{J_m}{dq} \exp(-\Gamma_R t) \sin(\Omega_{R'} t) \tag{6.55a}$$

$$\delta S(t) = \frac{\Gamma G_N S_0}{\Omega_R^2} \frac{J_m}{dq}\left[1 - \exp(-\Gamma_R t)\left\{\cos(\Omega_{R'} t) + \frac{\Gamma_R}{\Omega_{R'}} \sin(\Omega_{R'} t)\right\}\right] \tag{6.55b}$$

$$\Omega_{R'} = \left(\Omega_R^2 - \Gamma_R^2\right)^{1/2} \approx \Omega_R \tag{6.55c}$$

The optical output power is proportional to the photon density $S(t)$. Equation (6.55) indicates that $S(t)$ and $N(t)$ approaches asymptotically to a new steady state, after an evanescent oscillation of frequency Ω_R with damping factor Γ_R, as shown in Fig. 6.12. This oscillation is referred to as relaxation oscillation. An abrupt change in the injection current is followed by a gradual change in the carrier density, since the carrier is accumulated by the injection current density as indicated by the rate equation given by Eq. (6.28). The photon density cannot vary more rapidly than the photon

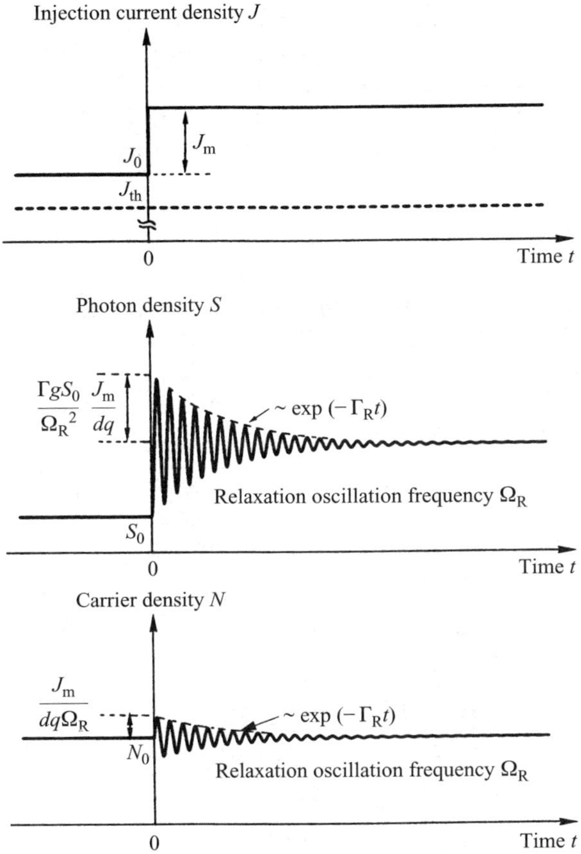

Figure 6.12 Transient response in direct modulation of a semiconductor laser.

lifetime, since the optical field propagates round in the resonator. Therefore, the increase in the injection current density is followed by an increase in the carrier density → increase in the photon density and the spontaneous emission → decrease in the carrier density → decrease in the photon density and the spontaneous emission, and this sequence is repeated with time delay to form an oscillation. Thus the relaxation oscillation results from the feedback with delay. The oscillation, however, attenuates in a time of the same order of magnitude as the carrier lifetime, as the carrier density approaches asymptotically the value that satisfies the oscillation condition for the new steady state. The relaxation oscillation frequency Ω_R and the damping factor Γ_R depend upon the laser structure and the choice of the operation point; the parameter dependence is given by Eqs (6.52) and (6.53). Omitting the gain saturation, which is small ($|G_S| \ll G_N$) in the vicinity of the threshold and at ordinary operation points, omitting the spontaneous emission coupling, and using an expression for the steady-state photon density $S_0 = \tau_{ph}(J - J_0)/dq$ derived from Eq. (6.34), we obtain an approximate expression for Ω_R:

$$\Omega_R^2 \approx \frac{\Gamma G_N S_0}{\tau_{ph}} = \frac{\Gamma G_N (J_0 - J_{th})}{dq} \tag{6.56}$$

The above expression shows that Ω_R is dominated by the effective differential gain ΓG_N and the volume density of the injection current. The gain saturation makes Ω_R higher. The damping factor Γ_R, on the other hand, is determined mainly by the carrier lifetime τ_s. When the output power increases, the attenuation becomes faster owing to the enhanced stimulated emission.

6.4.3 Sinusoidal Modulation and Frequency Response

We next consider a case where the injection current density is modulated by a small sinusoidal signal of frequency Ω. Using complex expressions for the modulation components, we put

$$J(t) = J_0 + \mathrm{Re}\{J_m \exp(-i\Omega t)\} \qquad (J_0 \gg |J_m|) \tag{6.57a}$$

$$N(t) = N_0 + \mathrm{Re}\{N_m \exp(-i\Omega t)\} \qquad (N_0 \gg |N_m|) \tag{6.57b}$$

$$S(t) = S_0 + \mathrm{Re}\{S_m \exp(-i\Omega t)\} \qquad (S_0 \gg |S_m|) \tag{6.57c}$$

Solving the linearized differential equations in Eqs (6.50) with Eqs (6.57) substituted for N_m and S_m, we obtain

$$N_m = \frac{-(i\Omega + \Gamma G_s S_0)}{\Omega_R^2 - \Omega^2 - 2i\Gamma_R \Omega} \frac{J_m}{dq} \tag{6.58a}$$

$$S_m = \frac{\Gamma G_N S_0}{\Omega_R^2 - \Omega^2 - 2i\Gamma_R \Omega} \frac{J_m}{dq} \tag{6.58b}$$

The spontaneous emission coupling was neglected by putting $C_s = 0$. Using Eq. (6.56), the complex amplitude of the modulation in the photon density for low-frequency modulation ($\Omega \to 0$) can be written as

$$S_m(0) = \frac{\Gamma G_N S_0}{\Omega_R^2} \frac{J_m}{dq} \approx \tau_{ph} \frac{J_m}{dq} \tag{6.59}$$

and that for modulation at frequency Ω can be written as

$$S_m(0) = S_m(0) H(\Omega) \tag{6.60}$$

where

$$H(\Omega) = \frac{\Omega_R^2}{\Omega_R^2 - \Omega^2 - 2i\Gamma_R\Omega} \tag{6.61}$$

The function $H(\Omega)$ is a normalized complex transfer function that describes the transfer of the modulation from the current to optical output. The frequency response of the intensity modulation is given by

$$|H(\Omega)| = \frac{\Omega_R^2}{\left[\left(\Omega_R^2 - \Omega^2\right)^2 + 4\Gamma_R^2\Omega^2\right]^{1/2}} \tag{6.62}$$

The frequency response is flat in the low-frequency region, has a peak in the vicinity of the relaxation oscillation frequency Ω_R, and decreases rapidly for frequency higher than Ω_R, as shown in Fig. 6.13. Equation (6.61) shows that, as the modulation frequency approaches the relaxation oscillation frequency, the modulation in the optical output starts to involve a phase retardation. The normalized complex transfer factor at $\Omega = \Omega_R$ giving the peak of the transfer function is

$$H(\Omega_R) = \frac{S_m(\Omega_R)}{S_m(0)} = \frac{i\Omega_R}{2\Gamma_R} \tag{6.63}$$

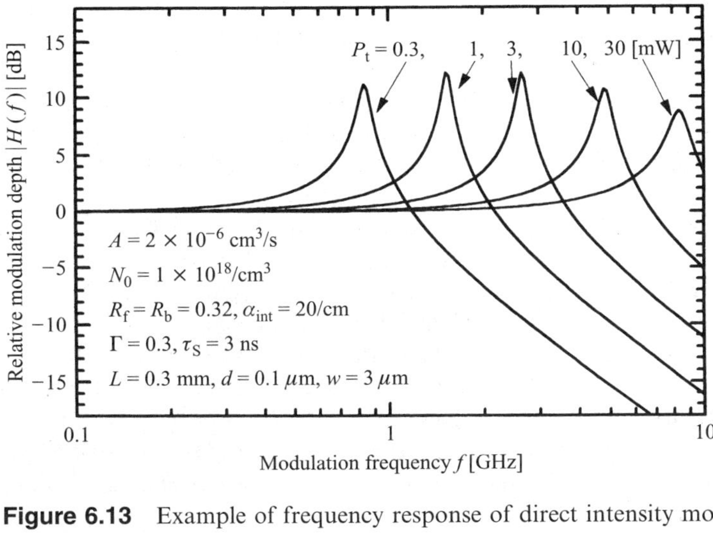

Figure 6.13 Example of frequency response of direct intensity modulation.

Although the above discussion considered modulation with a sinusoidal signal of a single frequency Ω, the modulation by an arbitrary small signal can be described by a superposition of the Fourier components, as Eqs (6.50) are linear equations. The above result shows that, whereas in the low-frequency range the output power is modulated with a high fidelity to the input signal, in the frequency range close to the relaxation frequency Ω_R, where the transfer function is not flat and phase retardation is involved, the modulated output suffers from distortion of the waveform. In the frequency range higher than Ω_R, the output is not modulated substantially. This means that the maximum frequency of practical direct modulation is limited by Ω_R. As we see from Eq. (6.56), Ω_R is determined mainly by the effective differential gain ΓG_N and the injection current density level. For a given laser, the larger the injection level is, the higher Ω_R becomes. Therefore, for high-speed modulation, the bias point must be set at a sufficiently large injection level. The relaxation frequency Ω_R of ordinary semiconductor lasers at a high injection level is several gigahertz.

In order to design a laser of large Ω_R for high-speed modulation, the product of the effective differential gain ΓG_N and the maximum volume density of the injection current should be maximized. Decreasing the active layer thickness d, within a constraint for avoiding large reduction in the confinement factor Γ, leads to a large maximum volume density of injection current, and therefore to a high Ω_R. As presented in Section 4.3, quantum well structures offer a differential gain G_N higher than that for bulk DH structures [8]. The small thickness d of quantum well structures is advantageous in obtaining a high volume density of injection current. In fact, in quantum well lasers using a SCH structure appropriately designed to assure a large confinement factor Γ, direct modulation of maximum frequency higher than that for bulk DH lasers can be obtained [8]. Although, from Eq. (6.56), Ω_R seems to be enhanced limitlessly by increasing the volume density of the injection current, this is a false result because the gain saturation was omitted. In fact, enhancement of Ω_R is limited by the gain saturation.

In the above discussion, the modulation characteristics determined by the intrinsic character of a semiconductor laser itself were considered. In practical laser devices, however, the high-speed modulation characteristics are affected by the junction capacitance, the electrode capacitance, and the inductance of the lead wire. There are many cases where the modulation speed is limited by such driving circuit problems [9]. Thus, implementation of semiconductor lasers for high-speed modulation requires careful design, fabrication, mounting, and packaging for minimizing the parasitic circuit elements. With such efforts, modulation bandwidths over 20 GHz or wider have been accomplished.

6.4.4 Frequency Chirping

The rate equation for the optical phase (Eq. (6.27)) indicates that the phase ϕ of the output wave changes when the refractive index changes through the variation in the carrier density caused by modulation. This means that the intensity modulation is associated with phase modulation [10]. To analyze the phase variation under small-signal modulation, we separate $\phi(t)$ into the value at the bias point and the modulation component and put

$$\phi(t) = \phi_0 + \delta\phi(t) \tag{6.64}$$

The first term ϕ_0 satisfies an equation for the steady state:

$$\frac{d}{dt}\phi_0 = \frac{\alpha_c}{2}\left(\Gamma G_0 - \frac{1}{\tau_{ph}}\right) \tag{6.65}$$

Substituting Eq. (6.64) into Eq. (6.27) and using Eq. (6.65), we obtain

$$\frac{d}{dt}\delta\phi = \frac{\alpha_c}{2}\Gamma G_N\,\delta N \tag{6.66}$$

The gain saturation has been omitted by putting $G_S = 0$. The phase modulation $\delta\phi$ is correlated with the modulation $\delta\omega$ in the instantaneous frequency ω by $\delta\omega(t) = d\delta\phi(t)/dt$.

Consider small-signal modulation by a sinusoidal wave of frequency Ω. Using $J(t)$ and $N(t)$ from Eq. (6.57), then from Eqs (6.58a) and (6.66) we obtain an expression for the frequency modulation $\delta\omega(t)$:

$$\delta\omega(t) = \frac{d}{dt}\delta\phi(t) = \frac{\alpha_c}{2}\Gamma G_N\,\mathrm{Re}\{N_m \exp(-i\Omega t)\} \tag{6.67a}$$

$$N_m = \frac{-(i\Omega + \Gamma G_s S_0)H(\Omega)}{\Omega_R^2}\frac{J_m}{dq} \tag{6.67b}$$

The normalized transfer function $H(\Omega)$ is given by Eq. (6.61). The above expression shows that the optical frequency of the output wave ω varies sinusoidally with a period $2\pi/\Omega$ (with an angular frequency Ω) [10]. Thus the output wave is frequency modulated. The modulation depth $\delta\omega_{max}$ is proportional to the antiguiding factor α_c, the effective differential gain ΓG_N and the modulation amplitude J_m of the driving current density. The frequency response is given by $[-(i\Omega + \Gamma G_S S_0)H(\Omega)]$; similarly to the intensity modulation, the depth of the frequency modulation (FM) increases in the vicinity of the relaxation oscillation frequency Ω_R and decreases rapidly for frequencies Ω higher than Ω_R. From Eqs (6.56), (6.61), and (6.67), the FM depth $\delta\omega_{max}$ for $\Omega \ll \Omega_R$ is given by

$$\delta\omega_{max} = \frac{\alpha_c \tau_{ph}}{2S_0}[\Omega^2 + (\Gamma G_s S_0)^2]^{1/2}\frac{J_m}{dq} \tag{6.68}$$

and, for a constant Ω, $\delta\omega_{max}$ decreases with increasing optical output power.

The time-averaged spectrum of the optical wave with sinusoidal frequency variation as given by Eq. (6.67a) is given by the Fourier transform of $\exp[-i\Omega t - i\delta\phi(t)]$. As is well known concerning the frequency modulation (FM) of a radio-frequency (RF) wave, the spectrum consists of side peaks at equidistant frequencies $\omega \pm m\Omega$ $(m=0, 1, 2, 3, \ldots)$ with the carrier frequency ω at the center. The magnitude of the mth side peak is given by $[J_m(\delta\omega_{max}/\Omega)]^2$, where J_m is the Bessel function of mth order. Observation of the spectrum of this output wave using a monochromator with resolution lower than the side wave separation Ω gives a distribution corresponding to the envelope of the side peaks. The width of the envelope is comparable with $\delta\omega_{max}$ given by Eq. (6.68). With increasing modulation amplitude J_m of the injection current density, the envelope spectrum evolves from a narrow single peak for $J_m=0$, through two-lobe to multilobe distributions with extended bandwidth, as shown in Fig. 6.14. For modulation of ordinary semiconductor lasers with Ω of the order of 100 MHz, the wavelength width of the spectrum can be as large as 1 Å or so [11].

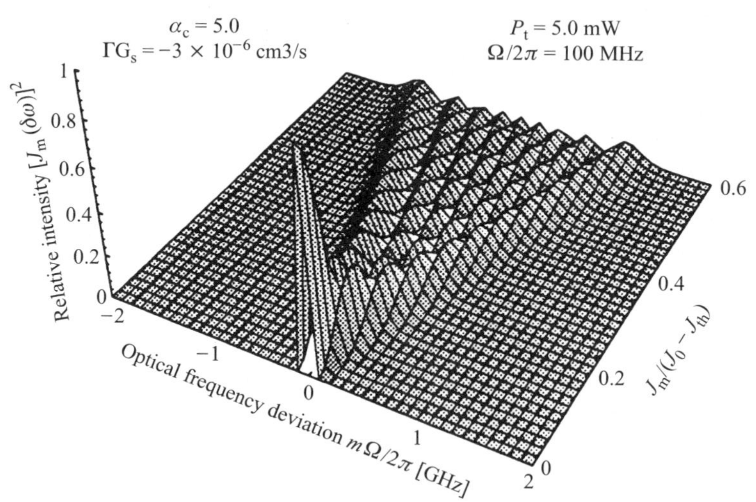

Figure 6.14 Variation in output light spectrum associated with direct intensity modulation.

The FM in the output wave due to the modulation in the driving current is referred to as frequency chirping. Frequency chirping can be utilized as a modulation of the optical frequency (or optical wavelength) [12], or for high-speed scanning of the optical frequency. In optical fiber communication applications, however, this phenomena is recognized as an unwanted optical frequency fluctuation caused by the intensity modulation for giving information [11]. If frequency chirping occurs, the spectrum width of the output wave is broadened to a width significantly larger than the bandwidth essential for the transmission of the signal information, giving rise to large distortion of the optical signal waveform after the transmission through an optical fiber that exhibits group velocity dispersion. Thus chirping prevents achievement of the ultimate transmission bandwidth and the transmission distance. To reduce the harmful chirping, the antiguiding factor α_c must be reduced. Quantum well structures, compared with bulk DH structures, offer a larger differential gain and accordingly a smaller α_c [8]. Therefore, quantum well lasers are advantageous for implementing a laser of small frequency chirping.

6.4.5 Large-Signal Transient Response

In the previous sections, characteristics of small-signal modulation were clarified under the assumption of single-mode oscillation. In many practical applications, modulation with a large amplitude is required. Although the results of the small-signal analysis give insights into understanding the outline of the large-signal modulation characteristics, the validity must be examined. To do this, general analysis should be done assuming multimode oscillation, rather then treating as single-mode oscillation.

The multimode rate equations, given by Eq. (6.29), are a set of nonlinear differential equations and therefore cannot be solved analytically for general cases including large-signal modulation. The analysis requires numerical calculation [13]. As a simple example, here we assume an index-guiding GaAs laser that oscillates in the quasisingle mode for constant-current driving as discussed in Section 6.3 and consider the modulation by driving it with a step current that rises from zero times to twice the threshold current I_{th} at time $t = 0$. The result of the calculation of the transient response is shown in Fig. 6.15. As for the longitudinal modes, the central mode at the gain maximum and the adjacent ±first and ±second side modes are considered, and the photon density for each mode is denoted by $S_m(t)$. The results indicates that, with the rise of the injection current, the carrier density $N(t)$ starts to increase and, at the instance when the increasing $N(t)$ reaches the threshold carrier density N_{th}, the photon densities $S_m(t)$ increase rapidly. Thus the oscillation starts with a delay from the rise of the driving current.

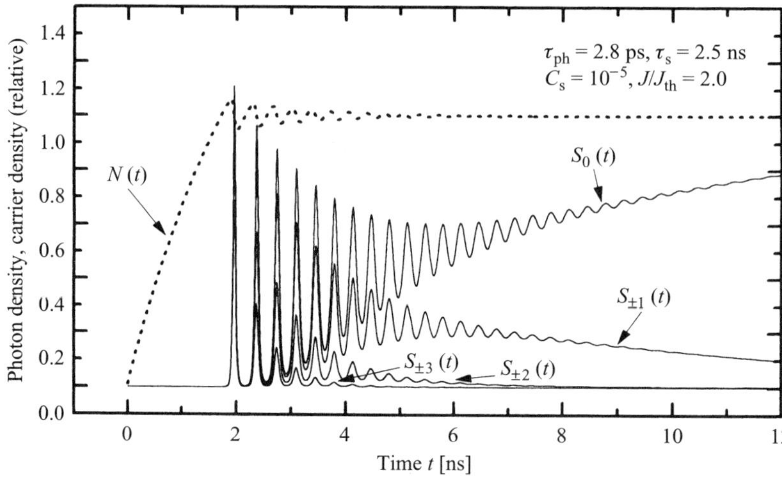

Figure 6.15 Calculated example of large-signal transient response.

The turn-on delay will be discussed in detail in Section 6.4.6. After the start of the oscillation, relaxation oscillation of large amplitudes occurs. The relaxation oscillation attenuates in several nanoseconds, and after that the laser reaches the steady state, maintaining the quasisingle-mode oscillation.

An important characteristic revealed by the result in Fig. 6.15 is that, during the large-amplitude relaxation oscillation, not only the central mode but also the side modes oscillate with large magnitudes associated with relaxation oscillation. This implies that even a steady-state quasisingle-mode laser oscillates in multiple longitudinal modes when it is modulated by a fast large-amplitude signal with frequencies of the order of 100 MHz or higher. Note that, for FP lasers, the photon density τ_{ph} does not depend substantially on the longitudinal mode, and that the gains for the side modes are very close to the gain for the central mode. In the transient state, the time derivatives as the right-hand sides of rate equations are not zero and, therefore, Eq. (6.42) does not hold. The absence of the condensation of the optical power into the central mode, i.e., the side mode suppression, described by Eq. (6.42), is the reason for the transient multimode oscillation.

Positive use of the multimode oscillation by large-signal fast modulation can be made as a means to avoid the noise problems due to mode hopping and external feedback, as will be discussed later. This method is actually employed in some optical disk memory systems. For high-speed long-distance optical fiber communication, the multimode oscillation is a harmful phenomenon. Note that the wavelength separation between adjacent longitudinal modes is usually a few ångströms, larger than the wavelength bandwidth broadened by the frequency chirping discussed in Section 6.4.4. If the laser operation becomes multimode oscillation, the wavelength bandwidth is broadened to a few nanometres to 10 nm, and then the transmission bandwidth and the transmission distance drastically deteriorate. To avoid this problem, modulation must be accomplished by using an external modulator rather than direct modulation, or dynamic single-mode lasers, that will be presented in Chap. 7, must be employed. For modulation at a relatively low frequency, on the other hand, the laser can be considered to be in the quasisteady state. Then the results obtained with the steady-state analysis, and the results using the single-mode model, discussed in the previous sections, apply well as approximate results. It should also be mentioned that, even for multimode oscillation under high-speed modulation, the behavior of the total optical power can be interpreted approximately on the basis of the single-mode model.

6.4.6 Turn-on Delay

As Fig. 6.15 shows, laser oscillation starts with a delay after the start of carrier injecton. This turn-on delay is undesirable phenomenon for applications of semiconductor lasers. The delay, however, can easily be avoided by giving a bias current close to the threshold, since the delay is no more than the time required for carrier accumulation up to the threshold. Then the extinction ratio of the modulated output may deteriorate. An appropriate bias level must be set in order to avoid substantially the delay and deterioration of the extinction ratio.

The turn-on delay provides a method for analyzing experimentally carrier recombination in a semiconductor laser. Since the turn-on delay is a phenomenon before the start of the oscillation, we can analyze it by using the rate equation of the carrier density where the photon density is set to zero:

$$\frac{\mathrm{d}}{\mathrm{d}t}N = -\frac{N}{\tau_\mathrm{s}} + \frac{J}{dq} \tag{6.69}$$

where the first term on the right-hand side, N/τ_s, represents the carrier recombination rate for unit volume. As described in Chap. 3, the recombination includes not only the spontaneous emission recombination but also nonradiative recombination at surfaces and defects, and Auger recombination. The carrier density dependences of the rates of these recombinations are given by $R_\mathrm{sp} = B_\mathrm{r}(P_0N + N^2) \approx B_\mathrm{r}N^2$, $R_\mathrm{sd} \approx A_\mathrm{sd}N$, from Eq. (3.43), and $R_\mathrm{A} \approx C_\mathrm{A}N^3$, from Eq. (3.46), respectively. Therefore, we can write

$$\frac{1}{\tau_\mathrm{s}(N)} \approx A_\mathrm{sd} + B_\mathrm{r}N + C_\mathrm{A}N^2 \tag{6.70}$$

In this chapter, thus far, the lifetime τ_s in the rate equations was treated as if it were a constant independent of N. When a laser oscillates and gives an optical output, the oscillation condition is satisfied at least approximately, and therefore N is close to the threshold carrier density N_{th}. Accordingly, the use of a constant τ_s given by $\tau_s(N_{th})$ has been an appropriate approximation. For the analysis of problems such as turn-on delay, where N varies in a wide range, Eq. (6.70) should be used to take the above dependence into account.

The turn-on delay time T_D for the case where the injection current density rises from J_0 up to J at time $t=0$ can be calculated using an expression which yields from separation of the variables and integration of Eq. (6.69):

$$T_D = \int_{N_0}^{N_{th}} \frac{dN}{J/dq - N/\tau_s(N)}$$
$$\frac{N_{th}}{\tau_s(N_{th})} = \frac{J_{th}}{dq}, \qquad \frac{N_0}{\tau_s(N_0)} = \frac{J_0}{dq} \tag{6.71}$$

where J_{th} is the threshold current density and N_0 the steady-state carrier density for injection current density J_0. For $J_0=0$ and $J \gg J_{th}$, from the above equation we obtain

$$T_D = \frac{dq}{J} N_{th} = \tau_s(N_{th}) \frac{J_{th}}{J} \tag{6.72}$$

Although Eq. (6.71) cannot be integrated analytically, T_D can be calculated easily by integrating numerically Eq. (6.71) with Eq. (6.70) substituted. The delay time T_D and J_{th}, J, and J_0 can easily be measured in experiment. Therefore, by comparing T_D measured under various conditions for a given laser with the theoretical values, one can determine the major component of the carrier recombination and the each coefficient in Eq. (6.70). The carrier recombination lifetime at the threshold $\tau_s(N_{th})$ can also be determined using Eq. (6.72).

6.5 NOISE CHARACTERISTICS

In this section we return to steady-state oscillation of semiconductor lasers and discuss the noise characteristics. As lasers, in general, are light-emitting devices based on the optical transition of electrons, they inevitably involve a quantum noise problem as described in Chap. 2. Semiconductor lasers, in particular, are characterized by a high gain, a wide gain bandwidth and a wide longitudinal mode separation. Ordinary FP semiconductor lasers use a resonator consisting of the facet mirrors with low reflectivities. Since the active region is injected with a high density of carriers, the fluctuation in the carrier density affects significantly the phase of the optical wave. Therefore, semiconductor lasers exhibit unique noise characteristics unlike those of other types of laser. Understanding and improving the noise characteristics are very important for all applications, especially for achieving the ultimate performances in long-distance high-speed optical data transmissions and precision optical measurements. Noises of the output light from practical laser are classified into intrinsic noises and extrinsic noises. In the following, the basic concepts for theoretical noise analysis are summarized, and then the quantum noises under the single-mode steady-state oscillation are discussed in terms of the intensity and frequency noises. Subsequently, noises for multimode operation, and noises caused by external feedback are discussed.

6.5.1 The Wiener–Khintchene Theorem and Shot Noise

Consider the frequency spectrum of a quantity $F(t)$ that fluctuates randomly with time as noises [14]. We assume that the time average of $F(t)$ is zero. Although $F(t)$ is a function for an infinite interval of time, for correlating the analysis with practical observation we cut the part in the finite time interval $-T < t < +T$ and put

$$F_T(t) = \begin{cases} F(t) & (|t| < T) \\ 0 & (|t| > T) \end{cases} \tag{6.73}$$

Let $F_T(\Omega) = \int F_T(t)\exp(\mathrm{i}\Omega_t)\,\mathrm{d}t$ be the Fourier transform of $F_T(t)$. Since $F(t)$ is real, we have $F_T(-\Omega) = F_T(\Omega)^*$. Since $\int F_T(t)^2\,\mathrm{d}t = \int |F_T(\Omega)|^2\,\mathrm{d}\Omega/2\pi$, the energy belonging to a frequency width $\mathrm{d}\Omega/2\pi$ is proportional to $|F_T(\Omega)|^2\mathrm{d}\Omega/2\pi$. We therefore define the power spectrum density of F by

$$S_F(\Omega) \equiv \lim_{T\to\infty}\left(\frac{\ll |F_T(\Omega)|^2 \gg}{2T}\right) \tag{6.74}$$

where $\ll\gg$ denotes ensemble average. Then $S_F(\Omega)\mathrm{d}\Omega/2\pi$ represents the power belonging to a frequency width $\mathrm{d}\Omega/2\pi$. Whereas the above $S_F(\Omega)$ is defined for the frequency range $-\infty < \Omega < +\infty$, the power spectrum density defined for positive frequency range is $2S_F(\Omega)$. If $F(t)$ is a electric signal of radio frequency (RF), $2S_F(\Omega)$ is obtained by a measurement using a RF spectrum analyzer. Next consider the autocorrelation function of $F_T(t)$ given by

$$\langle F(t+\tau)F(t)\rangle = \lim_{T\to\infty}\left[\left(\frac{1}{2T}\right)\int F_T(t+\tau)F_T(t)\,\mathrm{d}t\right] \tag{6.75}$$

and assume that the ensemble average equals the temporal average (assumption of ergodicity). Then the power spectrum and the autocorrelation function are correlated by

$$S_F(\Omega) = \int \langle F(t+\tau)F(t) \rangle \exp(\mathrm{i}\Omega\tau)\,\mathrm{d}\tau \tag{6.76}$$

This relation, referred to as the Wiener–Khintchene theorem, shows that the power spectrum is given by the Fourier transform of the autocorrelation function [14].

Wc next consider a stochastic signal, which is formed by quanta, such as electrons and photons, flowing into a detector or flowing out at a constant rate per unit time. Note that the flow is not of a continuous medium but of discrete quanta. Even though the average flow rate is a constant, the events are not a regular sequence of equal temporal separation, but a stochastic process randomly arranged on the time axis, as illustrated in Fig. 6.16. The intrinsic signal consists of random pulses of an infinitesimal width and a same height. By a realistic measurement system having a finite bandwidth, this signal is observed as a signal with a constant average value associated with fluctuation (quantum noise). Assuming that there is no correlation between quanta (Markov process), it can be shown that the statistical distribution of the number of events in a constant time width is Poissonion. The autocorrelation function of the signal is proportional to Dirac delta function $\delta(\tau)$, and the Fourier transform, i.e., the noise power spectrum, is flat independent of the frequency (white noise). It is known that, when the number of quanta per unit time is Q, the noise power spectrum equals Q [15]. These results were derived on the basis of the classical statistical theory for particles. The white noise is referred to as shot noise.

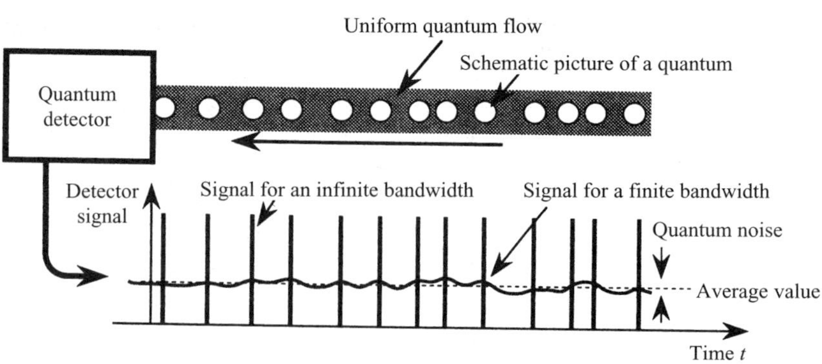

Figure 6.16 Schematic illustration of shot noise.

It has been verified that the same result is obtained on the basis of the quantum theory of photons. Thus, the quantum noise of a signal resulting from a quantum flow at a rate of Q quanta per unit time is a white noise of power spectrum density $S_F(\Omega)=Q$. For example, the noise power spectrum of a photocurrent resulting from detection of Q photons per unit time, $I=qQ$ [A], is $S_I(\Omega)=q^2Q=qI$ [A^2/Hz], if the spectrum is defined for $-\infty<\Omega<+\infty$, and $2qI$ [A^2/Hz], if defined for positive frequency range.

6.5.2 Langevin Noise Sources and Equations for Fluctuation

The rate equations for the carrier density N (=number of carriers in the active region/volume of the active region) and the photon density S (=number of photons in the resonator/volume of the active region) (Eqs (6.28)) can be rewritten as

$$\frac{d}{dt}N = -R_{st}S + R_{ab}S - \frac{N}{\tau_s} + \frac{J}{dq} \tag{6.77a}$$

$$\frac{d}{dt}S = +R_{st}S - R_{ab}S - \frac{S}{\tau_{ph}} + \frac{C_s N}{\tau_s} \tag{6.77b}$$

$$R_{st} - R_{ab} = \Gamma G \tag{6.77c}$$

by decomposing the net stimulated emission ΓGS into the stimulated emission $R_{st}S$ and the absorption $R_{ab}S$. Here, from the discussion in Appendix 5, we have the relation:

$$R_{st} = R_{sp}(\omega_m) = \frac{C_s V_a N}{\tau_s} \tag{6.78}$$

where V_a is the volume of the active region. Equations (6.77a) and (6.77b) represent the variations in N and S, respectively, per unit time, and each term on the right hand sides gives the specifications for each factor, i.e., stimulated emission, absorption, spontaneous emission, carrier injection, and photon number attenuation. These terms represent the time averages for the variation rates. Since each factor is an independent quantum process and therefore is associated with quantum fluctuation (quantum noise), each of these can be treated as shot noise. Therefore, we write the sum of the fluctuations for each term on the right-hand sides as $F_N(t)$ and $F_S(t)$ and append them to the right-hand sides of Eqs (6.77a) and (6.77b) to obtain rate equations with fluctuations considered:

$$\frac{\mathrm{d}}{\mathrm{d}t}N = -\Gamma GS - \frac{N}{\tau_s} + \frac{J}{dq} + F_N \tag{6.79a}$$

$$\frac{\mathrm{d}}{\mathrm{d}t}S = +\Gamma GS - \frac{S}{\tau_{ph}} + \frac{C_s N}{\tau_s} + F_s \tag{6.79b}$$

Similarly, the rate equation for the optical phase with the fluctuation considered can be written as

$$\frac{\mathrm{d}}{\mathrm{d}t}\phi = \frac{\alpha_c}{2}\left(\Gamma G - \frac{1}{\tau_{ph}}\right) + F_\phi \tag{6.80}$$

The terms $F_N(t)$, $F_S(t)$, and $F_\phi(t)$ appended to these equations represent phenomenologically the origins of the quantum noise and are referred to as Langevin noise sources [16]. Since they are random sources, the time averages are zero, i.e.,

$$\langle F_N(t)\rangle = \langle F_s(t)\rangle = \langle F_\phi(t)\rangle = 0 \tag{6.81}$$

We next represent the fluctuations in the carrier density, the photon density, and the optical phase associated with the quantum noise by δN, δS, and $\delta\phi$, respectively, and put

$$N(t) = N + \delta N(t) \qquad (|\delta N(t)| \ll N) \tag{6.82a}$$

$$S(t) = S + \delta S(t) \qquad (|\delta S(t)| \ll S) \tag{6.82b}$$

$$\phi(t) = \phi + \delta\phi(t) \qquad (|\delta\phi(t)| \ll 2\pi) \tag{6.82c}$$

Substituting these expressions into Eqs (6.79), in a similar manner to the analysis of the modulation characteristics, we obtain

$$\frac{\mathrm{d}}{\mathrm{d}t}\delta N = -\Gamma(G + G_s S)\,\delta S - \left(\Gamma G_N S + \frac{1}{\tau_s}\right)\delta N + F_N \tag{6.83}$$

$$\frac{\mathrm{d}}{\mathrm{d}t}\delta S = -\left(\frac{C_s N}{\tau_s S} - \Gamma G_s S\right)\delta S + \left(\Gamma G_N S + \frac{C_s}{\tau_s}\right)\delta N + F_s \tag{6.84}$$

$$\frac{\mathrm{d}}{\mathrm{d}t}\delta\phi = \frac{\alpha_c}{2}\Gamma G_N\,\delta N + F_\phi \tag{6.85}$$

The above equations are linearized rate equations for the fluctuations and can be used to analyze the noise characteristics of semiconductor lasers.

6.5.3 Correlation Functions and Power Spectra of Langevin Noise Sources [16–18]

The absolute value of each term on the right-hand sides of Eqs (6.77a) and (6.77b), multiplied by the volume V_a of the active region gives specifications for each origin of the fluctuation rate per unit time in the carrier number $V_a N$ and the photon number $V_a S$. Therefore, treating the fluctuations of these quantities as shot noises, their power spectra equal the absolute value of each term, multiplied by V_a. Since the noise sources are mutually independent stochastic processes, the power spectra of the total fluctuations $V_a F_N$ and $V_a F_S$ are given by the sum of the components for each source. Let $V_a^2\langle F_N F_N\rangle$ and $V_a^2\langle F_S F_S\rangle$ be the autocorrelation functions of $V_a F_N$ and $V_a F_S$. Then, since from the Wiener–Khintchene theorem the autocorrelation functions equal Fourier transforms of power spectra, we have

$$V_a^2\langle F_N F_N\rangle = \left(R_{st}S + R_{ab}S + \frac{N}{\tau_s} + \frac{J}{dq}\right) V_a \delta(\tau) \tag{6.86}$$

$$V_a^2\langle F_S F_S\rangle = \left(R_{st}S + R_{ab}S + \frac{S}{\tau_{ph}} + \frac{C_s N}{\tau_s}\right) V_a \delta(\tau) \tag{6.87}$$

As for the cross-correlation function between $V_a F_N$ and $V_a F_S$, $V_a^2\langle F_N F_S\rangle$, by taking the components that change N and S simultaneously, we can write

$$V_a^2\langle F_N F_S\rangle = -\left(R_{st}S + R_{ab}S + \frac{C_s N}{\tau_s}\right) V_a \delta(\tau) \tag{6.88}$$

where a minus sign is inserted in the right-hand side, since the correlation is negative, i.e., the increase in N is associated with a decrease in S and vice versa. Since we are considering a steady state, the rate equations, with the left-hand sides replaced by zero, hold. These equations (Eqs (6.77c) and (6.76)) and $V_a S \gg 1$ can be used to simplify the above expressions to yield

$$\langle F_N F_N\rangle = D_{NN}\delta(\tau), \qquad D_{NN} = \frac{2C_s NS}{\tau_s} + \frac{2N}{\tau_s V_a} \tag{6.89}$$

$$\langle F_S F_S\rangle = D_{SS}\delta(\tau), \qquad D_{SS} = \frac{2C_s NS}{\tau_s} \tag{6.90}$$

$$\langle F_N F_S\rangle = D_{NS}\delta(\tau), \qquad D_{NS} = -\frac{2C_s NS}{\tau_s} + \frac{S}{\tau_{ph} V_a} \tag{6.91}$$

Consider next the phase noise source F_ϕ. Let $\delta E(t)$ be a small fluctuation, which results from the Langevin noise sources, in the complex amplitude $E(t)$ of the optical field in the resonator. We write the complex amplitude $E(t)$ as

$$\begin{aligned} E(t) &= E + \delta E(t) \\ &= |E(t)|\ \exp[-\mathrm{i}\{\phi + \delta\phi(t)\}] \\ &= [|E| + \delta E_1(t) - \mathrm{i}\delta E_2(t)]\ \exp(-\mathrm{i}\phi) \end{aligned} \tag{6.92}$$

The fluctuation $\delta E(t)$ in the random phase can be decomposed into a component δE_1 of the same phase as E and another component δE_2

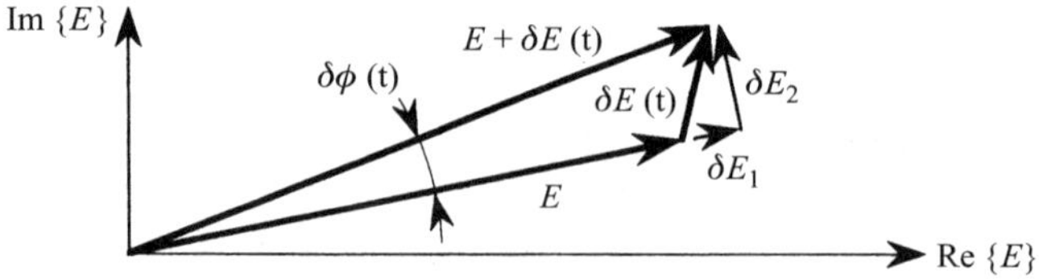

Figure 6.17 Relation between the amplitude fluctuation and phase fluctuation of an optical wave.

perpendicular to E, as shown in Fig. 6.17. Then the phase fluctuation can be written as

$$\delta\phi(t) = \frac{\delta E_2(t)}{|E|} \tag{6.93}$$

On the other hand, the photon density $S(t)$ and $E(t)$ are correlated by $\hbar\omega V_a S(t) = |E(t)|^2$, and from Eq. (6.92) we have $|E(t)|^2 - |E|^2 = 2|E|\,\delta E_1(t)$. Therefore, the Langevin noise source F_S for S, and the Langevin noise source F_{E_1} for δE_1 are correlated by $\hbar\omega V_a F_S = 2|E|F_{E_1}$. Let $\langle F_{E_1}F_{E_1}\rangle$ and $\langle F_{E_2}F_{E_2}\rangle$ be the autocorrelation functions of the noise sources F_{E_1} and F_{E_2} for δE_1 and δE_2, respectively. Then the autocorrelation function $\langle F_\phi F_\phi\rangle$ of F_ϕ, can be written as

$$\begin{aligned}\langle F_\phi F_\phi\rangle &= \frac{\langle F_{E_2}F_{E_2}\rangle}{|E|^2} \\ &= \frac{\langle F_{E_1}F_{E_1}\rangle}{|E|^2} \\ &= \frac{\langle F_S F_S\rangle}{4S^2} \\ &= D_{\phi\phi}\delta(\tau)\end{aligned} \tag{6.94}$$

$$D_{\phi\phi} = \frac{C_s N}{2\tau_s S}$$

Since $\delta E(t)$ has a random phase independent of the fluctuation δS in the photon density, there is no correlation between F_ϕ and F_S. Since the fluctuation in the optical phase ϕ resulting from the fluctuation δN in the carrier density has already been interpreted by the first term on the right-hand side of Eq. (6.85), there is no need to consider here the correlation between F_ϕ and F_N. Thus we can put

$$\langle F_\phi F_S\rangle = \langle F_\phi F_N\rangle = 0 \tag{6.95}$$

From the Wiener–Khintchene theorem, the power spectra (and the cross power spectra) $S_{ij}(\Omega)$ of the Langevin noise sources F_i ($i = N, S, \phi$) are given by the Fourier transforms of the correlation functions $\langle F_i F_j \rangle$ (calculated as above. Therefore, $S_{ij}(\Omega)$ are given by

$$S_{ij}(\Omega) = D_{ij} \qquad (i, j = N, S, \phi) \tag{6.96}$$

As Eqs (6.90) and (6.94) show, both S_{SS} and $S_{\phi\phi}$ are proportional to C_s, and $S_{SS} = S_{\phi\phi} = 0$ for $C_s = 0$. This implies that the ultimate origin of the optical noise is the coupling of the spontaneous emission to the laser oscillation mode.

6.5.4 Intensity Noise

Figure 6.18 illustrates the various kinds of noise of an output light from a semiconductor laser oscillating in a single mode. We first consider the intensity noise [19]. The output power of a single-mode laser is associated with fluctuation around the average value due to the quantum noise. Since the fluctuation is random, the intensity noise has a broad frequency spectrum. The noise power spectrum can be observed by using a wide-band photodetector to convert the optical intensity into a photocurrent, and using

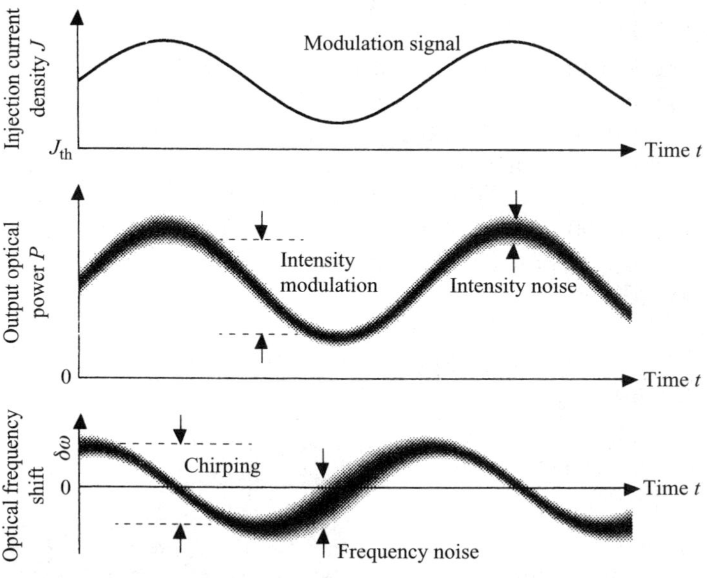

Figure 6.18 Schematic illustration of noises in laser output light.

a RF spectrum analyzer to measure and display the power spectrum of the photocurrent.

Noting that the optical output power is proportional to the photon density S, we first calculate the noise power spectrum for the photon density $|S_S(\Omega)|$ defined by putting $F(t) = \delta S(t)$ in Eq. (6.74). To solve the fluctuation equations (Eqs (6.83) and (6.84)), we use the Fourier transforms of the time functions $\delta N(t)$ and $F_N(t)$:

$$\delta N(\Omega) = \int \delta N(t) \exp(\mathrm{i}\Omega t)\, \mathrm{d}t \tag{6.97a}$$

$$F_N(\Omega) = \int F_N(t) \exp(\mathrm{i}\Omega t)\, \mathrm{d}t \tag{6.97b}$$

We also use similar Fourier transforms of $\delta S(t)$ and $F_S(t)$. Solving Eqs (6.83) and (6.84), with the inverse Fourier transform expressions for these quantities substituted, we obtain

$$\begin{aligned}\delta N(\Omega) = {} & [H(\Omega)\Omega_{\mathrm{R}}^2] \\ & \times \left[\left(\frac{C_{\mathrm{s}} N}{\tau_{\mathrm{s}} S} - \Gamma G_S S - \mathrm{i}\Omega \right) F_N(\Omega) - \Gamma (G + G_S S) F_{\mathrm{S}}(\Omega) \right]\end{aligned} \tag{6.98}$$

$$\begin{aligned}\delta S(\Omega) = {} & [H(\Omega)\Omega_{\mathrm{R}}^2] \\ & \times \left[\left(\Gamma G_N S + \frac{C_{\mathrm{s}}}{\tau_{\mathrm{s}}} \right) F_N(\Omega) + \left(\frac{1}{\tau_{\mathrm{s}}} + \Gamma G_N S - \mathrm{i}\Omega \right) F_S(\Omega) \right]\end{aligned} \tag{6.99}$$

where $H(\Omega)$ is the transfer function given by Eq. (6.61). The noise power spectrum for the photon density $S_S(\Omega)$ is obtained by calculating the autocorrelation function for $\delta S(t)$, the inverse Fourier transform of the above $\delta S(\Omega)$, and Fourier transforming the result based on the Wiener–Khintchene theorem. Using the correlation functions for the Langevin noise sources, Eqs (6.89), (6.90), and (6.91), the factors related to F_N and F_S in $S_S(\Omega)$ are replaced by the power spectra (and cross power spectra) $S_{ij}(\Omega) = D_{ij}$. Thus we obtain

$$\begin{aligned}S_S(\Omega) = {} & \frac{2\tau_{\mathrm{ph}}^2}{\Gamma^2 G_N^2 S} |H(\Omega)|^2 \left[\frac{C_{\mathrm{s}} N}{\tau_{\mathrm{s}}} \left(\Omega^2 + \frac{1}{\tau_{\mathrm{s}}^2} \right) \right. \\ & \left. + \frac{\Gamma G_N}{\tau_{\mathrm{ph}} \tau_{\mathrm{s}} V_{\mathrm{a}}} (1 + \Gamma G_N \tau_{\mathrm{ph}} N) S + \frac{\Gamma^2 G_N^2}{\tau_{\mathrm{ph}} V_{\mathrm{a}}} S^2 \right]\end{aligned} \tag{6.100}$$

In the above calculation, we used $C_{\mathrm{s}} \ll 1$, $NV_{\mathrm{a}} \gg 1$, and $\Gamma G_N \tau_{\mathrm{s}} / V_{\mathrm{a}} \ll 1$ and omitted the related small terms. Use has also been made of the approximate expression for Ω_{R} (Eq (6.56)). For laser operation near the threshold where

S is small, $S_S(\Omega)$ is dominated by the term proportional to $(C_s N/\tau_s)(\Omega^2 + 1/\tau_s^2)$ resulting from the spontaneous emission coupling and is proportional to S^{-1}. For high output operation (large S), on the other hand, $S_S(\Omega)$ is dominated by the final term in Eq. (6.100) and is proportional to S^{+1}. As can be seen from the fact that $H(\Omega)$ is included in Eq. (6.100), $S_S(\Omega)$ has a large peak at and near the relaxation oscillation frequency Ω_R.

From Eqs (6.52), (6.56), (6.62), and (6.100), $S_S(\Omega)$ for large S can be written approximately as

$$S_S(\Omega) \approx \frac{2S\tau_{ph}/V_a}{1 + \tau_{ph}^2 \Omega^2} \qquad (6.101)$$

The total noise power is calculated by integrating the above power density over all the frequency range to yield

$$\int \frac{S_S(\Omega)\,d\Omega}{2\pi} = \frac{S}{V_a} \qquad (6.102)$$

This result shows that the variance of the fluctuation in the photon number $V_a S$ in the laser is given by $V_a^2(S/V_a) = V_a S$. This is consistent with the theoretical result that the output light from a highly pumped laser is in the coherent state described in Chap. 2.

Although the above discussion is with the intensity noise in the photon density S, from a practical point of view, the important problem is the noise in the output light. As Eq. (6.35) shows, the output power P is proportional to S, and P in terms of the output photon number per unit time can be written as

$$P = KS, \qquad K \equiv \frac{F_{stm} V_a}{\tau_{ph}} \qquad (6.103)$$

where F_{stm} is the output coupling coefficient given by Eq. (6.36) and satisfies $0 < F_{stm} < 1$. Although a simple consideration would give a result that the noise power of P is the noise power of S multiplied by the square of the coefficient in the above relation, this is not correct. For the photons accumulated in the laser to leave the resonator to give an output, the photons must penetrate through the output facet mirror. However, because of the quantum nature of photons, the penetration and the reflection of the photons at the mirror are stochastic processes and give rise to additional noise [20]. Let a Langevin noise source F_P represent the photon partition noise, and put the fluctuation δP in the output power P as

$$\delta P(t) = K\,\delta S(t) + F_P(t) \qquad (6.104)$$

This noise source F_P has no correlation with the noise source for the carrier density F_N. However, since increase in P is associated with decrease in S, there is a negative correlation between F_P and F_S.

The correlation functions and the power spectra can be calculated by a similar method to that described in Section 6.5.3 as

$$\langle F_P F_P \rangle = D_{PP}\delta(\tau), \qquad D_{PP} = +P \tag{6.105}$$

$$\langle F_S F_P \rangle = D_{SP}\delta(\tau), \qquad D_{SP} = -\frac{P}{V_\mathrm{a}} \tag{6.106}$$

$$\langle F_N F_P \rangle = 0 \tag{6.107}$$

The noise power spectrum for the output optical power $S_P(\Omega)$ can then be calculated by a procedure similar to that for the calculation of the noise spectrum for the photon density.

As a convenient measure for representing the intensity noise and the spectrum, relative intensity noise (RIN) is often used. The RIN for the output power P is defined by

$$\mathrm{RIN}(\Omega) \equiv \frac{S_P(\Omega)}{P^2} \tag{6.108}$$

and the unit for the RIN is reciprocal Hertz. Calculating the noise power spectrum for the output power $S_P(\Omega)$, using Eqs (6.52), (6.56), (6.61), (6.99), (6.100), and (6.103)–(6.107), and substituting the result in Eq. (6.108) yields an expression for the RIN:

$$\begin{aligned}\mathrm{RIN}(\Omega) = \Bigg\{ & 2K\left(\frac{\tau_\mathrm{ph} K}{\Gamma G_N P}\right)^2 |H(\Omega)|^2 \\ & \times \left[\left(\frac{C_\mathrm{s} N}{\tau_\mathrm{s}} + \frac{\Gamma G_S P}{K V_\mathrm{a}}\right)\Omega^2 + \left(\frac{C_\mathrm{s} N}{\tau_\mathrm{s}^3}\right) \right. \\ & + \left(\frac{\Gamma^2 G_N^2 N}{\tau_\mathrm{s} V_\mathrm{a}} + \frac{\Gamma G_S}{\tau_\mathrm{s}^2 V_\mathrm{a}}\right)\frac{P}{K} \\ & \left. + \frac{\Gamma^2 G_N G_S}{\tau_\mathrm{s} V_\mathrm{a}}\left(\frac{P}{K}\right)^2 \right] + 1 \Bigg\} \frac{1}{P}\end{aligned} \tag{6.109}$$

The last term $1/P$ in the above expression corresponds to the shot noise, for the output consisting of P photons per unit time, in terms of the RIN. This quantity is referred to as the standard quantum limit (SQL) or shot noise limit. Equation (6.109) indicates that the intensity noise of the semiconductor laser output consists of the frequency-independent SQL noise and frequency-dependent excess noises. An example of the calculated RIN is shown in Fig. 6.19. The frequency dependence of the RIN is almost flat in the low-noise frequency range but exhibits a peak due to the resonance of the excess noises at and near the relaxation oscillation frequency Ω_R. For

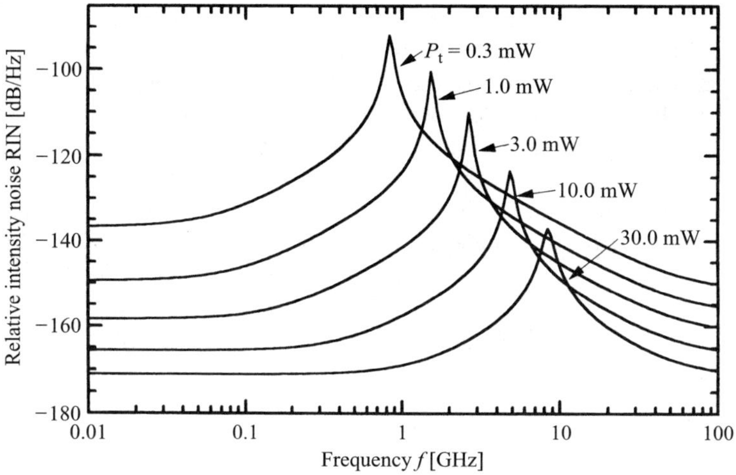

Figure 6.19 Example of the frequency spectrum of intensity noise.

frequency above Ω_R, the excess noises decrease rapidly and the noise level approaches asymptotically the SQL level. The RIN decreases with increasing output power P. From Eq. (6.109), for low frequencies and small S, we have the approximate expression

$$\mathrm{RIN} \approx \frac{2K(\tau_{\mathrm{ph}}K/\Gamma G_N P)^2 (C_s N/\tau_s^3) + 1}{P} \tag{6.110}$$

This expression shows that, with increasing P in the vicinity of the threshold, the RIN decreases in proportion to P^{-3}. With further increase in P, the term proportional to P^2 in Eq. (6.110) becomes the dominant term, and therefore the RIN decreases, with reduced speed, in proportion to P^{-1}.

6.5.5 Frequency Noise

In applications where the frequency of the laser light is modulated with a signal or measured, the fluctuation in the frequency is superimposed as a noise on the signal. Since the fluctuation in the instantaneous frequency due to the quantum noise is given by $\delta\omega(t) = \mathrm{d}\delta\phi(t)/\mathrm{d}t$, from Eq. (6.85) for the phase fluctuation $\delta\phi(t)$, $\delta\omega(t)$ can be written as

$$\delta\omega(\Omega) = \frac{\alpha_c}{2}\Gamma G_N\,\delta N(\Omega) + F_\phi(\Omega) \tag{6.111}$$

where $\delta\omega(\Omega)$ and $F_\phi(\Omega)$ are Fourier transforms of $\delta\omega(t)$ and $F_\phi(t)$, respectively. The above expression shows that the frequency fluctuation results from both the index fluctuation associated with the carrier fluctuation and the coupling of the spontaneous emission to the oscillation mode. Although the Fourier transform of the carried density fluctuation $\delta N(\Omega)$, given by Eq. (6.98), includes a term proportional to F_N and a term proportional to F_S, the latter is dominant and the former can be neglected. As Eqs (6.95) and (6.98) show, there is no correlation between δN and F_ϕ. In a similar manner to the calculation of $S_S(\Omega)$ in Section 6.5.4, calculation of the noise power spectrum $S_\omega(\Omega)$ using Eqs (6.90), (6.94), (6.98), and (6.111) results in

$$\begin{aligned} S_\omega(\Omega) &\approx \left(\frac{\alpha_c \Gamma G_N}{2}\right)^2 \left|\frac{H(\Omega)}{\Omega_R^2}\right|^2 \Gamma^2 G^2 D_{ss} + D_{\phi\phi} \\ &\approx \frac{C_s N}{2\tau_s S}\left[\alpha_c^2 |H(\Omega)|^2 + 1\right] \end{aligned} \tag{6.112}$$

where $H(\Omega)$ is the transfer function given by Eq. (6.61), and use has been made of Eq. (6.56) and $\Gamma G \approx 1/\tau_{ph}$ to obtain the second line.

The frequency noise power spectrum given by Eq. (6.112) describes the frequency Ω distribution of the time variation in the optical frequency fluctuation $\delta\omega$. Since the antiguiding factor (linewidth enhancement factor) α_c is usually considerably larger than unity, and $|H(\Omega)|$ is larger than unity for frequency Ω below the relaxation frequency Ω_R, the index fluctuation induced by the carrier density fluctuation dominates the frequency noise. The spectrum is dominated by $|H(\Omega)|^2$, flat for $\Omega < \Omega_R$, peaked at $\Omega \approx \Omega_R$, and decreases rapidly for $\Omega > \Omega_R$. It should be noted, however, that even for $\Omega > \Omega_R$ the frequency noise due to the spontaneous emission coupling to the oscillation mode remains. An example of calculated frequency noise power spectrum is shown in Fig. 6.20. We also see from Eq. (6.112) that, on reducing the frequency noise, C_s and α_c should be reduced and the laser should be operated with a high bias level.

6.5.6 Multimode Noises

Mode-Hopping Noise

Many of the FP semiconductor lasers oscillate in quasisingle mode in the high-output-power region. However, when the gain peak wavelength comes to the center between the two adjacent longitudinal modes, depending upon the operation conditions, alteration of the oscillation between the two modes takes place. This phenomenon, called mode hopping, causes a large intensity noise in the relatively low-frequency range lower than several tens of megahertz, since it is associated with not only a change in the oscillation

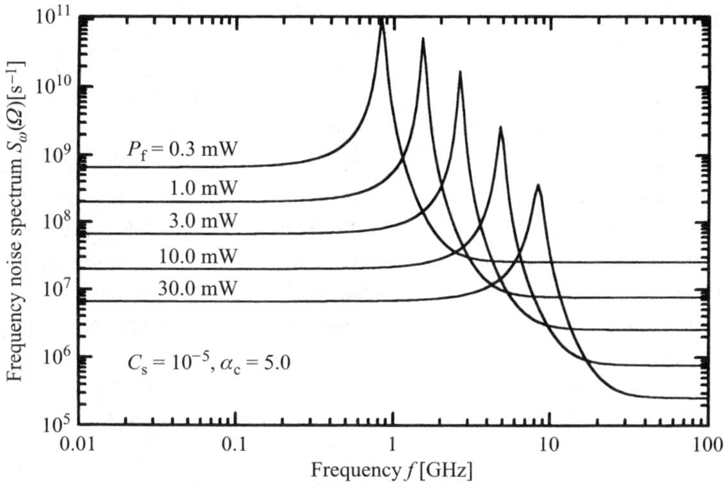

Figure 6.20 Example of the frequency spectrum of frequency noise.

wavelength but also a change in the oscillation intensity. The mode-hopping noise results from the alternate oscillation of the two modes with random durations. Stable simultaneous oscillation of both modes does not occur, since there are fluctuations due to the spontaneous emission and since oscillation of one of the two mode is suppressed by the cross gain saturation. Since the longitudinal mode wavelengths are sensitive to temperature change, in ordinary operating conditions it is difficult to avoid mode hopping by selecting the driving current level. Thus, mode hopping is a large problem for applications of FP semiconductor lasers.

For applications such as optical disk systems, the problem can be eliminated by various methods. One of these is to make the mode transition have a hysteresis characteristic [21], by high-density Te doping in the cladding layer to enhance the saturable absorption, or by enhancing the facet reflectivity to increase the photon density and to enhance the gain saturation. Then fast bidirectional mode transitions are avoided and therefore mode-hopping noise is eliminated. Another method is to implement a FP laser that performs stable pulse oscillation with a high repetition rate, by using the self-pulsation phenomenon discussed later. It is also possible to attain stable multimode oscillation by superimposing a RF signal of several hundred megahertz on the driving current [22]. Although then the output power varies at a fast speed, for detection of the total output of all modes with a narrow bandwidth around 10 MHz or lower, the noise is significantly reduced, enabling applications similar to those of lasers with stable continuous oscillation.

Mode Partition Noise

Even if a semiconductor laser is driven in a condition where mode hopping is avoided, it does not oscillate strictly in a single mode. The output includes not only the major oscillation mode but also side modes. Although the intensity noise measured for the total output power agrees fairly well with the result of single-mode analysis described in previous subsections, additional noise is involved for separate detection of each mode [23]. The additional noise results from random distribution of the photons produced by stimulated emission in each mode and is called the mode partition noise. The mode partition noise can be analyzed by using linearized multimode rate equations for fluctuations:

$$\frac{\mathrm{d}}{\mathrm{d}t}\delta N = -\sum_m \Gamma(G_m + G_s S_m)\,\delta S_m - \left(\Gamma G_N S_0 + \frac{1}{\tau_s}\right)\delta N + F_N$$
$$\frac{\mathrm{d}}{\mathrm{d}t}\,\delta S_m = +\left(\Gamma G_S S_m - \frac{C_s N}{\tau_s S_m}\right)\delta S_m + \left(\Gamma G_N S_m + \frac{C_s}{\tau_s}\right)\delta N + F_{Sm} \tag{6.113}$$

which is an extension of Eqs (6.83) and (6.84). In the above equations, S_0 denotes the photon density of the major oscillation mode. Here, the (cross) power spectra of the Langevin noise sources can be written as

$$D_{NN} = \frac{2C_s N}{\tau_s}\sum_m S_m + \frac{2N}{\tau_s V_a}$$
$$D_{SmSm} = \frac{2C_s N}{\tau_s} S_m, \qquad D_{SmSm'} = 0 \qquad (m \neq m') \tag{6.114}$$
$$D_{NSm} = -\frac{2C_s N}{\tau_s} S_m + \frac{S_m}{\tau_{phm} V_a}$$

since there is no correlation between noise sources for photon densities of different modes. Similarly to a single-mode case, we can calculate the noise power spectra for the photon densities $S_{Sm}(\Omega)$ and the relative intensity noise spectra RIN(Ω), by solving Eq. (6.113) by using Fourier transfoms and Eq. (6.114), and using the Wiener–Khintchene theorem. For the case where only the major mode and the two side modes adjacent to it are considered, an analytical solution can be obtained. Figure 6.21 shows the calculated RIN(Ω) with the side mode suppression ratio R_{sms} as a parameter. As the figure shows, the RIN coincides with the result of the single-mode analysis in the high-frequency range above the relaxation oscillation frequency. In the low frequency region, however, there is a significant mode partition noise component. Although the mode partition noise can be reduced by enhancing R_{sms}, it is still significant for

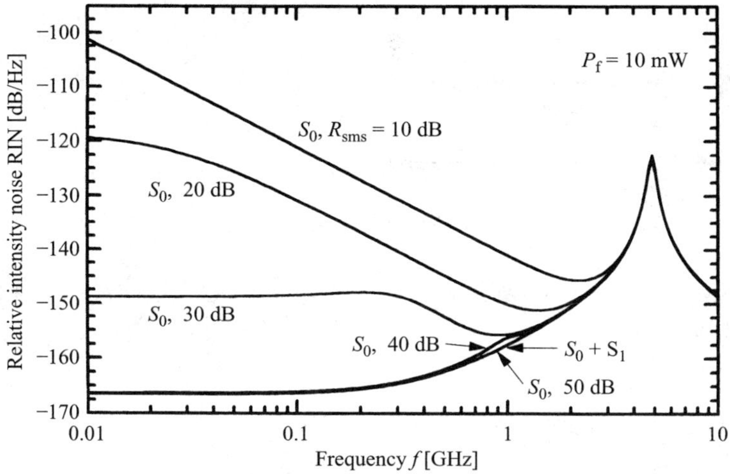

Figure 6.21 Example of the frequency spectrum of mode partition noise (S_0 and S_1 are the intensities of the major oscillation mode and the side mode, respectively).

$R_{sms} \sim 20$ dB, which can be achieved in ordinary FP lasers. Thus, the mode partition noise is an important noise that dominates the low-frequency RIN of the major mode output of FP lasers. The fact that significant noise remains in spite of the small side-mode powers is attributed to the fact that owing to the fluctuation the side-mode powers can be instantaneously much larger than the average power. In order to reduce the mode partition noise to a negligible level, an R_{sms} value as large as 40–50 dB is required. Such a high R_{sms} can be accomplished in distributed feedback (DFB) and distributed Bragg reflector (DBR) lasers presented in the next chapter.

6.5.7 Feedback-Induced Noise

Thus far, we have discussed the noise properties of semiconductor lasers operating independently. In many applications, however, some of the laser output light is reflected by optical elements and returns to the laser. In optical communications systems, reflection at the fiber input and output ports gives rise to optical feedback and, in optical disk systems, reflection at the disk also causes optical feedback. Semiconductor lasers under such operation conditions exhibit complicated behaviors, and the feedback significantly affects the noise characteristics. This section deals with feedback-induced noise. The external feedback and therefore feedback-induced noise can be eliminated by inserting a nonreciprocal isolator in the

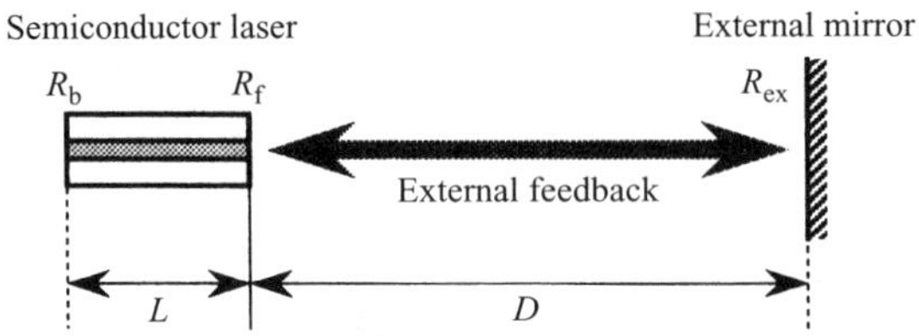

Figure 6.22 A model for the analysis of feedback-induced noise.

output beam path, since the isolator transmits the forward wave but blocks the reflected backward wave.

Model and Rate Equations

As a simple model for the analysis of feedback-induced noise, consider a system where a mirror is positioned at a distance D from a semiconductor laser, as shown in Fig. 6.22. The laser resonator and the external mirror constitute a composite resonator, and, in the steady state, the laser oscillates in the resonant mode(s) of the composite resonator. Although mathematical formulation of the composite resonator modes is possible, the formulation cannot enable analysis of laser dynamics to be made more rapidly than the round trip time $\tau = 2D/c$ for the distance D, which can generally be much longer than the laser resonator length L. Here we employ a method where the external feedback is treated as a perturbation and the combination of the laser output facet mirror and the external mirror is represented by an effective reflection coefficient [24]. Let R_f be the reflectivity of the output facet mirror, and R_{ex} be the effective reflectivity of the external mirror, including the optical losses during the round trip between the output facet and the external mirror and the loss in coupling back into the laser waveguide. Let $S(t)$ and $\phi(t)$ be the photon density and the optical phase, respectively, according to the definition in Section 6.2. Then the amplitude of the guided wave propagating in the laser towards the output facet is proportional to $[S(t)]^{1/2}\exp\{-[\omega_{th}t + \phi(t)]\}$, and the wave is reflected at the facet with the amplitude reflection coefficient $r_f = R_f^{1/2}$. The amplitude of the wave component that is fed back into the waveguide after a round trip to the external mirror, noting the double transmission and the retardation time τ, can be written as $(1 - R_f)[R_{ex}S(t-\tau)]^{1/2}\exp\{-i[\omega_{th}(t-\tau) + \phi(t-\tau)]\}$. Therefore, omitting the multiple reflection under the assumption $R_{ex} \ll 1$, the effective amplitude reflection coefficient for the combination of the output facet mirror and the external mirror can be written as

$$r_{feff} = r_f + \delta r_f \tag{6.115}$$

$$\delta r_{\mathrm{f}} = (1 - R_{\mathrm{f}})\left(\frac{R_{\mathrm{ex}} S(t-\tau)}{S(t)}\right)^{1/2} \exp\{\mathrm{i}[\omega_{\mathrm{th}}\tau + \phi(t) - \phi(t-\tau)]\}$$

The perturbations in the reflectivity and the reflection phase due to the external feedback are given by

$$\delta R_{\mathrm{f}} = 2r_{\mathrm{f}}\,\mathrm{Re}\{\delta r_{\mathrm{f}}\}, \qquad \delta\phi_{\mathrm{f}} = -\frac{\mathrm{Im}\{\delta r_{\mathrm{f}}\}}{r_{\mathrm{f}}} \tag{6.116}$$

We next consider how the rate equations are modified when the perturbation is given by the external feedback. The equation for the carrier density given by Eq. (6.17) should not be modified, since it is nothing to do directly with the optical feedback. Since the photon lifetime τ_{ph} included in the equation for the photon density given by Eq. (6.21) is a function of the reflectivity as shown by Eq. (6.20), the effective value under the perturbation is given by

$$\frac{1}{\tau_{\mathrm{phef}}} = \frac{1}{\tau_{\mathrm{ph}}} + \frac{\upsilon_{\mathrm{g}}}{2LR_{\mathrm{f}}}\,\delta R_{\mathrm{f}} \tag{6.117}$$

Thus the effective photon lifetime is not a constant any longer, and varies with time depending on the oscillation state. The longitudinal mode resonance condition (Eq. (6.23)) used in the derivation of the equation for the optical phase (Eq. (6.27)) should be rewritten, under the perturbation in the reflection phase, as

$$2L\left(\beta(\omega) + \frac{\Gamma\omega}{c}\Delta n\right) - \delta\phi_{\mathrm{f}} = 2\pi m \tag{6.118}$$

From Eqs (6.17), (6.21), (6.27), and (6.115)–(6.118), the rate equations for cases with the perturbation due to the external feedback are given by

$$\frac{\mathrm{d}}{\mathrm{d}t}N(t) = -\Gamma G(t)S(t) - \frac{N(t)}{\tau_{\mathrm{s}}} + \frac{J(t)}{dq} \tag{6.119a}$$

$$\frac{\mathrm{d}}{\mathrm{d}t}S(t) = \left(\Gamma G(t) - \frac{1}{\tau_{\mathrm{ph}}}\right)S(t) + \frac{C_{\mathrm{s}}N(t)}{\tau_{\mathrm{s}}} + 2\kappa[S(t)S(t-\tau)]^{1/2}\cos[\omega_{\mathrm{th}}\tau + \phi(t) - \phi(t-\tau)] \tag{6.119b}$$

$$\frac{\mathrm{d}}{\mathrm{d}t}\phi(t) = \frac{\alpha_{\mathrm{c}}}{2}\left(\Gamma G(t) - \frac{1}{\tau_{\mathrm{ph}}}\right) - \kappa\left(\frac{S(t-\tau)}{S(t)}\right)^{1/2}\sin[\omega_{\mathrm{th}}\tau + \phi(t) - \phi(t-\tau)] \tag{6.119c}$$

Note that the above equations are differential equations including time difference. The parameter κ represents the magnitude of the external feedback and is given by

$$\kappa = \frac{(1 - R_{\rm f})(R_{\rm ex}/R_{\rm f})^{1/2}}{\tau_L}, \qquad \tau_L = \frac{2L}{v_{\rm g}} \tag{6.120}$$

where $\tau_{\rm L}$ is the round-trip time in the laser resonator.

Steady-State Oscillation

Consider steady-state oscillation under constant external feedback. Assume that N, S, and G do not depend upon time, let ω be the oscillation frequency, and put $\phi(t) = (\omega - \omega_{\rm th})t$. Then from Eq. (6.119) we obtain

$$0 = -\Gamma GS - \frac{N}{\tau_{\rm s}} + \frac{J}{dq} \tag{6.121a}$$

$$S = \frac{C_{\rm s}N/\tau_{\rm s}}{\Gamma G - 1/\tau_{\rm ph} + 2\kappa\cos(\omega\tau)} \tag{6.121b}$$

$$\omega - \omega_{\rm th} = -\kappa[\sin(\omega\tau) + \alpha_{\rm c}\cos(\omega\tau)] - \frac{\alpha_{\rm c}C_{\rm s}N}{2S\tau_{\rm s}} \tag{6.121c}$$

The spontaneous emission coupling term at the end of the right-hand side of Eq. (6.121c) is small and can be neglected. Figure 6.23 illustrates the relation between the threshold oscillation frequency $\omega_{\rm th}$ without external feedback and the oscillation frequency ω with external feedback.

We define a parameter C by

$$C = \kappa\tau\left(1 + \alpha_{\rm c}^2\right)^{1/2} \tag{6.122}$$

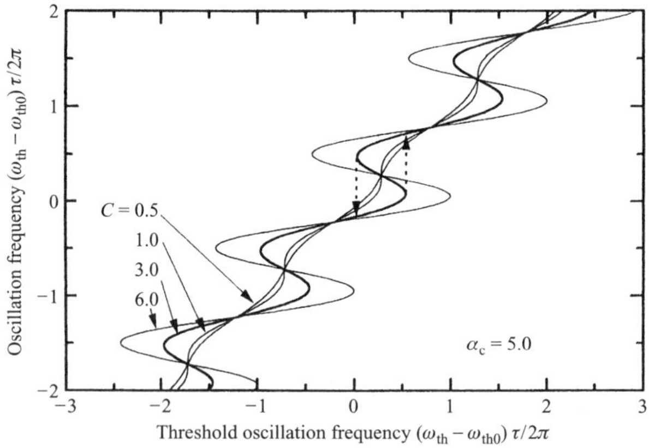

Figure 6.23 Relation between the threshold oscillation frequency without external feedback and the oscillation frequency with external feedback (The dotted arrows show mode hopping).

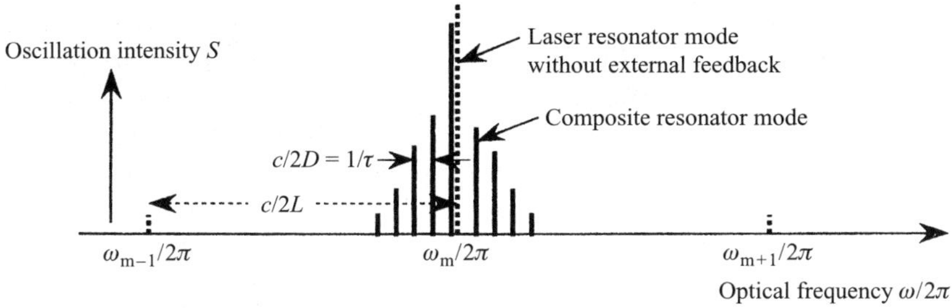

Figure 6.24 Composite resonance modes under the existence of external feedback.

and consider first a case where the external feedback is small so that $C < 1$. Then, there is only one ω value satisfying Eq. (6.121c); the laser oscillates in a single mode. It should be noted, however, that ω is slightly different from ω_{th}, and the difference $\omega - \omega_{th}$ varies periodically when τ varies. Equation (6.121b) shows that the output power also changes periodically when τ varies. In ordinary working conditions, changing the injection current density J gives rise to a change in ω_{th} through changes in the temperature and the carrier density. Therefore, even if τ is kept constant, ω varies periodically around ω_{th} when J changes. Then the output power also varies periodically around the output power for the laser without external feedback. This means that the graph of the injection current dependence of the output power (I–P characteristic) is a rippled line. This behavior results from the phase of the effective reflection coefficient r_{feff} and the dependence of the effective photon lifetime τ_{phef} on ω and τ.

We next consider a case where the external feedback is large so that $C > 1$. Then, Eq. (6.121c) is satisfied with a few values of ω, as shown in Fig. 6.23; the laser may oscillate in a few modes. This corresponds to the fact that several composite resonator modes are developed in the vicinity of a longitudinal mode of the laser resonator as shown in Fig. 6.24. The output powers in this situation can be calculated by using Eqs (6.121) extended to multimode oscillation. The result shows that the total output power for all composite resonator modes is close to the power in the oscillation without the external feedback. The frequency of each composite resonator mode and the power of each mode exhibit complicated dependences on τ and J.

Noise Characteristics

As described above, the magnitude of the external feedback can be specified by using the parameter C, given by Eqs (6.120) and (6.122), as a measure. Compared with other types of laser, the oscillation state of the semiconductor laser is affected strongly by external feedback, since they have a short resonator length L, a low mirror reflectivity R_f, and a large carrier-induced refractive index change represented by α_c. The noise characteristics, in particular, are strongly affected by external feedback, if any, and the oscillation may be completely different from that without external feedback. When the external feedback time t and/or injection current density J are changed in steady-state oscillation, hopping between composite resonator modes and hysteresis as shown by the arrows in Fig. 6.23 are observed. Mode hopping is associated with large noise. Depending upon the operation conditions, the slope $d\omega/d\omega_{th}$ of the curve in the figure may become infinitely large. In such situations, small quantum fluctuation results in large instability of the oscillation. As Eqs (6.119b) and (6.119c) show, the external feedback is associated with a delay. Because of these factors, under external feedback above a certain level, an abnormal increase in the noise and/or self-pulsation may take place. Quantitative analysis of the noise under existence of the external feedback can be made in a similar way to the cases without external feedback, by using linearized rate equations for the fluctuations which can be derived from Eq. (6.121) by adding the Langevin noise sources. The results of such analysis are consistent with the qualitative insights obtained by using Fig. 6.23. The laser behavior are classified into regions, denoted region I to region V. In the region where the feedback is small so that $C < 1$ (region I), when τ changes, the RIN level is enhanced or reduced in comparison with the level for cases without external feedback, depending upon the magnitude of $d\omega/d\omega_{th}$. In the region where the feedback is increased so that $C > 1$, in addition to the enhancement or reduction in RIN, splitting of the oscillation into two modes and hopping associated with increased low-frequency noise take place (region II). For larger C, the angular frequency separation between the adjacent composite resonator modes increases and approaches the upper limit $2\pi/\tau$. Then $d\omega/d\omega_{th}$ becomes small, the laser oscillates in a composite resonator mode with the lowest threshold, and the noise is reduced. This region (region III) is narrow. With further increase in C, many composite resonator modes oscillate simultaneously with mutual interaction, which may result in self-pulsation and unstable chaotic behavior (region IV). This phenomenon is referred to as coherence decay. When stable external feedback represented by an effective reflectivity comparable with the reflectivity of the output facet mirror is given, the laser oscillates in one of the composite resonator modes, which is selected depending upon the phase of the external feedback. In this situation, the oscillation is rather stable and the noise is reduced (region V). Thus the feedback-induced noise depends in a complex way upon various parameters. It should be noted that, in practical applications, τ and κ often fluctuate randomly. Unless special care is intentionally taken, external feedback generally increases the noise. Therefore, in applications where this kind of noise is harmful, use of an isolator is essential.

6.6 SINGLE-MODE SPECTRUM AND SPECTRUM LINEWIDTH

Even if a semiconductor laser is stably oscillating in a single mode, the optical frequency is not exactly a constant but is associated with fluctuation. Accordingly, the optical frequency spectrum is a sharp peak but is not a delta function. The spectrum has a finite linewidth, which is closely related to the frequency noise. The spectrum and the linewidth can be observed and measured using a high-resolution monochromator such as a FP spectrum analyzer. For applications such as coherent optical communications and optical interference measurements, where the coherence of laser light is utilized, it is very important to understand the characteristics related to the spectrum and the linewidth. In the following discussion, it is assumed that there is no external feedback.

6.6.1 The Schawlow–Townes Linewidth

We first consider the amplification of the stimulated emission in a laser resonator and discuss how the amplification and resonance change the emission spectrum to generate a coherent wave. As pointed out in Section 5.8, when an incoherent light source is put in a FP resonator, optical power at the longitudinal resonance frequency is accumulated effectively. Even if the spectrum of the source is wide, the output wave has a spectrum peaked at the longitudinal mode frequencies. In a passive FP resonator consisting of a waveguide of length L and propagation loss factor α, and facet mirrors of reflectivities R_f and R_b, the full width at half-maximum of the peaks is given from Eqs (5.135) and (5.137) by

$$\delta\omega = \frac{v_g}{L}\frac{1-R}{R^{1/2}}, \quad R = (R_f R_b)^{1/2}\exp(-\alpha L) \tag{6.123}$$

For a semiconductor laser injected with carriers, considering the mode gain Γg and the internal loss α_{int}, the loss factor α in the above expression should be replaced by an effective loss $\alpha_{\text{int}} - \Gamma g$. Then, for injection levels close to the threshold,

$$R = \exp\left[\left(\Gamma G - \frac{1}{\tau_{ph}}\right)\frac{L}{v_g}\right] \approx 1 + \left(\Gamma G - \frac{1}{\tau_{ph}}\right)\frac{L}{v_g} \tag{6.124}$$

and from Eqs (6.123) and (6.124) we obtain

$$\delta\omega \approx \frac{1}{\tau_{ph}} - \Gamma G = \frac{C_s N}{\tau_s S} \tag{6.125}$$

where use has been made of the steady-state rate equation (Eq. (6.31b)) to obtain the last expression. Whereas for the passive resonator the output spectrum width $\delta\omega$ is $1/\tau_{ph}$ determined by the photon lifetime τ_{ph}, the amplification gain extends the photon lifetime to an effective value $1/(1/\tau_{ph} - \Gamma G)$ and, as a result, $\delta\omega$ narrows. With increasing gain, the photon density S increases, while $\delta\omega$ narrows in inverse proportion to S and therefore to the output optical power. These results agree well with experimental results, for amplified spontaneous emission (ASE) below the oscillation threshold. For laser operation above the threshold, on the other hand, it has been shown by theoretical analysis that under oscillation the field fluctuation is stabilized by the nonlinear interaction and the spectrum width becomes half the value given by Eq. (6.125) [26]. The bandwidth $\delta\omega$ given by Eq. (6.125) is referred to as the Schawlow–Townes linewidth [27], and half of it as the modified Schawlow–Townes linewidth.

6.6.2 Spectrum Linewidth of Semiconductor Lasers

One of the important unique feature of semiconductor lasers is that there exist high-density carriers in the active region, and the fluctuation in the carrier density induces fluctuation in the refractive index, affecting significantly the oscillation spectrum. For this reason, the spectrum linewidth of semiconductor lasers cannot be described properly by the (modified) Schawlow–Townes formula, and an analysis with the index fluctuation taken into account is required.

The time-dependent amplitude of the laser output field can be written as

$$\mathrm{Re}\{E(t)\} = \mathrm{Re}\{|E(t)| \exp[-\mathrm{i}\omega_0 t - \mathrm{i}\,\delta\phi(t)]\} \tag{6.126}$$

where $E(t)$ is the complex amplitude and $\delta\phi(t)$ represents the phase fluctuation. Although the fluctuation is involved in the absolute amplitude $|E(t)|$, the effect on the frequency spectrum is minor. Therefore we omit the fluctuation in $|E(t)|$ and put $|E(t)|=E$. Then, from the Wiener–Khintchene theorem, the power spectrum $S_E(\omega)$ for the optical field amplitude is given by the Fourier transform of the autocorrelation function for $\mathrm{Re}\{E(t)\}$:

$$\begin{aligned} S_E(\omega) &= \tfrac{1}{2}\int \mathrm{Re}\big\{\langle E(t+\tau)^* E(t)\rangle\big\}\,\exp(\mathrm{i}\omega\tau)\,\mathrm{d}\tau \\ &= \frac{E^2}{2}\int \mathrm{Re}\big\{\exp(\mathrm{i}\omega_0\tau)\langle\exp[\mathrm{i}\,\Delta\phi(t,\tau)]\rangle\big\}\,\exp(\mathrm{i}\omega\tau)\,\mathrm{d}\tau \end{aligned} \tag{6.127}$$

$$\Delta\phi(t,\tau) = \delta\phi(t+\tau) - \delta\phi(t)$$

where ω is used for the optical frequency to distinguish from the frequency Ω in the RF region appearing in discussions in the previous sections.

The average phase factor $\langle \exp(\mathrm{i}\Delta\phi)\rangle$ is given by $\int f(\Delta\phi)\exp(\mathrm{i}\,\Delta\phi)\,\mathrm{d}(\Delta\phi)$ using the statistical distribution $f(\Delta\phi)$ of the time difference of the phase fluctuation $\Delta\phi(t,\tau)$. Assuming a Gaussian distribution for f, we obtain

$$\langle \exp(\mathrm{i}\,\Delta\phi(t,\tau))\rangle = \exp\left[-\frac{\langle \Delta\phi(t,\tau)^2\rangle}{2}\right] \tag{6.128}$$

Since, as we saw in Section 6.6.1, the fluctuation in the instantaneous optical frequency is given by $\delta\omega(t) = \mathrm{d}[\delta\phi(t)]/\mathrm{d}t$ using the phase fluctuation $\delta\phi(t)$, the Fourier transforms of $\delta\omega(t)$ and $\delta\phi(t)$ are correlated by $\delta\omega(\Omega) = -\mathrm{i}\Omega\,\delta\phi(\Omega)$. Therefore, the noise power spectra of ω and ϕ are correlated by $S_\omega(\Omega) = \Omega^2 S_\phi(\Omega)$. Using this relation and the Wiener–Khintchene theorem to calculate $\langle \Delta\phi^2\rangle$, we obtain

$$\begin{aligned}\langle \Delta\phi(t,\tau)^2\rangle &= \langle [\delta\phi(t+\tau) - \delta\phi(t)]^2\rangle \\ &= 2\langle [\delta\phi(t)]^2\rangle - 2\langle \delta\phi(t+\tau)\,\delta\phi(t)\rangle \\ &= \tau^2 \int S_\omega(\Omega)\left[\frac{\sin(\Omega\tau/2)}{\Omega\tau/2}\right]^2 \frac{\mathrm{d}\Omega}{2\pi}\end{aligned} \tag{6.129}$$

The frequency noise power spectrum $S_\omega(\Omega)$ included in the above integration is given approximately by Eq. (6.112). The phase fluctuations due to the spontaneous emission and the index fluctuation are included in $S_\omega(\Omega)$. Since the [ ]^2 factor in the above expression is small for large Ω, the integration is dominated by the value of $S\omega(\Omega)$ in the low-Ω region. Approximating $S_\omega(\Omega)$ by $S_\omega(0)$ for simplicity, we obtain

$$\begin{aligned}\langle \Delta\phi(t,\tau)^2\rangle &\approx \tau^2 S_\omega(0) \int \left[\frac{\sin(\Omega\tau/2)}{\Omega\tau(2)}\right]^2 \frac{\mathrm{d}\Omega}{2\pi} \\ &= |\tau| S_\omega(0)\end{aligned} \tag{6.130}$$

Substituting Eq. (6.130) into Eq. (6.128) and then into Eq. (6.127), and calculating the power spectrum $S_E(\omega)$ of the optical wave, we obtain

$$S_E(\omega) = \frac{S_\omega(0)E^2/4}{(\omega-\omega_0)^2 + [S_\omega(0)/2]^2} + \frac{S_\omega(0)E^2/4}{(\omega+\omega_0)^2 + [S_\omega(0)/2]^2} \tag{6.131}$$

This approximate expression shows that the optical spectrum is a Lorentzian distribution with peaks at the mode frequency $\pm\omega_0$. A more accurate expression can be obtained, through calculation of $\langle \Delta\phi^2\rangle$ by substitution of the frequency noise spectrum $S_\omega(\Omega)$ associated with the relaxation oscillation peaks, given by Eq. (6.112), into Eq. (6.129), and numerical integration of Eq. (6.127) using the result of $\langle \Delta\phi^2\rangle$ and Eq. (6.128).

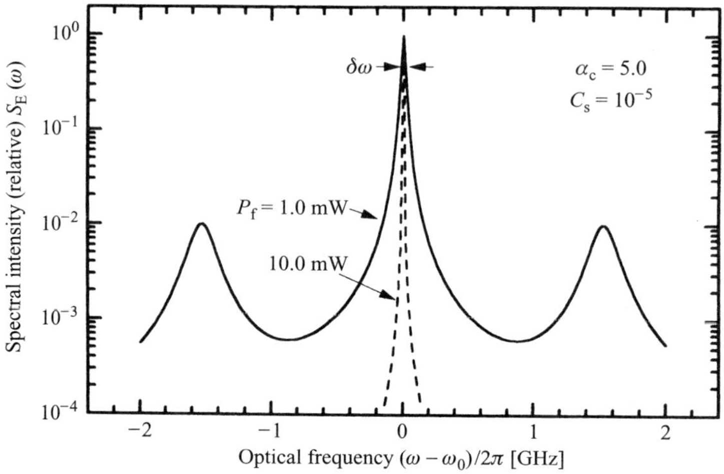

Figure 6.25 Calculated example of the frequency spectrum and the spectrum linewidth of the output light from a single-mode semiconductor laser.

The result is a spectrum with a nearly Lorentzian main peak at the frequency ω_0, (Fig. 6.25) as given by the approximate expression in Eq. (6.131), associated with subpeaks at frequencies $\omega_0 \pm \Omega_R$. The subpeaks result from the resonant enhancement in the frequency noise by the relaxation oscillation, but the intensity is low. Therefore, for evaluation of the full width at half-maximum of the main peak, we can use Eq. (6.131) to obtain [18, 28]

$$\delta\omega = S_\omega(0) = \frac{C_s N}{2\tau_s S}(1 + \alpha_c^2) \tag{6.132}$$

From Eq. (A5.7), the spontaneous emission coupling term can be written as $C_s/N\tau_s = R_{sp}(\omega_m)/V_a = n_{sp}\Gamma G/V_a$ using the population inversion factor n_{sp}. Using this relation and $\Gamma G \approx 1/\tau_{hp}$ for an oscillating laser, Eq. (6.132) can be rewritten as

$$\delta\omega = \frac{n_{sp}}{2\tau_{ph} V_a S}\left(1 + \alpha_c^2\right) \tag{6.133}$$

The above two expressions for $\delta\omega$ give a value $1 + \alpha_c^2$ times the value of the modified Schawlow–Townes linewidth. This implies that, for semiconductor lasers, the linewidth is increased by a value representing the spontaneous emission fluctuation effect, multiplied by α_c^2 representing the enhancement due to the index fluctuation. Therefore, α_c is often referred to as the linewidth enhancement factor. This factor is also denoted by α and is called the alpha parameter. Since α_c is usually larger than unity, the index fluctuation dominates the linewidth. The linewidth $\delta\omega$ narrows in inverse proportion to S, i.e., in inverse proportion to the output power. For many semiconductor lasers, a typical value of $\delta\omega$ at the maximum output is several tens of megahertz. The photon lifetime τ_{ph} can be increased by reducing the internal loss α_{int}, increasing the resonator length L, and enhancing the reflectivities R_f and R_b. Using quantum well structures, α_c smaller than that for DH structures can be obtained [8]. By increasing the threshold carrier density, n_{sp} and α_c can be reduced. By these methods, quantum well lasers of submegahertz linewidth have been implemented.

6.7 ULTRASHORT OPTICAL PULSE GENERATION

The temporal waveform and frequency spectrum of an optical pulse are correlated to each other by a Fourier transform. Therefore, the generation of short pulses requires a large spectrum width that is inversely proportional to the pulse width. Generation of ultrashort pulses is possible with semiconductor lasers, in principle, since they have a wide gain bandwidth. Optical pulses of picosecond width can be generated with relatively simple methods, and they offer many applications in ultrafast optical data transmissions, signal processing, and measurements. This section outlines the various methods for ultrashort pulse generation.

6.7.1 Self-Pulsation Oscillation

The relaxation oscillation discussed in Section 6.4.2 is a transient oscillation at a frequency Ω_R given by Eq. (6.53) decaying with the damping factor Γ_R given by Eq. (6.52). Usually the gain saturation coefficient G_S in Eq. (6.52) is negative and therefore Γ_R is positive. However, if G_S is positive, Γ_R can be negative. Then, even under constant-current driving, the relaxation oscillation does not attenuate but the amplitude increases with time. An increase in the stimulated emission is associated with a decrease in the carrier density, and the laser oscillation stops when the gain is reduced to below the threshold. Repetition of this sequence results in self-oscillation at frequencies of the order of several hundred megahertz, and optical pulses are periodically generated. This phenomenon is referred to as self-pulsation [29]. Many gain-guiding lasers oscillate in such a self-pulsation mode. In these lasers, nonuniformity in the current injection and the surface recombination of carriers in the vicinity of the facets give rise to formation of absorption regions in some parts within the resonator length, and the absorption tends to saturate with increase in the optical intensity. The mechanism of the self-pulsation may be attributed to such saturable absorption, which makes the effective G_S averaged over the resonator length negative. Index-guiding lasers, which oscillate in a similar self-pulsation mode, can also be implemented by inserting in the resonator a section without current injection, and/or designing the structure for weak index guiding, to incorporate intentionally a saturable absorption characteristic. Since with the saturable absorption the quality factor Q of the resonator is enhanced with increase in the optical intensity, the self-pulsation can be considered a kind of passive Q switching. Since the repetition period of the self-pulsation cannot be controlled externally, and the pulse width cannot be ultrashort, self-pulsation lasers are not very useful as a short optical pulse source. However, self-pulsation lasers are used in applications such as optical disk systems, as an effective method to avoid mode hopping and feedback-induced noises.

6.7.2 Gain Switching

As discussed in Section 6.4.5, when a semiconductor laser is driven with a step current of large step height, relaxation oscillation of large amplitude occurs at the start of the laser oscillation. If the current is kept constant, the relaxation oscillation attenuates, and the laser operation approaches steady-state oscillation. If the current injection is terminated at the instant when the first period of the relaxation oscillation is completed, a single short optical pulse having a width determined by the relaxation oscillation frequency is generated. Repeating such driving, periodic optical pulses can be generated. This technique, utilizing the rise and extinction of the laser oscillation through rapid changes in the gain by pulse driving, is referred to as gain switching. Advantages of the technique include the generation of ultrafast optical pulses using simple driving electronics, and the possibility of controlling the pulse width and the repetition rate by the driving current waveform. As the driving current, not only positive pulses, but also a large-amplitude sinusoidal wave superimposed on a negative bias current-can be used. In order to avoid the destruction of the laser diode, the backward voltage must be shunted by connecting a Shottky-barrier diode having a high-speed switching characteristic with inverse polarity in parallel with the laser diode.

Generation of ultrafast optical pulses by gain switching requires careful optimization of the driving current waveform so as to maintain the large gain only within one period of the relaxation oscillation. Under the optimum condition, the pulse width is dominated by the relaxation frequency. It should be noted that, even with a driving current pulse with widths of several hundred picoseconds, optical pulses of much shorter width, several tens of picoseconds, can be generated [30]. Using a quantum well laser with a differential gain larger than that for a bulk DH laser, the relaxation oscillation frequency can be enhanced, and shorter pulse widths can be achieved [31]. In quantum well DFB lasers, the differential gain and the relaxation frequency can be enhanced, and the pulse width can be shortened, by appropriate detuning of the Bragg wavelength towards a wavelength shorter than the gain peak wavelength. Through these optimization and improvements, pulse widths shorter than 10 ps were obtained. In gain-switching operation of FP lasers, the laser oscillates in multiple longitudinal modes, the spectrum bandwidth is remarkably broadened, and therefore the coherence degrades. In DFB and DBR lasers that maintain single-mode oscillation, the spectrum bandwidth is not greatly broadened. However, the frequency of the output pulse is chirped in the pulse duration, owing to the fast variation in the carrier density and the carrier-induced refractive index change. This chirping can be utilized for compressing the pulse width. The chirped pulses are transmitted through an element such as an optical fiber having an appropriately designed group delay dispersion, to realize the coincidence of the time of arrival for the leading and trailing edges of the pulse. For example, nearly Fourier-transform-limited ultrashort pulses of widths shorter than 3 ps have been generated by compressing gain-switching optical pulses of 10–20 ps widths [32].

6.7.3 Mode Locking

In ordinary multiple-longitudinal-mode oscillation of a FP semiconductor laser, the longitudinal modes have of frequencies nearly equal separations, but the mutual phases between modes are associated with random fluctuations. Accordingly, the frequency of the intensity variation resulting from the interference between modes (beat frequency) is associated with broadening and fluctuation. Therefore, the total output power fluctuates randomly like a noise in the frequency range of the same order of magnitude as the beat frequency, although the power averaged over a longer time is constant. However, if the mode separations are equalized and the relative phases of the modes are kept constant, as a result of coherent interference between modes, periodic optical pulses of a short pulse width are generated. This method of optical pulse generation is called mode locking. As a simple model for ideal mode locking, we consider an optical wave consisting of longitudinal modes arranged within a Gaussian envelope, as shown in Fig. 6.26. It is assumed that each mode has an infinitesimal linewidth, and all modes have a same phase at $t = 0$. Let $\Delta\omega$ be the frequency separation between adjacent modes, and $2\,\Delta\omega_e$ be the width of the envelope. Then the

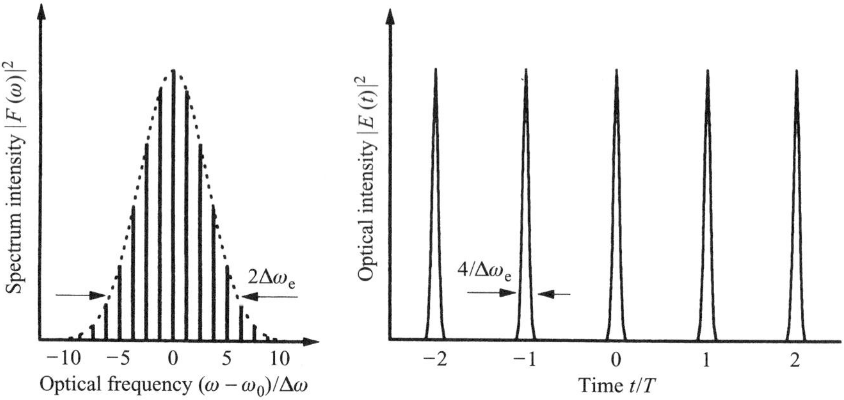

Figure 6.26 Frequency spectrum and temporal optical intensity waveform of mode-locked laser light.

frequency spectrum of the optical field can be written as

$$F(\omega) = \sum_m \exp\left[-\left(\frac{m\Delta\omega}{\Delta\omega_e}\right)^2\right]\delta(\omega - \omega_0 - m\,\Delta\omega) \tag{6.134}$$

The optical field amplitude, given by the Fourier transform of the above expression, is calculated as

$$E(t) = \frac{1}{2\pi^{1/2}}\frac{\Delta\omega_e}{\Delta\omega}\sum_n \exp\left[-\left(\frac{\Delta\omega_e}{2}\right)^2(t - nT)^2\right]\exp(i\omega_0 t) \tag{6.135}$$

$$T = \frac{2\pi}{\Delta\omega} = \frac{2LN_g}{c}$$

which represents periodic pulses with a $1/e^2$ full width of $4/\Delta\omega_e$ and a period T corresponding to the round-trip time for an optical wave in the resonator [33]. Since the broad gain bandwidth of semiconductor lasers allows a large value of $2\,\Delta\omega_e$, it is possible, in principle, to generate ultrashort pulses with a subpicosecond width.

Actual mode locking is classified into active mode locking and passive mode locking. Active mode locking of a semiconductor laser is accomplished by superimposing a RF wave of frequency Ω equal to the longitudinal mode separation $\Delta\omega$ on a constant driving current. Each mode is intensity modulated at a frequency Ω, and therefore, from a mode of frequency ω, optical wave components of frequency $\omega \pm \Omega$ are induced.

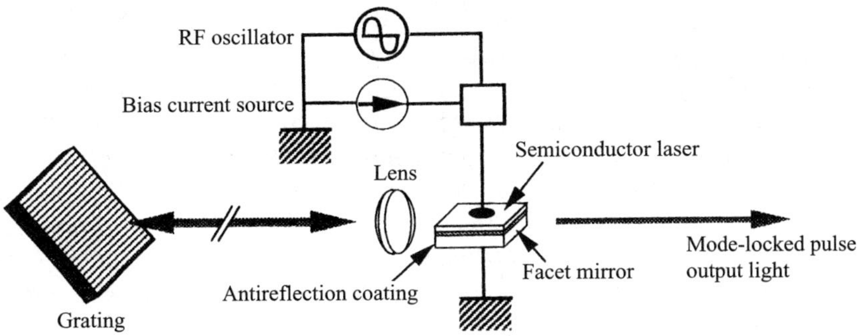

Figure 6.27 Example of the construction of a mode-locked semiconductor laser.

If Ω is equal or very close to $\Delta\omega$, the side-mode frequencies coincide (approximately) with the adjacent mode frequencies. As a result of coupling between adjacent modes, the laser operation is pulled into oscillation of equidistant modes with constant phase relation. Mode locking is thus accomplished. However, since ordinary FP lasers have a resonator length L of 1 mm or shorter, the longitudinal mode separation $\Delta\omega/2\pi = c/2N_eL$ is around 100 GHz. There is no practical RF source of such high frequency available. To solve the problem, an extended cavity laser with an effective resonator length L around 10 cm is constructed by combining a FP laser with one of the facets antireflection (AR) coated and an external mirror or grating. Then, the mode separation is as low as 1 GHz or so, and appropriate RF sources are available. An example of such a configuration is shown in Fig. 6.27. There exist many longitudinal modes within the gain bandwidth, and they are arranged with a wavelength separation. The separation, however, is not exactly a single value, since the group index of the laser waveguide is wavelength dependent. If a large number of modes are to be mode locked, only some of them are locked well, and the oscillations in other modes give rise to deterioration in the pulse shape. It is therefore necessary to limit the number of oscillation modes by using a grating or an etalon to reduce the feedback bandwidth and to optimize the driving condition. Under such optimization, ultrashort optical pulses of width from a few picoseconds to a few tens of picoseconds can be generated [34].

Passive mode locking is accomplished by inserting a saturable absorber in the laser resonator. Even if mode locking is not accomplished, the optical intensity distributed in the resonator is not uniform. In the saturable absorber, the optical wave is less attenuated and is amplified more in places where the intensity is high. Therefore the circulating optical wave is amplified more in positions where the intensity is higher, and the wave approaches the mode-locked pulses described by Eq. (6.125) [33]. A large saturation and a fast response are required for the saturable absorber for effective mode locking. Multiple quantum well (MQW) structures bombarded with protons to improve the temporal response are often used as a saturable absorber. Mode-locked semiconductor lasers were implemented by inserting a MQW saturable absorber in an extended cavity, or by monolithic integration of a laser and a saturable absorber. Ultrafast optical pulses of subpicosecond widths to widths of a few picoseconds have been generated by passive mode locking or by hybrid (active and passive) mode locking [35].

REFERENCES

1. Y. Suematsu and K. Furuya, *Electron. Commun. Jpn.*, **E-60**, 467 (1977).
2. K. Petermann, *IEEE J. Quantum Electron.*, **QE-15**, 566 (1979).
3. Y. Suematsu, S. Akiba, and T. Hong, *IEEE J. Quantum Electron.*, **QE-13**, 596 (1977).
4. T. P. Lee, C. A. Burrus, J. A. Copeland, A. G. Dentai, and D. Marcuse, *IEEE J. Quantum Electron.*, **QE-18**, 1101 (1982).
5. W. Streifer, D. B. Scifres, and R. D. Burnham, *Appl. Phys. Lett.*, **40**, 305 (1982).
6. M. Yamada, *IEEE J. Quantum Electron.*, **QE-19**, 1365 (1983).
7. T. Ikegami and Y. Suematsu, *Proc. IEEE*, **55**, 122 (1967).
8. Y. Arakawa and A. Yariv, *IEEE J. Quantum Electron.*, **QE-21**, 1666 (1985).
9. R. S. Tucker and I. P. Kaminov, *J. Lightwave Technol.*, **LT-2**, 385 (1984).
10. K. Kishino, S. Aoki, and Y. Suematsu, *IEEE J. Quantum Electron.*, **QE-18**, 343 (1982).
11. N. K. Dutta, N. A. Olsson, L. A. Koszi, P. Besoni, and R. B. Wilson, *J. Appl. Phys.*, **56**, 2167 (1984).
12. S. Kobayashi, Y. Yamamoto, M. Ito, and T. Kimura, *IEEE J. Quantum Electron.*, **QE-18**, 582 (1982).
13. D. Marcuse and T. P. Lee, *IEEE J. Quantum Electron.*, **QE-19**, 1397 (1983).
14. M. Born and E. Wolf, *Principles of Optics*, Pergamon, Oxford (1970).
15. A. Yariv, *Introduction to Optical Electronics*, Holt, Rinehart and Winston, New York (1971).
16. M. Lax and W. H. Louisell, *Phys. Rev.*, **185**, 568 (1969).
17. C. H. Henry, *IEEE J. Quantum Electron.*, **QE-18**, 259 (1982).
18. C. H. Henry, *IEEE J. Quantum Electron.*, **QE-19**, 1391 (1983).
19. Y. Yamamoto, S. Saito, and T. Mukai, *IEEE J. Quantum Electron.*, **QE-19**, 47 (1983).
20. Y. Yamamoto and N. Imoto, *IEEE J. Quantum Electron.*, **QE-22**, 2032 (1986).
21. N. Chinone, T. Kuroda, T. Ohtoshi, T. Takahashi, and T. Kajimura, *IEEE J. Quantum Electron.*, **QE-21**, 1264 (1985).

22. A. Oishi, N. Chinone, M. Ojima, and A. Arimoto, *Electron. Lett.*, **20**, 821 (1984).
23. G. P. Agrawal, *Phys. Rev.*, A, **37**, 2488 (1988).
24. R. Lang and K. Kobayashi, *IEEE J. Quantum Electron.*, **QE-16**, 347 (1980).
25. R. W. Tkach and A. R. Chraplyvy, *J. Lightwave Technol.*, **LT-4**, 1655 (1986).
26. M. Lax, *Phys. Rev.*, **160**, 290 (1967).
27. A. L. Schawlow and C. H. Townes, *Phys. Rev.*, **112**, 1940 (1958).
28. K. Vahala and A. Yariv, *IEEE J. Quantum Electron.*, **QE-19**, 1102 (1983).
29. J. Z. Van der Ziel, J. L. Mertz, and T. L. Paoli, *J. Appl. Phys.*, **50**, 4620 (1979).
30. H. F. Liu, M. Fukazawa, Y. Kawai, and T. Kamiya, *IEEE J. Quantum Electron.*, **QE-25**, 1417 (1989).
31. Y. Arakawa, T. Sogawa, M. Nishioka, M. Tanaka, and H. Sakaki, *Appl. Phys. Lett.*, **51**, 1295 (1987).
32. H. F. Liu, Y. Ogawa, S. Oshiba, and T. Nonaka, *IEEE J. Quantum Electron.*, **QE-27**, 1655 (1991).
33. A. E. Soegman, *Lasers*, University Science Books, Mill Valley, California (1986).
34. M. Schell, A. G. Weber, E. Schoell, and D. Bimberg, *IEEE J. Quantum Electron.*, **QE-27**, 1661 (1991).
35. P. A. Morton, J. E. Bowers, L. A. Koszi, M. Zoler, and D. P. Wilt, *Appl. Phys. Lett.*, **56**, 111 (1990).

7
Distributed Feedback Lasers

This chapter presents distributed feedback (DFB) and distributed Bragg reflector (DBR) lasers, which maintain stable single-mode oscillation even under high-speed modulation. They are suitable for monolithic integration and accomplishing advanced functions and performances. After the concepts of the dynamic single-mode lasers are outlined, the fundamental theory based on coupled-mode equations is explained. Then, characteristics of DBR and DFB lasers are discussed in some detail.

7.1 DYNAMIC SINGLE-MODE LASERS

In many applications, such as optical communications and measurements, semiconductor lasers that maintain stable and pure single-mode oscillation even under high-speed modulation are required. Fabry–Perot (FP) lasers cannot satisfy this requirement. Therefore, various types of dynamic single-mode semiconductor laser have been developed.

Dynamic single-mode oscillation can be achieved by giving the oscillation threshold gain a sharp mode selectivity to prevent oscillation of modes except for a single mode. One of the simplest methods is to combine a FP laser with an external mirror, thereby constructing of a composite resonator consisting of the two facet mirrors and the third external mirror, as shown in Fig. 7.1(a). One of the laser waveguide FP modes that coincides with one of the resonance frequencies for the facet and the external mirror oscillates [1]. Figure 7.1(b) shows a coupled-cleaved-cavity (C^3) laser [2]. A composite resonator is constructed by coaxial alignment of two FP lasers of different waveguide lengths with an appropriate gap. Single-longitudinal-mode operation is achieved by adjusting the injection currents for both sections to obtain coincidence of the resonance modes for the two FP resonators. The external grating resonator laser, shown in Fig. 7.1(c), uses a semiconductor laser with one of the facet mirrors antireflection (AR) coated

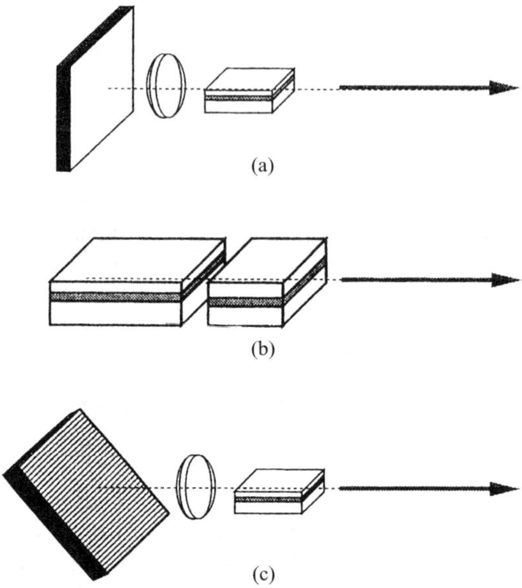

Figure 7.1 Constructions of dynamic single-mode lasers: (a) external mirror composite reasonator laser; (b) coupled cleaved cavity (C^3) laser; (c) external grating composite reasonator laser.

to suppress the FP mode, and an external grating with Littrow configuration for wavelength-selective feedback. The longitudinal modes are defined by the round-trip phase shift, and the laser oscillates with a mode that is fed back most efficiently. The wavelength of the strongest feedback can be tuned by changing the grating angle, and the longitudinal mode wavelengths shift with the change in the distance of the grating from the laser. Therefore, a semiconductor laser capable of continuous tuning of the oscillation wavelength can be implemented by using a mechanism for changing the angle and the shift of the grating with appropriate correlation.

If a grating having sharp wavelength selectivity is integrated as a feedback element in a semiconductor laser waveguide, dynamic single-mode lasers can be implemented without spoiling the compactness of semiconductor lasers [3–5]. This type of semiconductor laser is classified as a distributed feedback (DFB) laser using a grating within the active section (carrier injection section), and a distributed Bragg reflector (DBR) laser using a grating outside the active section. Figure 7.2 illustrates the DFB and DBR laser structures. Although the fabrication of these lasers requires advanced techniques, they can offer excellent performances including dynamic single-mode oscillation. The DFB and DBR configurations are

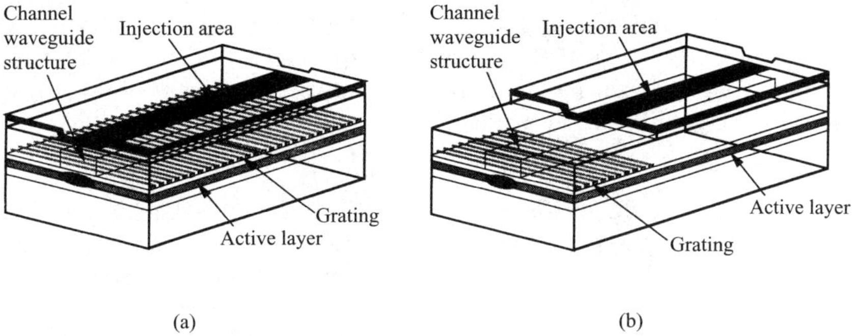

Figure 7.2 Schematic illustrations of (a) DFB and (b) DBR lasers.

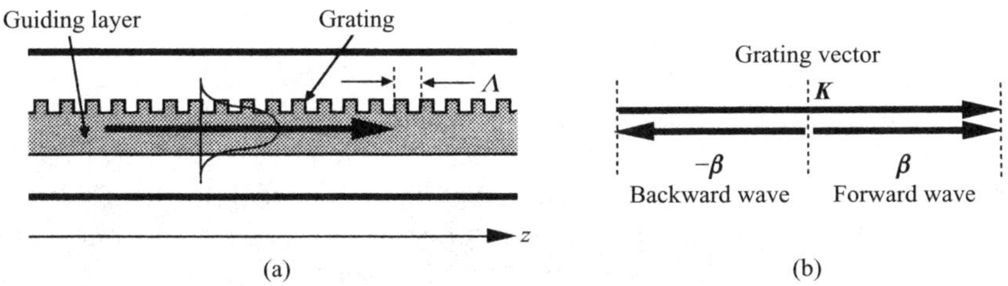

Figure 7.3 Waveguide with a grating: (a) cross-sectional structure; (b) wave vector diagram.

suitable for implementing advanced devices such as wavelength-tunable lasers. They do not require facet mirrors, and this unique feature facilitates monolithic integration of lasers and other optical elements, such as photodetectors and passive elements, in a semiconductor waveguide. With these advantages, DFB and DBR lasers are the most important components for implementation of various photonic integrated circuits.

7.2 COUPLED-MODE EQUATIONS

7.2.1 Wave Vector Diagram and Bragg Condition

Figure 7.3(a) shows schematically the cross section of a typical waveguide grating structure used in DFB and DBR lasers. As a generalized structure, consider a periodic structure, where the optical constants, such as the refractive index, of the medium are spatially modulated with a period Λ [6]. The grating structure is represented by a grating vector $\boldsymbol{K}$ ($|\boldsymbol{K}| = K = 2\pi/\Lambda$) having a magnitude of $2\pi/\Lambda$ and an orientation along the direction of the period. When an optical wave represented by a wave vector $\boldsymbol{\beta}$ propagates in this structure, spatial harmonic waves represented by wave vectors $\boldsymbol{\beta}-q\boldsymbol{K}$ ($q=1,2,\ldots$) are induced. If $\boldsymbol{\beta}-q\boldsymbol{K}$ equals a wave vector of an optical wave that can propagate in the medium, the incident wave couples with this wave. This is the Bragg diffeaction of the qth order. When

$$\boldsymbol{\beta} - q\boldsymbol{K} = -\boldsymbol{\beta} \qquad (q = 1, 2, \ldots) \tag{7.1}$$

holds, in particular, the incident wave of $\boldsymbol{\beta}$ couples with a backward wave of $-\boldsymbol{\beta}$. This is the Bragg reflection. For guided waves in a waveguide, the Bragg condition is given by $|\boldsymbol{\beta}| = \beta = 2\pi N_e/\lambda$, where λ is the wavelength and N_e the effective index of refraction. The wave vector diagram for $\boldsymbol{\beta}$ and $\boldsymbol{K}$ is shown in Fig. 7.3(b), and the Bragg condition given by Eq. (7.1) can be written as

$$\beta = \frac{qK}{2} \tag{7.2}$$

The above relation, called the phase-matching condition, indicates that only an optical wave with a wavelength that satisfies $\lambda = 2\Lambda N_e/q$ is effectively reflected. Effective reflection for a wave at wavelength λ can be attained by using a grating with a period $\Lambda = q\lambda/2N_e$. Although the best performances are obtained with a grating of fundamental order (first order). Second- and third-order gratings are also used because of easier fabrication.

7.2.2 Mathematical Expression for Gratings

Consider a canonical waveguide lossless and uniform in the direction of wave propagation (z direction). The structure is represented by a distribution of relative dielectric constant $\varepsilon(x, y)$. The amplification gain and the index change induced by the carrier injection and the scattering loss are represented by a distribution of complex dielectric constant $\delta\varepsilon(x, y)$. Let $\Delta\varepsilon(x, y, z)$ be the perturbation in the distribution of (complex) dielectric constants representing a grating structure formed in the waveguide. Then the waveguide grating structure shown in Fig. 7.3(a) is expressed by $\varepsilon(x, y) + \delta\varepsilon(x, y) + \Delta\varepsilon(x, y, z)$. Since the grating is a structure that is periodic in the z direction, $\Delta\varepsilon(x, y, z)$ can be written as a Fourier expansion:

$$\Delta\varepsilon(x, y, z) = \sum_q \Delta\varepsilon_q(x, y) \exp(\mathrm{i}qKz) \tag{7.3}$$

where $\Delta\varepsilon_q(x, y)$ is the Fourier amplitude of the qth order. If the grating is in the form of refractive index modulation, $\Delta\varepsilon(x, y, z)$ is real and, therefore, we have $\Delta\varepsilon_{-q}(x, y) = \Delta\varepsilon_q(x, y)^*$. If the grating is symmetrical with respect to the sign inversion of the z axis, $\Delta\varepsilon(x, y, -z) = \Delta\varepsilon(x, y, z)$, the relation $\Delta\varepsilon_{-q}(x, y) = \Delta\varepsilon_q(x, y)$ holds and, therefore, $\Delta\varepsilon_q(x, y)$ and $\Delta\varepsilon_{-q}(x, y)$ are real. A grating that results when the above grating is shifted by s in the z

direction is expressed by $\Delta\varepsilon(x, y, z - s) = \sum_q \Delta\varepsilon_q(x, y)\exp(-iqKs)\exp(iqKz)$. If $\Delta\varepsilon(x, y, z)$ is complex, we put $\Delta\varepsilon = |\Delta\varepsilon|\exp(i\theta)$. Then, by using the $\Delta\varepsilon_q(x, y)$ value for the symmetrical index-modulated grating, the general grating can be expressed as

$$\Delta\varepsilon(x, y, z) = \sum_q \exp(i\theta - iqKs)\ \Delta\varepsilon_q(x, y)\exp(iqKz) \tag{7.4}$$

$$\Delta\varepsilon_q(x, y) = \Delta\varepsilon_q(x, y)^* = \Delta\varepsilon_{-q}(x, y) = \Delta\varepsilon_{-q}(x, y)^*$$

7.2.3 Coupled-Mode Equations [6,7]

Let $\boldsymbol{E}$ and $\boldsymbol{H}$ be the optical electromagnetic fields in the waveguide with a grating, and $\boldsymbol{E}^0$ and $\boldsymbol{H}^0$ be the fields in the canonical waveguide. Then, from the Maxwell equations

$$\Delta \times \boldsymbol{E} = i\omega\mu_0\boldsymbol{H}, \quad \Delta \times \boldsymbol{H} = -i\omega\varepsilon_0(\varepsilon + \delta\varepsilon + \Delta\varepsilon)\boldsymbol{E} \tag{7.5}$$

$$\Delta \times \boldsymbol{E}^0 = i\omega\mu_0\boldsymbol{H}^0, \quad \Delta \times \boldsymbol{H}^0 = -i\omega\varepsilon_0\varepsilon\boldsymbol{E}^0 \tag{7.6}$$

with the use of a vector formula, in the same manner as the derivation of Eq. (5.89), we obtain

$$\Delta[\boldsymbol{E} \times \boldsymbol{H}^{0*} + \boldsymbol{E}^{0*} \times \boldsymbol{H}] = i\omega\varepsilon_0\boldsymbol{E}^{0*}(\delta\varepsilon + \Delta\varepsilon)\boldsymbol{E} \tag{7.7}$$

Integration of both sides of the above equation in a region having an infinitely large area perpendicular to the waveguide z axis and an infinitesimal thickness yields

$$\frac{\partial}{\partial z}\iint[\boldsymbol{E} \times \boldsymbol{H}^{0*} + \boldsymbol{E}^{0*} \times \boldsymbol{H}]_z\,dx\,dy = i\omega\varepsilon_0\iint \boldsymbol{E}^{0*}(\delta\varepsilon + \Delta\varepsilon)\boldsymbol{E}\,dx\,dy \tag{7.8}$$

We here identify $\boldsymbol{E}^0$ and $\boldsymbol{H}^0$ as a guided mode in the canonical waveguide and put

$$\boldsymbol{E}^0 = \boldsymbol{E}_m(x, y)\exp(+i\beta_m z), \qquad \boldsymbol{H}^0 = \boldsymbol{H}_m(x, y)\exp(+i\beta_m z) \tag{7.9}$$

The field $\boldsymbol{E}(x, y, z)$ in the waveguide with a grating can be expanded with the canonical waveguide modes $\boldsymbol{E}_m(x, y)$ as

$$\boldsymbol{E}(x, y, z) = \sum_m A_m(z)\boldsymbol{E}_m(x, y)\exp(+i\tilde{\beta}_m z) \tag{7.10}$$

where $\tilde{\beta}_m = \beta_m + \delta\beta_m$ is a complex propagation constant, and the perturbation term $\delta\beta$ representing the effects of the perturbation $\delta\varepsilon$ is given by Eq. (5.92). Substituting Eqs (7.9) and (7.10) into Eq. (7.8), and using the orthonormal relation of the modes given by Eq. (5.29), we obtain

$$\pm\frac{\mathrm{d}}{\mathrm{d}z}A_m(z) = \mathrm{i}\sum_n \kappa_{mn}(z)A_n(z) \tag{7.11}$$

$$\kappa_{mn}(z) = \frac{\omega\varepsilon_0}{4}\iint \boldsymbol{E}_m^* \cdot \Delta\varepsilon\, \boldsymbol{E}_n\, \mathrm{d}x\,\mathrm{d}y \exp[\mathrm{i}(\tilde{\beta}_n - \tilde{\beta}_m)z] \tag{7.12}$$

Since we used $\delta\beta$ determined by Eq. (5.92), the $\delta\varepsilon$ term has been canceled out. On the left-hand side of Eq. (7.11), the plus sign is taken for modes m propagating towards the $+z$ direction, and the minus sign is taken for modes m propagating towards the $-z$ direction. Equation (7.11) represents a group of equations that describe generally the variation in the mode amplitudes A_m under propagation along the z direction in the waveguide. They are called coupled-mode equations.

Consider next a case where the qth order Bragg condition $2\beta = qK$ is satisfied at least approximately. A single-lateral-mode waveguide is assumed. Then Eq. (7.10) reduces to a sum of the forward and backward waves of fundamental mode:

$$\boldsymbol{E}(x, y, z) = \boldsymbol{E}(x, y)E(z) \tag{7.13a}$$

$$E(z) = A_\mathrm{f}(z)\exp(+\mathrm{i}\tilde{\beta}z) + A_\mathrm{b}(z)\exp(-\mathrm{i}\tilde{\beta}z) \tag{7.13b}$$

The complex propagation constant $\tilde{\beta}$ can be written as

$$\tilde{\beta} = \frac{\mathrm{i}\alpha}{2} + \beta = -\frac{\mathrm{i}g}{2} + \beta \tag{7.14}$$

where α is the mode attenuation constant and g is the mode gain. The function $\kappa(z)$, calculated by substituting Eq. (7.3) into Eq. (7.12), includes components that oscillates rapidly with respect to z. The oscillation components, however, do not contribute substantially to the optical coupling. Therefore, omitting the oscillation components, we obtain coupled-mode equations describing the interaction between the forward and backward waves:

$$+\frac{\mathrm{d}}{\mathrm{d}z}A_\mathrm{f}(z) = \mathrm{i}\kappa_{\mathrm{fb}}A_\mathrm{b}(z)\exp(-\mathrm{i}2\tilde{\Delta}z) \tag{7.15a}$$

$$-\frac{\mathrm{d}}{\mathrm{d}z}A_\mathrm{b}(z) = \mathrm{i}\kappa_{\mathrm{bf}}A_\mathrm{f}(z)\exp(+\mathrm{i}\tilde{\Delta}z) \tag{7.15b}$$

where $2\tilde{\Delta}$ is a complex parameter defined by

$$2\tilde{\Delta} = 2\tilde{\beta} - qK = \mathrm{i}\alpha + 2\Delta, \qquad 2\Delta = 2\beta - qK \tag{7.16}$$

and the real part $2\varDelta$ represents the deviation from the Bragg condition. The quantities κ_{fb} and κ_{bf} are coupling coefficients defined by

$$\kappa_{\mathrm{fb}} = \kappa \exp(\mathrm{i}\theta - \mathrm{i}qKs), \qquad \kappa_{\mathrm{bf}} = \kappa \exp(\mathrm{i}\theta + \mathrm{i}qKs) \tag{7.17}$$

$$\begin{aligned} \kappa &= \kappa_q \\ &= \frac{\omega\varepsilon_0}{4}\iint \Delta\varepsilon_q(x,y)\,|\boldsymbol{E}(x,y)|^2 \mathrm{d}x\,\mathrm{d}y && (7.18\mathrm{a}) \\ &= \frac{k^2}{2\beta}\frac{\iint \Delta\varepsilon_q(x,y)\,|\boldsymbol{E}(x,y)|^2\,\mathrm{d}x\,\mathrm{d}y}{\iint |\boldsymbol{E}(x,y)|^2\,\mathrm{d}x\,\mathrm{d}y} && (7.18\mathrm{b}) \end{aligned}$$

where κ is a real constant. Although Eq. (7.18a) applies for the mode field $\boldsymbol{E}(x, y)$ normalized so as to satisfy Eq. (5.29), Eq. (7.18b) can be used for an unnormalized field.

7.2.4 Solution of Coupled-Mode Equations and Normal Modes

To solve the coupled-mode equations in Eqs (7.15), we assume a solution in the form

$$A_{\mathrm{f}}(z) = F\exp(-\mathrm{i}\tilde{\varDelta}z)\exp(\mathrm{i}\gamma z) \tag{7.19a}$$

$$A_{\mathrm{b}}(z) = B\exp(+\mathrm{i}\tilde{\varDelta}z)\exp(\mathrm{i}\gamma z) \tag{7.19b}$$

and substitute them into Eqs (7.15). Then, we obtain a characteristic equation to determine γ:

$$(\gamma - \tilde{\varDelta})F - \kappa_{\mathrm{fb}}B = 0 \tag{7.20a}$$

$$\kappa_{\mathrm{bf}}F + (\gamma + \tilde{\varDelta})B = 0 \tag{7.20b}$$

For Eq. (7.20) to have solutions other than the trivial solution $F = B = 0$, the determinant of the coefficient matrix must vanish. We therefore obtain $\gamma^2 = \tilde{\varDelta}^2 - \kappa_{\mathrm{fb}}\kappa_{\mathrm{bf}}$. We put

$$\gamma \equiv \pm\left(\tilde{\varDelta}^2 - \kappa_{\mathrm{fb}}\kappa_{\mathrm{bf}}\right)^{1/2} = \pm\left(\tilde{\varDelta}^2 - \exp(2\mathrm{i}\theta)\kappa^2\right)^{1/2} \tag{7.21}$$

where, on the right-hand side, the sign is selected so as to have the same sign for $\mathrm{Im}\{\gamma\}$ and $\mathrm{Im}\{\tilde{\varDelta}\} = \alpha/2$. Putting

$$\gamma_+ \equiv \frac{\gamma - \tilde{\varDelta}}{\kappa_{\mathrm{fb}}} = -\frac{\kappa_{\mathrm{bf}}}{\gamma + \tilde{\varDelta}} \tag{7.22a}$$

$$\gamma_- \equiv \frac{\gamma - \tilde{\varDelta}}{\kappa_{\mathrm{bf}}} = -\frac{\kappa_{\mathrm{fb}}}{\gamma + \tilde{\varDelta}} \tag{7.22b}$$

the general solution of the coupled-mode equations in Eqs (7.15) can be written as

$$\begin{aligned} A_{\mathrm{f}}(z) &= F\exp(-\mathrm{i}\tilde{\Delta}z)\exp(+\mathrm{i}\gamma z) + r_{-}B\exp(-\mathrm{i}\tilde{\Delta}z)\exp(-\mathrm{i}\gamma z) \\ A_{\mathrm{b}}(z) &= r_{+}F\exp(+\mathrm{i}\tilde{\Delta}z)\exp(+\mathrm{i}\gamma z) + B\exp(+\mathrm{i}\tilde{\Delta}z)\exp(-\mathrm{i}\gamma z) \end{aligned} \tag{7.23}$$

Substituting Eq. (7.23) into Eq. (7.13b), we see that the z dependence of the optical field amplitude can be written as

$$\begin{aligned} E(z) = {} & F\exp(+\mathrm{i}\gamma z)\left[\exp\left(+\frac{\mathrm{i}qKz}{2}\right) + r_{+}\exp\left(-\frac{\mathrm{i}qK_z}{2}\right)\right] \\ & + B\exp(-\mathrm{i}\gamma z)\left[r_{-}\exp\left(+\frac{\mathrm{i}qKz}{2}\right) + \exp\left(-\frac{\mathrm{i}qKz}{2}\right)\right] \end{aligned} \tag{7.24}$$

The first and second lines of the above equation express modes of an optical wave that can exist independently in the waveguide with a grating and are called normal modes [5,6]. As this expression shows, a normal mode is a combination of the forward wave and the backward wave. Without the grating, $\kappa = 0$ and $r_{+} = r_{-} = 0$, and the normal modes reduce to a forward wave and a backward wave.

To consider the characteristics of normal modes, we assume a case where there is no gain nor loss ($\alpha = 0$). Then the normal mode propagation constant β_{nm} is given by $\pm qK/2 \pm \gamma = \pm qK/2 \pm (\Delta^2 - \exp(2\mathrm{i}\theta)\kappa^2)^{1/2}$, where $\Delta = \beta - qK/2$. Figure 7.4 illustrates the relation between the normal mode propagation constant β_{nm} and the wave number k ($= 2\pi/\lambda = \omega/c$). This is called the Brillouin diagram. In the frequency region far away from the Bragg condition where $|\Delta|$ is large, the normal mode propagation constant is close to the propagation constants of the forward- and backward-propagating spatial harmonics $\pm\beta$ or $\pm\beta \pm qK$. In the vicinity of the Bragg condition, however, the normal modes exhibit a dispersion different from that of single forward or backward wave. For a refractive index grating ($\theta = 0$), the normal mode propagation constant takes a complex value in the region of $|\Delta| < |\kappa|$, as shown in Fig. 7.4(a). This implies that the optical wave undergos Bragg reflection and cannot propagate for a long distance. This frequency range is referred to as a stop band. For a grating in the form of periodic modulation in loss or gain ($\theta = \pm\pi/2$), on the other hand, the normal mode propagation constant for $\Delta \approx 0$ is approximately $\pm\beta \pm \kappa$, implying that the forward wave couples with the backward wave without the appearance of a stop band.

7.2.5 Coupling Coefficient and Radiation Decay Factor

The coupling coefficient κ, which is an important parameter for analysis of a waveguide with a grating, is given by Eq. (7.18). The value can be calculated

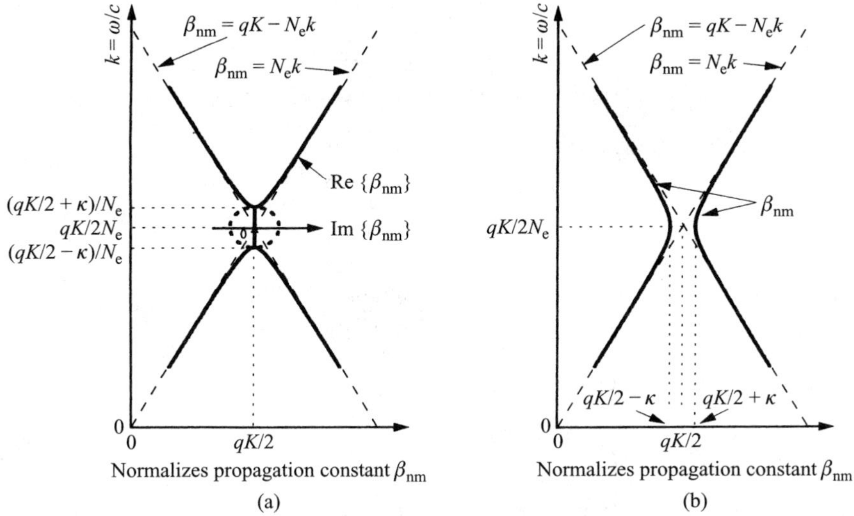

Figure 7.4 Brillouin diagram describing the dispersion characteristic of waveguide gratings (the region where $\beta_{nm} < 0$ is given by the mirror image with respect to the ordinate): (a) index grating; (b) gain grating.

through integration by substituting in this expression the Fourier amplitude $\Delta\omega_q(x, y)$ of the grating and the lateral mode function $\boldsymbol{E}(x, y)$ [6,8]. The magnitude depends upon the optical constants and structure parameters describing the waveguide and the grating. The grating is usually formed by periodic etching of the cladding layer, or a boundary of the separate-confinement heterostructure (SCH). As a typical example, a grating of periodic rectangular cross section is shown in Fig. 7.5. The Fourier amplitude describing this structure is

$$\begin{aligned} \Delta\varepsilon_q(x) &= \frac{1}{\Lambda}\int \Delta\varepsilon(x, y)\exp(-\mathrm{i}qKz)\,\mathrm{d}z \\ &= \left(n_2^2 - n_1^2\right)\frac{\sin(qa\pi)}{q\pi} \end{aligned} \tag{7.25}$$

in the grating layer, and it is zero outside the grating layer. Here, n_1 and n_2 are the refractive indexes of the media above and below the corrugated boundary, respectively, and a denotes the ratio of the width of the grating tooth to the grating period Λ. Exact calculation of the coupling coefficient κ can be made by using the lateral mode expression. However, since the grating groove depth d_g of gratings used in semiconductor lasers is much smaller than the lateral expanse of the lateral mode, from Eq. (7.18), κ can

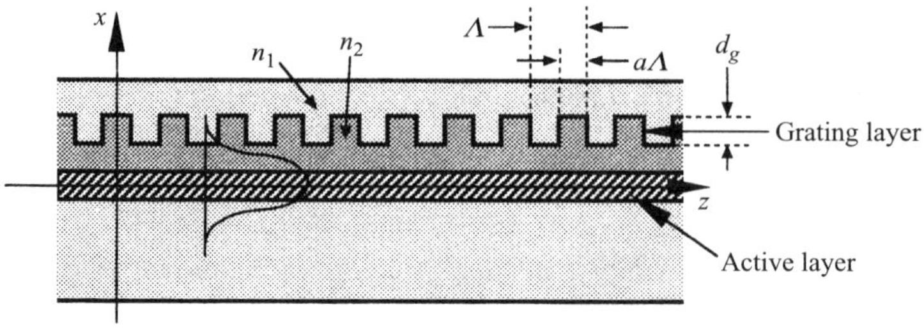

Figure 7.5 Cross section of a waveguide grating structure.

be written approximately as

$$\kappa_q = \frac{k^2}{2\beta}\left(n_2^2 - n_1^2\right)\Gamma_g \frac{\sin(qa\pi)}{q\pi} \tag{7.26}$$

where Γ_g is the ratio of the optical power propagating in the grating layer to the total power in the waveguide. Calculation of Γ_g can be made in a similar manner to the calculation of the confinement factor in the active region, and Γ_g is proportional approximately to d_g. The above expression shows that κ is also proportional to the difference in the squared refractive index. For the fundamental order, κ_1 is a maximum at $a = \frac{1}{2}$. Note that, for a symmetrical grating with $a = \frac{1}{2}$, the κ_q values for even q are zero. The second-order coupling coefficient κ_2 or third-order coupling coefficient, κ_3 utilized in the second-order or third-order grating, respectively, is maximized at $a = \frac{1}{4}$ and $\frac{3}{4}$, or at $a = \frac{1}{2}$, respectively. The magnitude of κ_2 or κ_3 is smaller than the value obtained by a grating of fundamental order.

In a structure with a grating of second or higher order, in addition to the coupling between the forward and the backward waves, coupling between the guided mode and radiation modes takes place owing to the lower-order Fourier component $\Delta\varepsilon_q$. In a second-order grating, for example, optical waves propagating along directions perpendicular to the waveguide axis are induced. The coupling to radiation modes gives rise to attenuation of the guided mode power in the form of $\exp(-a_r z)$ or $\exp(+\alpha_r z)$. The radiation decay factor α_r is given by $\alpha_r = \pi|k_r|^2$, where the coupling coefficient k_r is calculated by Eq. (7.18a) with $|\boldsymbol{E}(x, y)|^2$ replaced by the product $\boldsymbol{E}_r^*(x, y) \cdot \boldsymbol{E}(x, y)$ of the radiation mode $\boldsymbol{E}_r(x, y)$ and the guided mode $\boldsymbol{E}(x, y)$. Other methods for calculation of the radiation decay factor α_r include series expansion based on the Floquet–Bloch theorem, and use of equivalent electric circuits [6]. The radiation decay factor is approximately proportional to $|\Delta\varepsilon_q|^2$ with the order q of the radiation coupling, and the square d_g^2 of the grating groove depth. The effect of the radiation decay factor α_r on the laser oscillation is similar to that of the internal loss α_{int} of the waveguide; both of these give rise to an increase in the oscillation threshold and reduction in the efficiency. In the analysis, the radiation loss effect can be taken into account by simply including α_r in α_{int}.

7.3 DISTRIBUTED FEEDBACK LASERS

7.3.1 Oscillation Condition of Distributed Feedback Lasers

The propagation of optical waves in a DFB laser is expressed by Eq. (7.24), which can be rewritten in a form of the sum of the forward- and backward-propagating components as

$$E(z) = \exp\left(+\frac{iqKz}{2}\right)[F\exp(+i\gamma z) + r_- B\exp(-i\gamma z)] + \exp\left(-\frac{iqK_z}{2}\right)[r_+ F\exp(+i\gamma z) + B\exp(-i\gamma z)] \tag{7.27}$$

Consider a waveguide with a grating that is terminated with facet mirrors of reflectivities R_f and R_b at positions $z = L$ (>0) and $z = 0$, respectively. Then from the boundary conditions

$$R_f^{1/2}\exp\left(+\frac{iqKL}{2}\right)[F\exp(+i\gamma L) + r_- B\exp(-i\gamma L)] = \exp\left(-\frac{iqKL}{2}\right)[r_+ F\exp(+i\gamma L) + B\exp(-i\gamma L)] \tag{7.28a}$$

$$R_b^{1/2}(r_+ F + B) = (F + r_- B) \tag{7.28b}$$

we obtain

$$(r_+ - r_f)\exp(+2i\gamma L)F + (1 - r_f r_-)B = 0 \tag{7.29a}$$

$$(1 - r_b r_+)F + (r_- - r_b)B = 0 \tag{7.29b}$$

with

$$r_f = R_f^{1/2}\exp(+iqKL), \qquad r_b = R_b^{1/2}$$

The oscillation condition for the DFB laser is that the above equations have a solution other than the trivial solution $F = B = 0$. From the determinant of the coefficient matrix, we obtain the oscillation condition

$$\frac{(r_+ - r_f)(r_- - r_b)}{(1 - r_b r_+)(1 - r_f r_-)}\exp(i2\gamma L) = 1 \tag{7.30}$$

For the case where there is no grating present ($\kappa = 0$), the above equation coincides with the oscillation condition of FP lasers. For cases where the facet reflectivities are zero, $r_{\mathrm{f}} = r_{\mathrm{b}} = 0$, the oscillation condition reduces to the simple form

$$r_+ r_- \exp(\mathrm{i}2\gamma L) = 1 \tag{7.31}$$

This equation is in a form similar to the oscillation condition of FP lasers and show that a DFB laser is represented by the effective reflectivities r_+ and r_- and the effective gain $\mathrm{i}\gamma$. It should be noted, however, that, unlike FP lasers, r_+ and r_- depend not only on the coupling coefficient κ but also on the deviation Δ from the Bragg condition and the gain γ, and the effective gain $\mathrm{i}\gamma$ depends not only on g but also on κ and Δ. Increasing the gain by injecting carriers into a DFB laser, oscillation starts when Eq. (7.30) or (7.31) is satisfied. In the following, the oscillation wavelength and the threshold gain are discussed for various kinds of DFB laser.

7.3.2 Index-Coupled Distributed Feedback Lasers

The relief-type grating shown in Fig. 7.5 is in the form of periodic modulation in the refractive index and is expressed by Eq. (7.4) with $\theta = 0$ substituted. Consider the oscillation condition for the case where the facets are free from reflection. At the wavelength that satisfies the Bragg condition $\Delta = 0$, $\tilde{\Delta} = -\mathrm{i}g/2$, $\gamma = -\mathrm{i}(g^2/4 + \kappa^2)^{1/2}$ and therefore both $\tilde{\Delta}$ and γ are pure imaginary. Although $\exp(\mathrm{i}2\gamma L)$ in Eq. (7.31) can be increased by increasing g, Eq. (7.31) cannot be satisfied since $\gamma_+\gamma_- = (\gamma - \tilde{\Delta})^2/\kappa^2 < 0$. This implies that oscillation does not take place at the Bragg wavelength, since the phase condition cannot be satisfied. To clarify the oscillation wavelength and threshold, we next simplify Eq. (7.31). From Eq. (7.22), $\gamma_+\gamma_- = (\gamma - \tilde{\Delta})^2/\kappa^2 = \kappa^2/(\gamma + \tilde{\Delta})^2$, and Eq. (7.31) is rewritten as $[(\gamma - \tilde{\Delta})/\kappa] \times \exp(+\mathrm{i}\gamma L) = \pm 1$ or $-[\kappa/(\gamma + \tilde{\Delta})]\exp(+\mathrm{i}\gamma L) = \pm 1$. Eliminating $\tilde{\Delta}$ from this expression, the oscillation condition can be rewritten as

$$\kappa L \sinh(\mathrm{i}\gamma L) = \pm\gamma L, \qquad \gamma = -\mathrm{i}\left[\left(\frac{g}{2} + \mathrm{i}\Delta\right)^2 + \kappa^2\right]^{1/2} \tag{7.32}$$

The above equation shows that the oscillation condition is satisfied at wavelengths corresponding to positive and negative values with the same absolute value of Δ. The solution of the above equations cannot be obtained analytically. In the limiting case when $\kappa \to 0$, however, the above equation can be rewritten approximately as $(\kappa L)^2 \exp(gL + \mathrm{i}2\Delta L) = -(gL + \mathrm{i}2\Delta L)^2$

and can then be decomposed into the amplitude and phase conditions

$$(\kappa L)^2 \exp(gL) = (gL)^2 + (2\,\Delta L)^2 \tag{7.33a}$$

$$\Delta L = \left(m - \frac{1}{2}\right)\pi + \tan^{-1}\left(\frac{2\,\Delta L}{gL}\right) \tag{7.33b}$$

where $m = 1, 2, \ldots$ is the order of the DFB modes. Omitting the $\tan^{-1}$ term in Eq. (7.33b), from $\Delta L = \pi$ we see that the DFB mode separation is given approximately by $\Delta\omega = v_g\pi/L$, which is approximately the same as the longitudinal mode separation of FP lasers. Equation (7.33a) shows that the threshold gain depends upon the mode and is higher for a higher mode. The oscillation wavelength and threshold gain satisfying Eq. (7.32) can be calculated by iterative calculation with the result for weak coupling as the first approximation. The calculated dependences of the normalized deviation ΔL of the oscillation wavelength from the Bragg wavelength and the normalized threshold gain gL on the normalized coupling coefficient κL are shown in Fig. 7.6.

The result in Fig. 7.6 is consistent with the result obtained under the weak-coupling approximation. It shows that the threshold gain for a mode decreases with increasing κL, and that the deviation ΔL increases with increasing κL. Because of the significant mode dependence of the threshold gain, actual oscillation takes place with the fundamental mode $m = 1$. However, as pointed out above, the oscillation condition is satisfied simultaneously at wavelengths corresponding to positive and negative

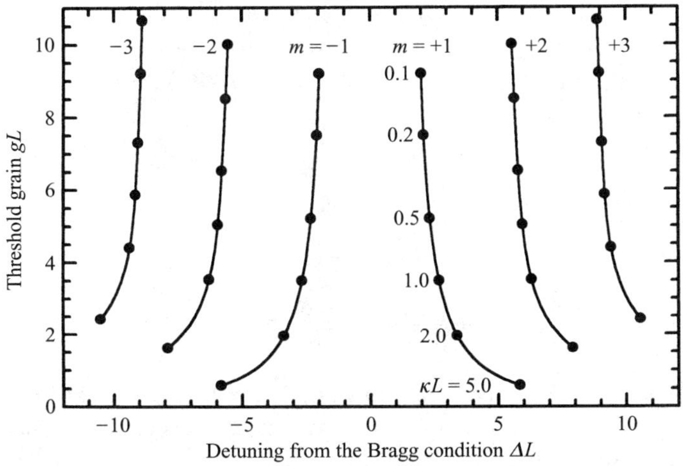

Figure 7.6 Oscillation threshold gain and oscillation wavelength detuning of an index-coupled DFB laser.

values of Δ, and therefore the two modes oscillate simultaneously. In fact, such two-mode oscillation is experimentally observed. The above result was obtained under the assumption of reflectionless facets. For realistic lasers, the facet reflection cannot be neglected, unless the facets are antireflection coated. By reducing the reflectivity for the forward facet and enhancing it for the back facet, the output power can be enhanced, and single-mode oscillation may be achieved. For this case, the exact oscillation condition is given by Eq. (7.30) and is affected by the reflection phase determined by the position(s) of the facet(s) relative to the grating phase [9]. Since the facet is formed by cleaving, it is almost impossible to control precisely the facet position for the optimization of reflection phase that requires submicron accuracy. Accordingly, some fabricated lasers oscillate in two modes, and some of them oscillate in a single mode. There are significant sample-to-sample variation in the oscillation wavelength and the threshold current. Thus, the DFB laser of this type suffers from the rather low yield of devices with good performances. This is the outstanding drawback of index-coupling DFB lasers.

7.3.3 Phase-Shifted Index-Coupled Distributed Feedback Lasers

The drawback pointed out in Section 7.3.2 comes from the fact the DFB lasers do not oscillate at the Bragg wavelength. To eliminate the problem, the device structure should be modified to satisfy the phase condition at the Bragg wavelength. From the boundary condition for Eq. (7.27) in the case where the reflectivity is zero at $z = L$, we obtain the relation $Fr_+ \exp(+2igL) + B = 0$. From this, the ratio of the amplitude of the backward wave to that of the forward wave at $z = 0$, i.e., the complex reflection coefficient R, is given by

$$\begin{aligned} R &= \frac{r_+\{1 - \exp(2i\gamma L)\}}{1 - r_+ r_- \exp(+2i\gamma L)} \\ &= \frac{\gamma - \tilde{\Delta}}{\kappa} \exp(iqKs) \\ &\quad \times \frac{1 - \exp(+2i\gamma L)}{1 - [(\gamma - \tilde{\Delta})^2/\kappa^2] \exp(+2i\gamma L)} \end{aligned} \tag{7.34}$$

where use has been made of Eqs (7.17) and (7.22) for the modification to obtain the second line, and $\theta = 0$ was used to consider an index grating. Since, in index gratings, at the Bragg wavelength, $\tilde{\Delta} = -ig/2$, $\gamma = -i(g^2/4 + \kappa^2)^{1/2}$, and both $\tilde{\Delta}$ and γ are pure imaginary, the factor $[1 - \exp(+2irL)]/\{1 - [(r - \tilde{\Delta})^2/\kappa^2] \exp(+zirL)\}$ in the above expression is real. Therefore,

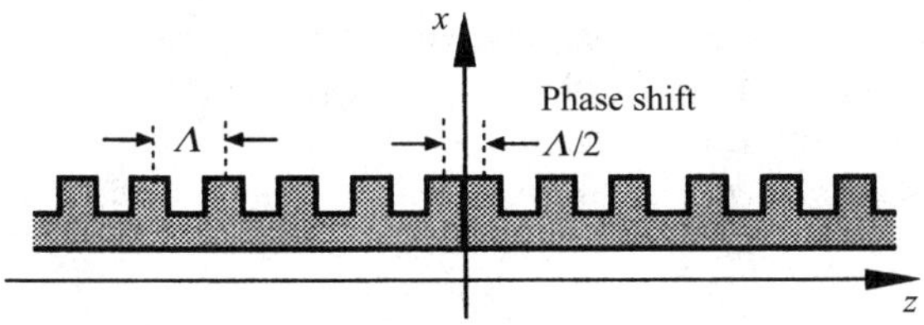

Figure 7.7 $\lambda/4$ phase shift grating.

the phase of R equals the phase of $\pm i\exp(iqKs)$. Thus the phase of R is determined by the grating phase s. The phase factor of R is $\pm i$ for $s=0$ and is ± 1 for $qKs=\pi/2$ or $s=\Lambda/4q$. In a DFB laser constructed from a combination of a grating section with $s=\Lambda/4q$ in $0<z<L$ and a grating section in $-L<z<0$ corresponding to its mirror image, as shown in Fig. 7.7, the phase factor corresponding to a round trip of the optical wave is $(\pm 1)^2=1$, showing that the phase oscillation condition is satisfied [10]. The phase shift between the right- and left-hand sections of the grating is $2s=\Lambda/2q=\lambda/4N_e$ (for $q=1$), i.e., a quarter of the optical wavelength in the waveguide. Thus, the DFB laser using a $\lambda/4$ phase-shifted index grating can oscillate at the Bragg wavelength. The simple index grating DFB laser presented in Section 7.3.2 can be considered a structure consisting of a grating section of $s=0$ and its mirror image. The phase factor corresponding to a round trip is $(\pm i)^2=-1$, showing that the feedback is negative and therefore oscillation does not occur at the Bragg wavelength.

The oscillation condition for DFB lasers of structures having mirror symmetry with respect to $z=0$, as described above, is the condition for the optical wave to have the same complex amplitude after a round-trip propagation in the sections in $z>0$ and $z<0$ as the initial amplitude. It can be written as $R^2=1$ or $R=\pm 1$ using R given by Eq. (7.34). For a DFB laser using a $\lambda/4$ phase-shifted index-coupled grating, $\theta=0$, $s=\Lambda/4q$, the oscillation condition can be written as

$$\exp(i\gamma L)=\frac{\pm i\kappa[\pm\kappa-i(\gamma-\tilde{\Delta})]}{(\gamma-\tilde{\Delta})[\pm\kappa+i(\gamma-\tilde{\Delta})]} \tag{7.35}$$

$$\tilde{\Delta}=-\frac{ig}{2}+\Delta, \qquad \gamma=-i\left[\left(\frac{g}{2}+i\Delta\right)^2+\kappa^2\right]^{1/2}$$

where the total resonator length $2L$, with use of Eq. (7.34), has been rewritten as L. The coupling coefficient κ is real, and both $\tilde{\Delta}$ and γ are pure imaginary at the Bragg wavelength. The above equation shows that the oscillation condition is satisfied at wavelengths corresponding to the positive and negative values with the same absolute value of Δ. In the limiting case of weak coupling ($\kappa \to 0$), the above equation reduces to $\pm\kappa L \exp(gL/2 + \mathrm{i}\,\Delta L) = gL + 2\mathrm{i}\,\Delta L$, which can be decomposed into amplitude and phase conditions:

$$(\kappa L)^2 \exp(gL) = (gL)^2 + (2\,\Delta L)^2 \tag{7.36a}$$

$$\Delta L = m\pi + \tan^{-1}\left(\frac{2\,\Delta L}{gL}\right) \tag{7.36b}$$

where $m = 1, 2, \ldots$ is the order of the DFB modes. The above expression shows hat the fundamental DBR coincides with the Bragg wavelength corresponding to $\Delta = 0$. The DBR mode separation equals approximately the separation of the FP mode. Since the threshold gain is higher for a higher-order DFB mode, the laser oscillates in the single fundamental mode. It should also be noted that the threshold gain for the fundamental mode is lower than that for the DFB laser without the phase shift. Because of these advantages, at present the $\lambda/4$ phase-shifted DFB laser is the major DFB laser [11]. The calculated dependences of the normalized deviation ΔL of the oscillation wavelength from the Bragg wavelength and the normalized threshold gain gL on the normalized coupling coefficient κL, calculated from Eq. (7.35), are shown in Fig. 7.8. The result is consistent with the result obtained using the weak-coupling approximation. Although the facet reflection is omitted here again, the threshold gain is affected by the facet

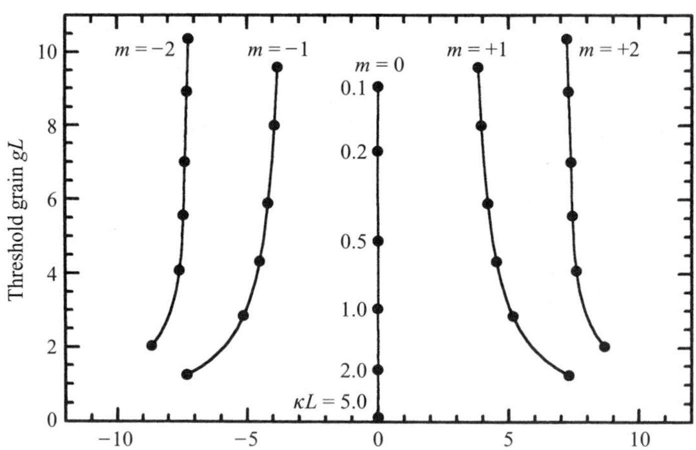

Figure 7.8 Oscillation threshold gain and oscillation wavelength detuning of a $\lambda/4$ phase-shifted index-coupled DFB laser.

reflection, in a manner similar to that presented in Section 7.3.3. Therefore the facet should be antireflection coated to achieve a good performance.

7.3.4 Gain-Coupled Distributed Feedback Lasers

Another method to attain laser oscillation at the Bragg wavelength is to employ a grating structure in the form of periodic modulation in the gain (or loss) [3]. The grating can be realized by several methods, such as periodic etching of an absorption layer inserted in the laser waveguide, and formation of a structure including the active and cladding layers with periodically modulated thickness by crystal regrowth on a corrugated substrate [12]. Although the gratings fabricated by these methods involve periodic modulation in both refractive index and gain or loss, it is possible to implement pure gain gratings by appropriate design of the cross-sectional structure for canceling out the index grating.

A simple gain (loss) grating can be expressed by Eq. (7.4) with $\theta = \pm\pi/2$. Then, from Eqs (5.93), (7.4), and (7.18) we see that the gain factor varies in the form $g + 4\kappa\cos(qKz)$. At the Bragg wavelength satisfying $\Delta = 0$, $\tilde{\Delta} = -ig/2$, $\gamma = -i(g^2/4 - \kappa^2)^{1/2}$, and therefore, for large g, both $\tilde{\Delta}$ and γ are pure imaginary. In the oscillation condition in Eq. (7.31), r_+r_- satisfies $r_+r_- = -(\gamma - \tilde{\Delta})^2/\kappa^2 > 0$ and, with increasing $g\exp(i2gL)$ in Eq. (7.31) r_+r_- increases. Therefore, the phase and amplitude oscillation conditions can be satisfied, and the laser oscillates at the Bragg wavelength. Since, from Eq. (7.22), $r_+r_- = -(\gamma - \tilde{\Delta})^2/\kappa^2 = -\kappa^2/(\lambda + \tilde{\Delta})^2$, Eq. (7.31) can be rewritten as $[(\gamma - \tilde{\Delta})/\kappa]\exp(+i\gamma L) = \pm i$ or $[\kappa/(\gamma + \tilde{\Delta})]\exp(+i\gamma L) = \pm i$. Eliminating $\tilde{\Delta}$, the oscillation condition can be written as

$$\kappa L \sinh(+i\gamma L) = \pm i\gamma L, \quad \gamma = -i\left[\left(\frac{g}{2} + i\Delta\right)^2 - \kappa^2\right]^{1/2} \tag{7.37}$$

For the limiting case of weak coupling ($\kappa \to 0$), the above equation can be rewritten approximately as $(\kappa L)^2\exp(gL + i2\Delta L) = (gL + i2\Delta L)^2$. Decomposing this equation into the amplitude and the phase, we obtain the same result as Eq. (7.36) in Section 7.3.4. This means that gain-coupled DFB lasers oscillate in a single fundamental mode, in a similar manner to phase-shifted index-coupled DFB lasers. The dependences of the normalized deviation ΔL of the oscillation wavelength from the Bragg wavelength and the normalized threshold gain gL on the normalized coupling coefficient κL, calculated from Eq. (7.37), are shown in Fig. 7.9. The result is consistent with the result obtained by using the weak-coupling approximation, and analogous to the result shown in Fig. 7.8. Thus gain-coupled DFB lasers have the same advantage as phase-shifted index-coupled

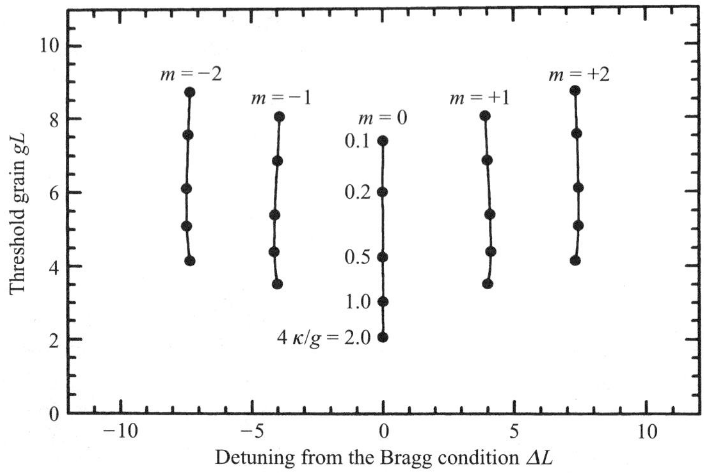

Figure 7.9 Oscillation threshold gain and oscillation wavelength detuning of a gain-coupled DFB laser.

DFB lasers. Another advantage of the gain-coupled DFB lasers is that they are less sensitive to the facet reflectivities, and good performance can be obtained without an antireflection coating on the facets.

7.3.5 Photon Lifetime and Rate Equations

In order to apply the rate equation analysis to DFB lasers, the photon lifetime τ_{ph}, the photon density S, and output power P must be expressed by using the device parameters of DFB lasers. The threshold gain $g = g_{th}$ of a DFB laser calculated in the previous sections is given in the form of the net gain for the guided mode. The threshold gain G_{th} for use in the rate equations is expressed in the form of a temporal material gain G, and g_{th} and G_{th} are correlated to each other by

$$g_{th} = \frac{\Gamma G_{th}}{v_g} - \alpha_{int} \tag{7.38}$$

where Γ is the confinement factor and α_{int} the internal loss. The oscillation condition is satisfied when the complex propagation constant is $\tilde{\beta} = \beta - ig_{th}/2$. Equation (7.27) with this $\tilde{\beta}$ substituted gives the DFB resonance mode function satisfying the wave equation and the boundary conditions. According to the definition of the photon lifetime τ_{ph} described in Section 5.8.3, the temporal attenuation of the optical power in a resonator without gain is in the form $\exp(-t/\tau_{\mathrm{ph}})$. This means that the optical wave of a complex frequency $\omega - \mathrm{i}/2\tau_{\mathrm{ph}}$ satisfies the wave equation and the boundary conditions in the resonator without gain. From Eq. (5.141), the complex propagation constant of this wave is given by $\tilde{\beta} = \tilde{\beta}(\omega - \mathrm{i}/2\tau_{\mathrm{ph}}) = \beta + \mathrm{i}\alpha_{\mathrm{int}} - \mathrm{i}/2v_{\mathrm{g}}\tau_{\mathrm{ph}}$. By equating this complex propagation constant to $\beta - ig_{\mathrm{th}}/2$, we obtain an expression for the photon lifetime for DFB lasers:

$$\frac{1}{\tau_{\mathrm{ph}}} = v_{\mathrm{g}}(\alpha_{\mathrm{int}} + g_{\mathrm{th}}) \tag{7.39}$$

For DFB lasers using a higher-order grating, the radiation decay factor α_{r} should be included in the internal loss α_{int}. The oscillation condition is expressed in the form of the rate equation is $\Gamma G_{\mathrm{th}} = 1/\tau_{\mathrm{ph}}$, and this equation with Eq. (7.39) substituted coincides with Eq. (7.38).

The multimode rate equations for the carrier density and the photon densities for a FP laser are given by Eq. (6.29). For a DFB laser, they are given in a similar form:

$$\frac{\mathrm{d}}{\mathrm{d}t}N = -\sum_{m} \Gamma_m G_m S_m - \frac{N}{\tau_{\mathrm{s}}} + \frac{J}{dq} \tag{7.40a}$$

$$\frac{\mathrm{d}}{\mathrm{d}t}S_m = \Gamma_m G_m S_m - \frac{S_m}{\tau_{\mathrm{ph}m}} + \frac{C_{\mathrm{s}}N}{\tau_{\mathrm{s}}} \tag{7.40b}$$

where the gain saturation is omitted. The outstanding difference from FP lasers is that, since the threshold gain g_{th} of a DFB laser depends upon the DFB mode order m, the photon lifetime τ_{ph} given by Eq. (7.39) exhibits a remarkable dependence on the mode. To clarify this, in the above equations, τ_{ph} is denoted as $\tau_{\mathrm{ph}m}$. The fundamental mode has the largest lifetime $\tau_{\mathrm{ph}m}$. On the other hand, the mode dependences of the gain G and the confinement factor Γ are not so strong as to that of τ_{ph}, and therefore the values for the side modes can be approximated by the value for the fundamental mode. For DFB lasers that oscillate in the fundamental mode at the Bragg wavelength, from the Bragg condition $\beta = qK/2$, Eq. (6.24a) with the right-hand side replaced by $qK/2$ and Eq. (6.24b) hold. Therefore, similarly to FP lasers, the rate equation for the optical phase for the major oscillation mode, i.e., the fundamental mode, is given by

$$\frac{\mathrm{d}}{\mathrm{d}t}\phi = \frac{\alpha_{\mathrm{c}}}{2}\left(\Gamma G - \frac{1}{\tau_{\mathrm{ph}}}\right) \tag{7.41}$$

Consider next the optical power distribution of the DBR resonant mode function. When the lateral mode function $\boldsymbol{E}(x, y)$ is normalized, the power flows of the forward and backward waves, denoted by P^{+} and P^{-}, respectively, are given by the absolute squares of the first and second lines respectively, of Eq. (7.27). For a DFB laser without a phase shift in $0 < z < L$ and without reflection at the facets, ($r_f = r_b = 0$), we can write

$$P^+(z) = |2F \sinh(\mathrm{i}\gamma z)|^2 \tag{7.42a}$$

$$P^-(z) = |2F \sinh[\mathrm{i}\gamma(L - z)]|^2 \tag{7.42b}$$

where use has been made of a relation $F + r_- B = 0$, obtained from Eq. (7.29), and the oscillation condition in Eq. (7.31). The forward output power is given by

$$P_f = P^+(L) = |2F \sinh(\mathrm{i}\gamma L)|^2 \tag{7.43}$$

Optical energy stored in the resonator is given by the sum of P^+/v_g and P^-/v_g (v_g is the group velocity) integrated over the waveguide resonator in $0 < z < L$. Since this equals the product of the photon density S, the volume of the active region V_a, and $\hbar\omega$, we have

$$v_g \hbar\omega V_a S = \int [P^+(z) + P^-(z)]\mathrm{d}z \tag{7.44}$$

and, from Eqs (4.42)–(4.44), we obtain a relation between the optical output power P_f and the photon density S:

$$P_f = S\frac{v_g \hbar\omega V_a}{L} \times |\sinh(\mathrm{i}\gamma L)|^2 \left[\frac{\sinh[2\,\mathrm{Im}\{\gamma\}\,L]}{2\,\mathrm{Im}\{\gamma\}\,L} - \frac{\sin[2\,\mathrm{Re}\{\gamma\}\,L]}{2\,\mathrm{Re}\{\gamma\}\,L}\right]^{-1} \tag{7.45}$$

For a $\lambda/4$ phase-shifted index-coupled DFB laser in $-L/2 < z < +L/2$, the power flow for the fundamental mode can be calculated by using Eq. (7.27) as

$$P^+(z) = |F\,[\exp(+\mathrm{i}\gamma|z|) + |r|^2 \exp(\mathrm{i}\gamma L)\exp(-\mathrm{i}\gamma|z|)]|^2$$

$$P^-(z) = \left|2Fr\exp\left(+\frac{\mathrm{i}\gamma L}{2}\right)\sinh\left[+\mathrm{i}\gamma\left(\frac{L}{2} - |z|\right)\right]\right|^2$$

$$|r|^2 = \frac{\{[(g/2)^2 + \kappa^2]^{1/2} - g/2\}^2}{\kappa^2} \tag{7.46}$$

where $\mathrm{i}\gamma$ takes a positive value. The distribution of the power flow described by the above expression is shown in Fig. 7.10. From the above equation, the

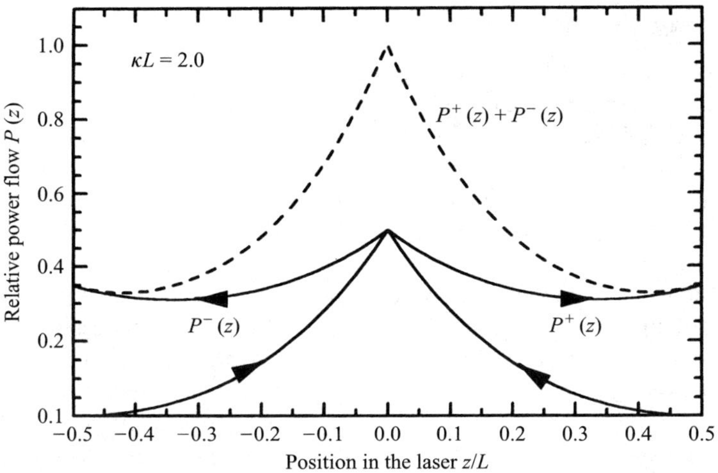

Figure 7.10 Power flow distribution in a $\lambda/4$ phase-shifted index-coupled DFB laser.

forward output power is given by

$$P_{\mathrm{f}} = P^{+}(L/2) = \left| F \exp\left(+ \frac{\mathrm{i}\gamma L}{2} \right) (1 + |r|^2) \right|^2 \tag{7.47}$$

and the output power P_{f} and the photon density S are correlated to each other by

$$P_{\mathrm{f}} = \frac{\mathrm{i}\gamma v_{\mathrm{g}} \hbar\omega V_{\mathrm{a}} (1 + |r|^2)}{[1 + |r|^2 \exp(\mathrm{i}\gamma L)]\,[1 - \exp(-\mathrm{i}\gamma L)]} S \tag{7.48}$$

As we see in Fig. 7.10, the optical intensity in a DFB laser is not uniform. Therefore, the validity of the rate equation analysis, where the intensity is considered constant and represented by the average value, is not necessarily guaranteed. In many cases, the accuracy of the rate equation analysis applied to DFB lasers is lower than that for FP lasers. The intensity distribution should be examined to check the validity. If the nonuniformity of the intensity is large, spatial hole burning occurs at positions where the intensity is large: it gives rise to nonuniformity in the refractive index and leads to deterioration of the laser performances. The nonuniformity can be reduced by optimizing the phase shift in the grating, appropriate design of the injection current distribution, and/or use of a phase-shifted grating having more than two sections. Improvement can also be made by using quantum wells with a small linewidth enhancement factor α_{c}.

7.3.6 Operating Characteristics

DFB lasers, as well as FP lasers, are implemented by using a bulk double heterostructure (DH) or quantum well structure as an active region, various types of channel waveguide structure, and a structure for current injection. In this structure, a grating of the Bragg wavelength that coincides with the gain peak wavelength is fabricated. The grating should have sufficient spatial overlap with the guided mode. Direct formation of the grating on the surface of the active layer, however, involves deterioration of the light emission characteristics due to the damage induced by the processing. To avoid this problem, in many cases the grating is formed by patterning a grating layer inserted at an appropriate distance from the active layer. The grating pattern is formed with holographic interference recording or electron-beam writing, and the pattern is transferred to a semiconductor layer by chemical etching or ion etching. Although ion etching is advantageous to the resolution and the controllability, the apparatus should be carefully selected and the etching conditions should be optimized to avoid degradation of the light-emitting characteristics due to the damage that may be caused by the processing. After the grating formation, the upper cladding is deposited by crystal regrowth. This means that fabrication of a DFB laser generally requires two epitaxal growth steps. At least one of the two waveguide ends should be cleaved for output coupling, and the cleaved facet is often antireflection coated to improve the performance.

DFB lasers have specified device parameters more than those of FP lasers, and the operating characteristics have complex dependences on them. As presented above, the oscillation condition for DFB lasers is significantly different from that of FP lasers. They have unique longitudinal mode characteristics and oscillate in a single longitudinal mode. However, many of the operating characteristics are common to those of FP lasers, as they are described by rate equations similar to those of FP lasers. Although care must be taken for the hole-burning problem pointed out in Section 7.3.5, the steady-state oscillation characteristics, modulation characteristics, and noise characteristics are interpreted in terms of the results obtained upon the basis of a single-mode analysis, which was discussed in detail with FP lasers in Chap. 6. In the following, characteristics peculiar to DFB lasers, except for the oscillation condition discussed already, are presented.

Oscillation Wavelength

The oscillation wavelength of a DFB laser is determined by the Bragg wavelength of the grating. The Bragg wavelength must be within the gain bandwidth. Although the highest output power is obtained for the case where the Bragg wavelength coincides with the gain peak wavelength, it is not necessary for them to coincide exactly each other. Even if a slight detuning is given intentionally between them, the laser oscillates at the Bragg wavelength. This is in contrast with FP lasers, where oscillation takes place with the major longitudinal mode closest to the gain peak. The oscillation wavelength can be fine controlled by only a slight change in the grating period in lasers of the same material and structure. This feature allows production of DFB lasers belonging to several groups for different wavelengths from a single epitaxal structure, and implementation of a monolithic array of DFB lasers that emit light of various wavelengths. However, while the temperature dependence of the Bragg wavelength is dominated by the temperature coefficient of the refractive index, that of the gain peak wavelength is dominated by the temperature coefficient of the bandgap energy. Because the latter is approximately one order of magnitude larger than the former, a large change in ambient temperature gives rise to large detuning and the oscillation stops. Therefore it is necessary to optimize the relation between the bandgap energy and the grating period, so that the detuning is small at the prescribed operation condition.

Polarization Characteristic

As presented before, the material gain G of a bulk semiconductor does not depend upon polarization. The confinement factors Γ for the active layer are slightly different for transverse electric (TE) and transverse magnetic (TM) modes; it is larger for a TE mode that is better guided. The mode gain ΓG is therefore slightly larger for a TE mode, but the difference is not significant. The Bragg wavelengths for a TE and TM modes, however, are different from each other, since the mode indexes are different. The mode gain at the Bragg wavelength may be larger for a TE mode or for a TM mode, depending upon the detuning of the grating. The TE and TM mode gains are also affected by the strain imposed in the active region. The coupling coefficient κ, on the other hand, is approximately proportional to the overlap of the mode profile with the grating layer as shown by Eq. (7.26) and is larger for a TE mode with a larger mode size in many cases. The radiation decay factor a_r for a second-or third-order grating is larger and enhances more the oscillation threshold for a TE mode where electric fields of the guided and radiation modes are parallel to each other, in comparison with that for a TM mode where the fields are perpendicular to each other. Thus, for bulk DH DFB lasers, the oscillation polarization is determined depending upon many factors, and some lasers oscillate in a TE mode and some lasers in a TM mode. A laser may oscillate in both TE and TM modes, and on changing the injection current the oscillation polarization may alternate between TE and TM modes because of the change in detuning through the change in the temperature. This characteristic of DH

DFB lasers is in contrast with that of FP lasers, which oscillate in a TE mode without exception. Although the possibility of TM mode oscillation may be an advantage, the oscillation polarization is not fixed and a high extinction ratio is not assured automatically. Optimization of device parameters is required to obtain a significant difference between the thresholds of the desired polarization and the other polarization. These drawbacks and problems are eliminated if quantum well structures are used. As presented in Chap. 4, the gain of the quantum well exhibits a strong polarization dependence. For ordinary single-quantum-well and multiple-quantum-well structures, the TE gain is significantly larger than the TM gain. Strained quantum wells can be designed so as to have a larger gain for a TE mode or a TM mode. Therefore, DFB lasers with a high polarization extinction ratio can be implemented by using a SCH structure with a quantum well active region and an appropriately designed grating.

Side-Mode Suppression Ratio

Consider the (longitudinal) mode spectrum of the output light from a DFB laser under steady-state oscillation. Putting the left-hand side of Eq. (7.40b) as zero, we obtain

$$S_m = \frac{\tau_{\mathrm{ph}m} C_{\mathrm{s}} N / \tau_{\mathrm{s}}}{1 - \tau_{\mathrm{ph}m} \Gamma G} \tag{7.49}$$

where m is the DFB mode order. The m dependence of the mode gain ΓG and the spontaneous emission coupling coefficient C_{s} are neglected. The photon lifetime $\tau_{\mathrm{ph}m}$ are correlated with the threshold gain $g_{\mathrm{th}m}$ by Eq. (7.39). For the major fundamental oscillation mode ($m=0$), Eq. (7.49) reduces to $S_0 = (\tau_{\mathrm{ph}0} C_{\mathrm{s}} N / \tau_{\mathrm{s}})/(1 - \tau_{\mathrm{ph}0} \Gamma G)$. Therefore, the photon density for the adjacent side-mode relative to that for the major oscillation mode, i.e., the side-mode suppression ratio, is given by

$$\begin{aligned} R_{\mathrm{sms}} &= \frac{S_0}{S_1} \\ &= 1 + \left(\frac{1}{\tau_{\mathrm{ph}1}} - \frac{1}{\tau_{\mathrm{ph}0}} \right) \frac{S_0 \tau_{\mathrm{s}}}{C_{\mathrm{s}} N_{\mathrm{th}0}} \\ &= 1 + \Delta G_{\mathrm{th}} \frac{S_0 \tau_{\mathrm{s}}}{C_{\mathrm{s}} N_{\mathrm{th}0}} \end{aligned} \tag{7.50}$$

$$\Delta G_{\mathrm{th}} = v_{\mathrm{g}} \, \Delta g_{\mathrm{th}}, \qquad \Delta g_{\mathrm{th}} = g_{\mathrm{th}1} - g_{\mathrm{th}0}$$

The output power $P_{\mathrm{f}0}$ of the major oscillation mode is proportional to the photon density S_0, and they are correlated by Eq. (7.45) or (7.48). The suppression ratio R_{sms} increases with increasing output power. As DFB lasers use an index waveguide, C_s is small, of the order of 10^{-5}–10^{-4}. The threshold gain margin Δg_{th} is large; $\Delta g_{th}L$ is roughly of the order of unity. Therefore, a very high R_{sms} can be achieved in the high-output region. While in FP lasers, where the side mode is suppressed by the slight gain margin, the suppression ratio R_{sms} given by Eq. (6.45) is 100 or less even in the high-output region, in DFB lasers R_{sms} larger than 1000 (30 dB) can easily be obtained. In quantum well DFB lasers, in particular, R_{sms} as large as 40–50 dB has been obtained [13]. The high side-mode suppression ratio R_{sms} and the small linewidth enhancement factor α_c enable high-purity single-mode oscillation of sub-megahertz spectrum linewidth to be attained [14]. The very high side-mode suppression ratio leads to excellent noise characteristics. There is no problem of mode-hopping noise. Mode partition noise is also very low. Therefore, the result of the noise analysis based on the single-mode model presented in Chap. 6 applies with good accuracy.

High-Speed Modulation Characteristics

Under high-speed modulation or in transient state, the side-mode suppression ratio cannot be defined simply, since the intensities of each mode vary with time. However, from the rate equation in Eq. (7.40b), omitting the mode dependence of the gain ΓG, we obtain the relation

$$S_m = \frac{C_s N/\tau_s}{(\mathrm{d}S_m/\mathrm{d}t)/S_m - \Gamma G + 1/\tau_{phm}} \tag{7.51}$$

and from this we obtain a formal expression for the ratio of the photon density of the major mode to the photon density of the side mode:

$$\frac{S_0}{S_1} = 1 + \left(\Delta G_{th} + \frac{\mathrm{d}S_1/\mathrm{d}t}{S_1} - \frac{\mathrm{d}S_0/\mathrm{d}t}{S_0}\right)\frac{S_0\tau_s}{C_s N} \tag{7.52}$$

$$\Delta G_{th} = v_g\,\Delta g_{th}, \qquad \Delta g_{th} = g_{th1} - g_{th0}$$

This ratio, which can be considered a time-dependent side-mode suppression ratio, is in the form of the suppression ratio for the steady state given by Eq. (7.50) to which $(\mathrm{d}S_1/\mathrm{d}t)/S_1 - (\mathrm{d}S_0/\mathrm{d}t)/S_0$ has been added. This added term varies with time. Assuming that the laser is modulated with a signal of frequency Ω superimposed on an appropriate bias current, $|(\mathrm{d}S_1/\mathrm{d}t)/S_1 - (\mathrm{d}S_0/\mathrm{d}t)/S_0|$ can be estimated at most as values of the same order of magnitude as Ω. On the other hand, for appropriately designed DFB lasers, $\Delta g_{th}L$ is the order of unity, and therefore $\Delta G_{th} = (v_g/L)\,\Delta g_{th}\,L$ is of the order of $10^{10}\,\mathrm{s}^{-1}$ or larger. Therefore, the factor $\Delta G_{th} + (\mathrm{d}S_1/\mathrm{d}t)/S_1 - (\mathrm{d}S_0/\mathrm{d}t)/S_0$ in Eq. (7.52) remains at values comparable with $\Delta G_{\rm th}$ even under gigahertz modulation. This implies that the side-mode suppression ratio is not greatly degraded even under high-speed modulation. More precise analysis of the high-speed modulation characteristics can be made, by solving the multimode rate equation in Eq. (7.40) numerically. It has been shown that appropriately designed DFB lasers maintain a side-mode suppression ratio higher than 30 dB even under large-signal pulse modulation in the gigahertz band, provided that the laser is driven with an appropriate bias current. For small-signal modulation, the results of the single-mode analysis discussed in Chap. 6 apply with good accuracy including the high-frequency range. The frequency chirping characteristic is similar to that of FP lasers, as seen from the fact that the rate equation for the optical phase (Eq. (7.41)) is common. In quantum well DFB lasers having a high relaxation oscillation frequency and a small linewidth enhancement factor α_c, in particular, single-mode high-speed modulation above 10 GHz is possible [13], and frequency chirping is small. Because of these excellent characteristics, quantum well DFB lasers are most suitable as a light source for high-speed optical fiber communications.

7.4 DISTRIBUTED BRAGG REFLECTOR LASERS

7.4.1 Distributed Bragg Reflector

A DBR is a structure where an index grating of a period corresponding to the Bragg condition is formed in a passive waveguide. The optical wave in this waveguide can be expressed, similarly to Eq. (7.27), as a sum of the forward and backward wave components:

$$\begin{aligned} E(z) &= \exp\left(+\frac{\mathrm{i}qKz}{2}\right)[F\exp(+\mathrm{i}\gamma z) + rB\exp(-\mathrm{i}\gamma z)] \\ &\quad + \exp\left(-\frac{\mathrm{i}qKz}{2}\right)[\mathrm{r}F\exp(+\mathrm{i}\gamma z) + B\exp(-\mathrm{i}\gamma z)] \end{aligned} \tag{7.53}$$

$$\begin{aligned} &\gamma = \mathrm{i}\left[\left(\frac{\alpha}{2} - \mathrm{i}\Delta\right)^2 + \kappa^2\right]^{1/2}, \qquad r = \frac{[\gamma - (\mathrm{i}\alpha/2 + \Delta)]}{\kappa} \\ &2\Delta = 2\beta - qK \end{aligned} \tag{7.54}$$

where α is the propagation loss factor. For higher-order gratings, α includes the radiation decay factor α_r representing the radiation loss due to lower-order diffraction. We assume a grating of index modulation type ($\theta = 0$), and put $s = 0$ for simplicity. When the grating is terminated by a facet of a reflectivity

R_f at $z = L$, a boundary condition

$$R_f^{1/2} \exp\left(+\frac{iqKL}{2}\right) [F \exp(+i\gamma L) + rB \exp(-i\gamma L)]$$
$$= \exp\left(-\frac{iqKL}{2}\right) [rF \exp(+i\gamma L) + B \exp(-i\gamma L)] \qquad (7.55)$$

holds. If the reflection at the grating end is negligible, the above equation reduces to

$$B = -rF \exp(2i\gamma L) \qquad (7.56)$$

Using Eqs (7.54) and (7.56), Eq. (7.53) can be rewritten as

$$E(z) = A \exp\left(+\frac{iqKz}{2}\right)\left[\gamma \cosh[i\gamma(L - z)] - \left(\frac{i\alpha}{2} + \Delta\right) \sinh[i\gamma(L - z)]\right] + A \exp\left(-\frac{iqKz}{2}\right) \kappa \sinh[i\gamma(L - z)] \qquad (7.57)$$

$$A = -\frac{2rF \exp(i\gamma L)}{\kappa}$$

Let $P^+(z)$ and $P^-(z)$ be the power flows of the forward and backward waves, respectively. From the above equations, the relative spatial distributions of $P^+(z)$ and $P^-(z)$ can be written as

$$\frac{P^+(z)}{P^+(0)} = \frac{|\gamma \cosh[i\gamma(L - z)] - (i\alpha/2 + \Delta) \sinh[i\gamma(L - z)]|^2}{|\gamma \cosh(i\gamma L) - (i\alpha/2 + \Delta) \sinh(i\gamma L)|^2} \qquad (7.58a)$$

$$\frac{P^-(z)}{P^+(0)} = \frac{|\kappa \sinh[i\gamma(L - z)]|^2}{|\gamma \cosh(i\gamma L) - (i\alpha/2 + \Delta) \sinh(i\gamma L)|^2} \qquad (7.58b)$$

Note that, at the Bragg wavelength, $i\gamma$ is negative. The power flow distributions are shown in Fig. 7.11 for cases where the Bragg condition is satisfied. Here a lossless waveguide ($\alpha = 0$) is assumed. The figure shows that, during the propagation, the forward wave power is transferred to the backward power owing to the distributed Bragg reflection.

From Eq. (7.57), the ratio of the amplitude of the backward wave to that of the forward wave, i.e., the complex amplitude reflection coefficient,

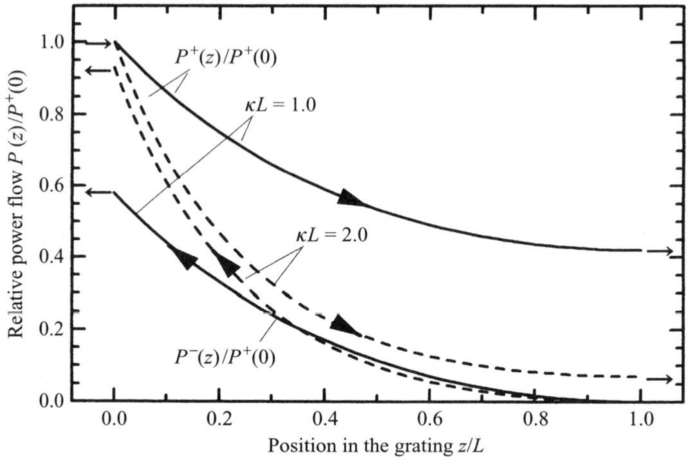

Figure 7.11 Power flow distribution for the forward and backward waves in a DBR.

can be written as

$$r_{\mathrm{DBR}} = \frac{\kappa \sinh(\mathrm{i}\gamma L)}{\gamma \cosh(\mathrm{i}\gamma L) - (\mathrm{i}\alpha/2 + \Delta)\sinh(\mathrm{i}\gamma L)} \tag{7.59}$$

$$\gamma = \mathrm{i}\left[\left(\frac{\alpha}{2} - \mathrm{i}\Delta\right)^2 + \kappa^2\right]^{1/2}, \qquad 2\Delta = 2\beta - qK$$

From Eq. (7.58b) or (7.59), the power reflectivity is given by

$$\begin{aligned} R_{\mathrm{DBR}} &= |r_{\mathrm{DBR}}|^2 \\ &= \frac{P^-(0)}{P^+(0)} \\ &= \frac{|\kappa \sinh(\mathrm{i}\gamma L)|^2}{|\gamma \cosh(\mathrm{i}\gamma L) - (\mathrm{i}\alpha/2 + \Delta)\sinh(\mathrm{i}\gamma L)|^2} \end{aligned} \tag{7.60}$$

The reflectivity at the Bragg wavelength ($\Delta = 0$) in a lossless waveguide ($\alpha = 0$) is given by $R_{\mathrm{DBR}} = \tanh^2(\kappa L)$. The power transmissivity is given by

$$\begin{aligned} T_{\mathrm{DBR}} &= \frac{P^+(L)}{P^+(0)} \\ &= \frac{|\gamma|^2}{|\gamma \cosh(\mathrm{i}\gamma L) - (\mathrm{i}\alpha/2 + \Delta)\sinh(\mathrm{i}\gamma L)|^2} \end{aligned} \tag{7.61}$$

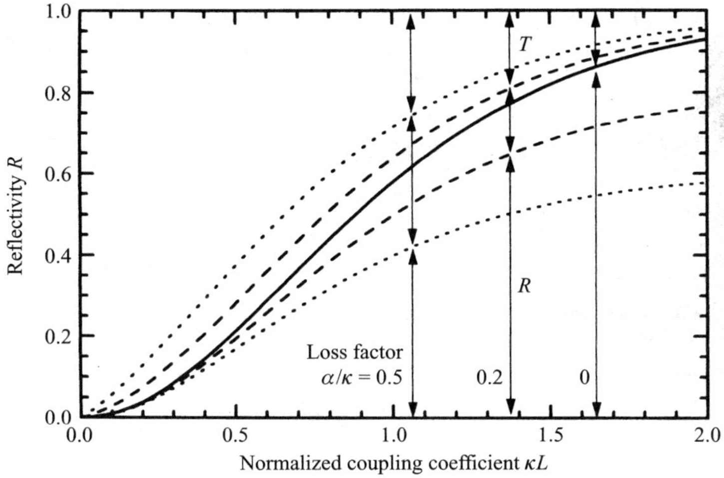

Figure 7.12 Power reflectivity and transmissivity of a DBR at the Bragg wavelength.

Obviously, the relation $R_{\mathrm{DBR}} + T_{\mathrm{DBR}} = 1$ holds in lossless waveguides ($\alpha = 0$), although $R_{\mathrm{DBR}} + T_{\mathrm{DBR}} < 1$ in waveguides with loss. The dependences of R_{DBR} and T_{DBR} at the Bragg wavelength on a normalized grating length κL is shown in Fig. 7.12. The reflectivity decreases with increasing detuning from the Bragg wavelength. The dependences of the power reflectivity $R_{\mathrm{DBR}} = |r_{\mathrm{DBR}}|^2$ and the reflection phase $\phi_{\mathrm{DBR}} = \arg\{r_{\mathrm{DBR}}\}$ on the normalized detuning ΔL ($\Delta = \beta - qK/2$) are shown in Fig. 7.13. The graph of the reflectivity shows that a DBR grating functions as a reflector with a sharp wavelength selectivity. The Bragg reflection bandwidth can be represented by $2(\Delta L)$, the full width of ΔL for the half-maximum of R_{DBR}. When $\kappa L = 1$ and $\alpha = 0$, the bandwidth is given approximately by $2\Delta L \approx 4$. The bandwidth $2\Delta L$ is broadened with increasing κL and α. The reflection phase is $\phi_{\mathrm{DBR}} = \pi/2$ at the Bragg wavelength. For a grating with $\kappa L < 1$, the dependence of the phase ϕ_{DBR} upon detuning is given approximately by $\phi_{\mathrm{DBR}} \approx \pi/2 + \Delta L$, within the Bragg bandwidth, since the effective center of reflection is at $z \approx L/2$.

7.4.2 Oscillation Condition of Distributed Bragg Reflector Lasers

A DBR laser is implemented by replacing both or one of the facet mirrors of a FP laser by DBR grating(s). Accordingly, the mathematical expressions for a DBR laser are obtained by substituting the DBR reflectivity, instead of

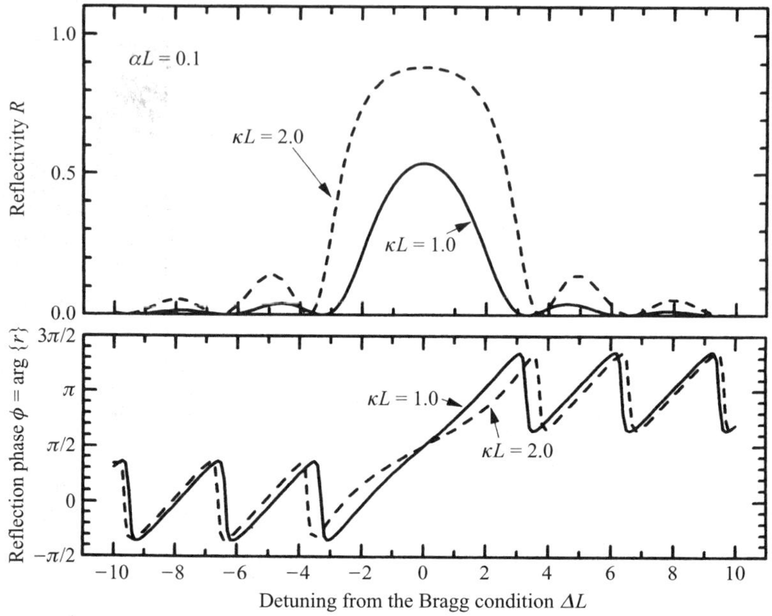

Figure 7.13 Dependence of the reflectivity and reflection phase of a DBR on detuning from the Bragg condition.

the facet reflectivity, into the expressions for a FP laser [15]. The oscillation condition for a DBR laser that results when both facets are replaced by symmetrical gratings, i.e., the condition for the optical wave to have the same complex amplitude after a round trip, can be written as

$$r_{\mathrm{DBR}}^2 \exp[(\Gamma g - \alpha_{\mathrm{int}} + 2\mathrm{i}\beta)L_{\mathrm{act}}] = 1 \tag{7.62}$$

where L_{act} is the length of the active section, ΓG the mode gain, and α_{int} the internal loss. The above equation can be rewritten as separate amplitude and phase conditions:

$$\Gamma g = \alpha_{\mathrm{int}} + \alpha_{\mathrm{DBR}}, \quad \alpha_{\mathrm{DBR}} = \frac{1}{L_{\mathrm{act}}} \ln\left(\frac{1}{R_{\mathrm{DBR}}}\right) \tag{7.63}$$

$$\phi_{\mathrm{DBR}} = m\pi - \beta L_{\mathrm{act}} = m\pi - \left(\Delta + \frac{qK}{2}\right) L_{\mathrm{act}} \tag{7.64}$$

where α_{DBR} represents the reflectivity of the DBR grating converted into a form of loss and corresponds to the mirror loss for a FP laser. The detuning Δ corresponding to the oscillation wavelength that satisfies Eq. (7.64) can be indicated by the crossing points of the curves for both sides of Eq. (7.64), as

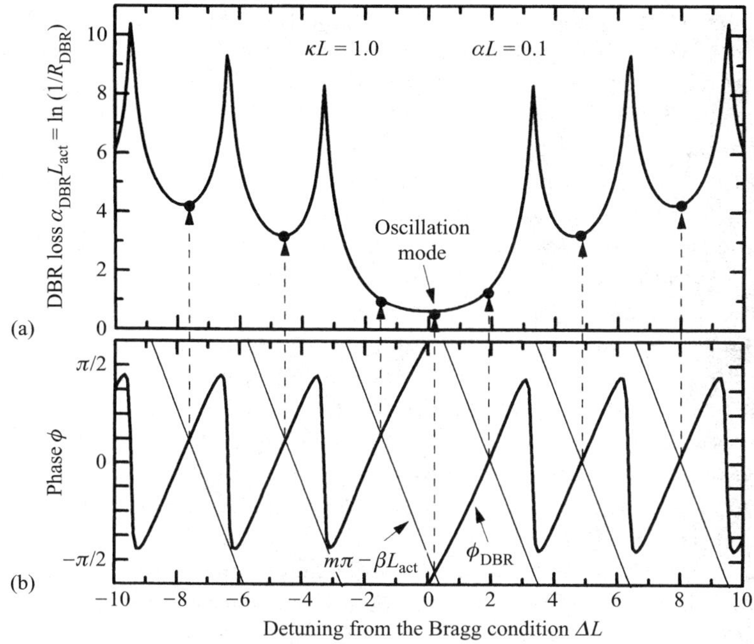

Figure 7.14 Oscillation condition of a DBR laser.

shown in Fig. 7.14(b). Each crossing point represents a DBR longitudinal mode. The values of the DBR loss α_{DBR} given by Eq. (7.63) for each mode are indicated by full circles on the curve shown in Fig. 7.14(a). They have strong mode dependence. Accordingly, the threshold gains $\Gamma g_{thm} = \alpha_{int} + \alpha_{DBRm}$ for each mode exhibit strong mode dependence. Noting that $\phi_{DBR} \approx \pi/2 + \Delta L_{DBR}$ within the Bragg bandwidth, and using $\Delta = \beta - qK/2$, Eq. (7.64) can be approximated as $(L_{act} + L_{DBR})\Delta = (m-1/2)\pi - qKL_{act}/2$. Therefore, the separation between the major oscillation mode and the adjacent side is given by

$$\Delta\omega = \frac{2\pi c}{2N_g(L_{act} + L_{DBR})} \tag{7.65}$$

which is similar to the expression for a FP laser.

7.4.3 Implementation and Operation Characteristics of Distributed Bragg Reflector Lasers

DBR lasers are fabricated by forming a grating structure, similar to that of DFB lasers, at a position adjacent to the active section of a channel waveguide structure including a bulk DH or quantum well structure similar to that of FP lasers. An outstanding feature of DBR lasers, unlike DFB lasers, is that the structure with a DBR grating formed independently out of the active section offers large design and fabrication flexibilities. Since the DBR section does not require an amplification function, the possible degradation of the light emission characteristic due to damage by grating fabrication process is not a concern, and there is no need to form a p–n junction and an electrode for carrier injection in the DBR section. Therefore, fabrication of DBR lasers does not necessarily require crystal rcgrowth on the grating. On the other hand, the propagation loss in the DBR section without carrier injection can be a large problem. For bulk DH structures, an optical wave with a wavelength at or near the gain peak wavelength for an active section suffers from a large absorption loss in a passive waveguide of the same material and structure as the active section. To realize a high-reflectivity DBR grating, a passive waveguide of different composition must be connected to the active section. Therefore, the device fabrication is complicated, and a connection loss may be involved. As pointed out in Section 4.3.8, for quantum wells and strained quantum wells, the absorption loss in the passive waveguide for the gain peak wavelength is significantly smaller than that for bulk DH structures. By detuning the Bragg wavelength to a wavelength appropriately longer than the gain peak wavelength, the propagation loss can be reduced. Careful design enables implementation of a high-reflectivity DBR grating and a high-performance DBR laser. Moreover, it is possible to reduce further the passive waveguide loss by the area-selective disordering of quantum well in the DBR section. Long-optical-cavity DBR lasers, which are advantageous for achieving a very narrow spectrum linewidth, have also been implemented by inserting a passive waveguide section with loss reduced by disordering between the active and DBR sections.

The operation characteristics of a DBR laser can be analyzed using rate equations, in a similar manner to those for FP and DFB lasers. The photon lifetime $\tau_{\mathrm{ph}m}$ is given by

$$\frac{1}{\tau_{\mathrm{ph}m}} = \Gamma g_{\mathrm{th}m} = \alpha_{\mathrm{int}} + \alpha_{\mathrm{DBR}m} \tag{7.66}$$

where $\alpha_{\mathrm{DBR}m}$ is the DBR loss given by Eq. (7.63). The relation between the photon density and the optical output power is given by the expression for FP lasers, with the reflectivity R and the transmissivity $1-R$ of the facet mirror(s) replaced by the reflectivity $R_{\mathrm{DBR}m}$ and the transmissivity $T_{\mathrm{DBR}m}$ of the DBR grating(s). Since the photon lifetime $\tau_{\mathrm{ph}m}$ exhibits strong mode dependence, DBR lasers as well as DFB lasers oscillate in a single mode with

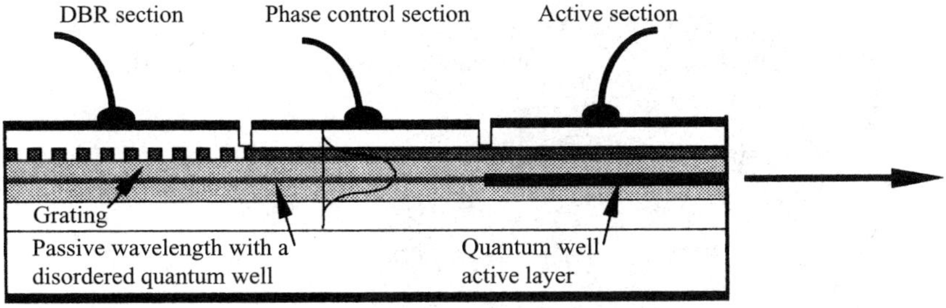

Figure 7.15 Cross-sectional structure of a three-section-type wavelength-tunable DBR laser.

a high side-mode suppression ratio. A laser using two DBRs at both ends of the active section has a side-mode suppression ratio higher than that of a laser using a single DBR and a facet mirror. The high side-mode suppression ratio is maintained also under high-speed modulaton. It should be noted that, as we see from Fig. 7.15, two different modes close to the Bragg wavelength may have values of $\tau_{\mathrm{ph}m}$ the same or very close to each other, depending upon the relation between the resonator length and the Bragg wavelength. This problem can be avoided by appropriate choice of the operating point, since the effective optical length of the resonator depends upon the carrier density and the ambient temperature. The characteristics of DBR lasers concerning the oscillation wavelength, polarization mode, high-speed modulation, noise, and spectrum linewidth are similar to those of DFB lasers. Although the quantitative characteristics of DBR lasers depend upon many device parameters in a complicated manner, performances as high as, or in part even higher than, those of DFB lasers have been obtained, particularly in quantum well DBR lasers.

7.4.4 Wavelength-Tunable Distributed Bragg Reflector Lasers

Semiconductor lasers that oscillate in a high-purity single mode at a wavelength continuously tunable electrically are required in many applications such as wavelength-division-multiplexed optical communication, coherent optical communications, and precision optical measurements. Compact lasers satisfying this requirement can be implemented by modifying a DBR laser [16,17]. An example is schematically illustrated in Fig. 7.15 [17]. The device, called a three-section-type wavelength-tunable DBR laser, is constructed with an active section, a passive waveguide phase control section, and a DBR section, monolithically integrated in a continuous-channel waveguide structure. Each section has an independent electrode. Let N_{eact}, N_{epc}, and N_{eDBR} be the effective refractive indexes for each section. Then the Bragg wavelength and the longitudinal mode resonance condition are given by

$$\lambda_{\text{B}} = \frac{2\Lambda N_{\text{eDBR}}}{q} \tag{7.67}$$

$$2\frac{2\pi}{\lambda}(N_{\text{eact}}L_{\text{act}} + N_{\text{epc}}L_{\text{pc}}) + \phi_{\text{DBR}} = 2m\pi \tag{7.68}$$

The DBR reflection phase is $\phi_{\text{DBR}} = \pi/2$ for the Bragg wavelength λ_{B}. Carrier injection in the phase control section or in the DBR section gives rise to an increase in the carrier density. Then the refractive index decreases through the band-filling effect. When N_{epc} is adjusted so that Eq. (7.68) holds for $\lambda = \lambda_{\text{B}}$, by control of the current injection in the phase control section, the laser oscillates at λ_{B}. The Bragg wavelength λ_{B} can also be changed through change in N_{eDBR} by control of the current injection in the DBR section. Therefore, by changing the currents injected in both sections jointly to maintain an appropriate condition, continuous tuning of the oscillation wavelength can be achieved. The refractive index can also be changed by applying a backward bias voltage to the electrode, instead of current injection. This effect can be utilized to increase the tuning range. In two-section-type devices without the phase control section, the oscillation wavelength can be tuned, but a mode jump takes place when the injection current in the DBR section is changed over a wide range. The three-section-type tunable lasers have been implemented by using quantum wall structures. Tuning ranges of 2–10 nm, high-speed tuning of the order of nanoseconds, and submegahertz spectrum linewidths have been obtained.

REFERENCES

1. K. Y. Lion, C. A. Burrus, R. A. Linke, I. P. Kaminov, S. W. Granlund, C. B. Swan, and P. Besomi: *Appl. Phys. Lett.*, **45**, 729 (1984).
2. T. W. Tsang, N. A. Olsson, and R. A. Logan, *Appl. Phys. Lett.*, **42**, 650 (1983).
3. H. Kogelnik and C. V. Shank, *J. Appl. Phys.*, **43**, 2327 (1972).
4. M. Nakamura, A. Yariv, H. W. Yen, S. Someck, and H. L. Garvin, *Appl. Phys. Lett.*, **23**, 224 (1973).
5. S. Wang, *IEEE J. Quantum Electron.*, **QE-10**, 413 (1974).
6. H. Nishihara, M. Haruna, and T. Suhara, *Integrated Optical Circuits*, McGraw-Hill, New York (1989) (*Integrated Optical Circuits* (Japanese version), Ohmsha, Tokyo (1993).
7. A. Yariv, *IEEE J. Quantum Electron*, **QE-9**, 919 (1973).

8. W. Streifer, D. R. Scifres, and R. D. Burnham, *IEEE J. Quantum Electron.*, **QE-12**, 74 (1976).
9. W. Streifer, R. D. Burnham and D. R. Scifres, *IEEE J. Quantum Electron.*, **QE-11**, 154 (1975).
10. H. A. Haus and C. V. Shank, *IEEE J. Quantum Electron.*, **QE-12**, 532 (1976).
11. H. Soda, Y. Kotaki, H. Ishikawa, S. Yamakoshi, and H. Imai, *IEEE J. Quantum Electron.*, **QE-23**, 804 (1987).
12. Y. Luo, Y. Nakano, K. Tada, T. Inoue, H. Hosomatsu, and H. Iwaoka, *IEEE J. Quantum Electron.*, **QE-27**, 1725 (1991).
13. K. Uomi, T. Tuchiya, H. Nakano, M. Aoki, M. Suzuki, and N. Chinone, *IEEE J. Quantum Electron.*, **QE-27**, 1705 (1991).
14. M. Okai, T. Tuchiya, K. Uomi, N. Chinone, and T. Harada, *IEEE J. Quantum Electron.*, **QE-27**, 1767 (1991).
15. Y. Suematsu, S. Arai, and K. Kishino, *J. Lightwave Techol.*, **LT-1**, 161 (1983).
16. T. L. Koch and U. Koren, *J. Lightwave Techol.*, **LT-8**, 274 (1990).
17. T. Hirata, M. Maeda, M. Suehiro, and H. Hosomatsu, *IEEE J. Quantum Electron.*, **QE-27**, 1609 (1991).

8
Semiconductor Laser Amplifiers

Chapters 6 and 7 described semiconductor laser oscillators. Another important function of semiconductor lasers is an optical amplifier. Semiconductor optical amplifiers (SOAs) offer many applications in optical communication systems, including direct amplification of optical signals such as optical power boosting in a transmitter, amplification in a repeater, and preamplification in a receiver, and also applications as a power amplifier to implement a high-power laser source. SOAs can also be used to realize other functions such as optical wavelength conversion, wavelength filtering, and optical phase conjugation. This chapter presents the fundamental characteristics of semiconductor laser amplifiers.

8.1 GAIN SPECTRUM AND GAIN SATURATION

Let us consider the optical amplification gain in a carrier-injected active region from the viewpoint of application as an amplifier. The gain factor g can be given approximately by a linear function of the carrier density as in Eq. (3.40). Approximating the gain spectrum by a Lorentz function of width $2\,\Delta\omega$, the dependences of g on the optical frequency and the carrier density can be written approximately as

$$g(\omega, N) = \frac{A_g(N - N_0)}{1 + [(\omega - \omega_0)/\Delta\omega]^2} \tag{8.1}$$

The carrier density N in the active region, where the optical wave is amplified by stimulated emission, depends not only on the injection current density but also on the optical intensity. Therefore, it is convenient to rewrite the above expression as a function of the injection current density and the optical power density. In the analysis of optical amplifiers, in general, it is not appropriate to employ the rate equations used in the analysis of laser oscillators in the same form, since the carrier density N and the photon density S in an amplifier, depending upon the injection current density J and the input optical power, have distributions that are nonuniform in the direction of the optical axis. However, so far as the carrier density $N(z)$ is concerned, the rate equation

$$\frac{\mathrm{d}}{\mathrm{d}t}N = -v_g \Gamma g S - \frac{N}{\tau_s} + \frac{J}{dq} \tag{8.2}$$

holds locally at each position in the amplifier. The local photon density $S(z)$ in a laser amplifier is correlated with the local optical power flow $P(z)$ by

$$P = \hbar\omega d W v_g S \tag{8.3}$$

where d and W are the thickness and the width, respectively, of the active channel. Putting $\mathrm{d}/\mathrm{d}t = 0$ in Eq. (8.2) for the steady state, solving for N by substituting Eq. (8.1) into Eq. (8.2), and substituting the result into Eq. (8.1) again yield

$$g(\omega, J, P) = \frac{B_g(J - J_0)}{1 + [(\omega - \omega_0)/\Delta\omega]^2 + P/P_{sat}} \tag{8.4}$$

$$B_g = \frac{A_g \tau_s}{dq}, \qquad J_0 = \frac{dqN_0}{\tau_s} \tag{8.5}$$

$$P_{sat} = \frac{\hbar\omega d W}{\Gamma \tau_s A_g} = \frac{\hbar\omega}{q\Gamma B_g} W \tag{8.6}$$

where J_0 is the current density at transparency. Here use has been made of Eq. (8.3) to convert S into P. The effective gain factor for a guided mode is given by $\Gamma g - \alpha_{int}$, where g is the material gain given by the above expression, Γ the confinement factor, and α_{int} the internal loss.

Equation (8.4) indicates that the gain g depends not only on the injection current density J but also on the optical power P, as shown in Fig. 8.1. In the region where the optical power is small, most injected carriers are consumed by spontaneous emission recombination. Here, as Eq. (8.2) shows, $N\tau_s J/dq$ and, with increasing injection current density, the carrier density and the gain increase. The small-signal gain in this region does not depend significantly on the optical power. In the region where the optical power is high, on the other hand, the large value of S limits g to a small value, as seen from Eq. (8.2). Here, most of the injection current density J over the J_0 required for maintaining transparency is consumed by stimulated emission recombination. Thus the carrier density is limited to

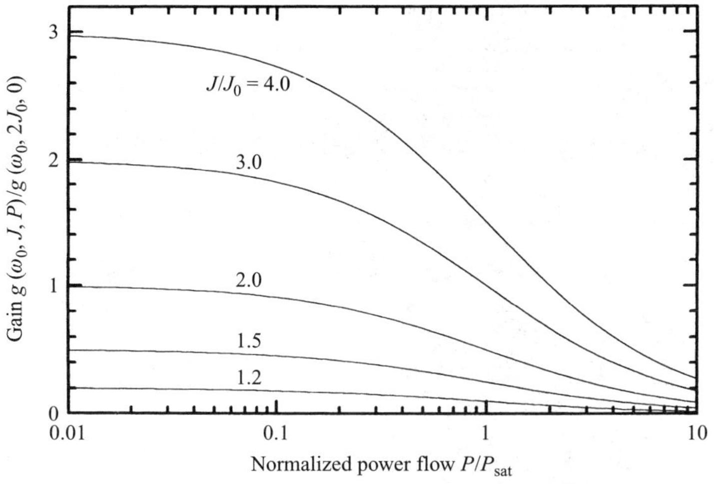

Figure 8.1 Characteristic of saturation in a semiconductor laser amplifier.

a value close to the transparency value N_0, and the gain is limited. In this region, the increment in the optical power, $dP/dz = \Gamma gP$, equals the stimulated recombination carrier number per unit length in the z direction per unit time, $(J - J_0)W/q$, multiplied by the photon energy $\hbar\omega$. Therefore, $g \approx \hbar\omega W(J - J_0)/q\Gamma P$, and the gain is inversely proportional to P. Thus, the gain saturates with increase in the optical power. Equation (8.4) gives a convenient approximate expression for the gain including the saturation characteristic. In the expression, P_{sat} represents the optical power level where the gain starts to be involved with substantial saturation and is given by Eq. (8.6). For ordinary index-guiding lasers, P_{sat} is 1 w to several tens of milliwatts.

8.2 RESONANT LASER AMPLIFIERS

A Fabry–Perot (FP) laser can be used as an optical amplifier, as shown in Fig. 8.2, when driven by an injection current below the threshold level, by coupling an optical signal from one of the facets and taking the output through another facet [1]. This type of amplifier is called a resonant amplifier or FP-type amplifier. We assume for simplicity that the carrier and photon densities are approximately uniform in the resonator, as in the analysis of oscillators. Then, by replacing the loss factor α in the expression for the transmissivity of a passive FP resonator (Eq. (5.134)) by the effective

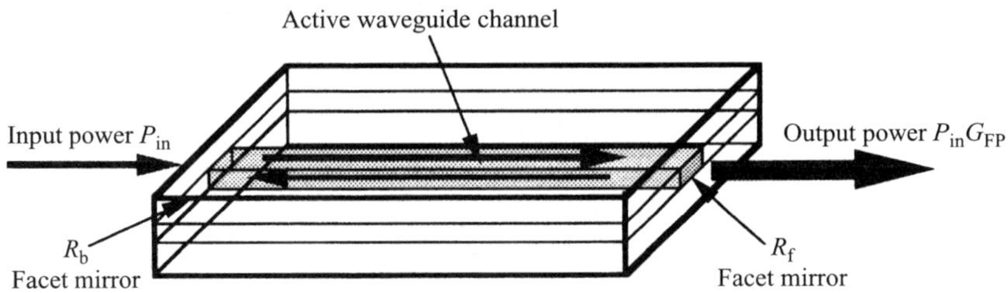

Figure 8.2 Resonant semiconductor laser amplifier.

gain factor $\Gamma g - \alpha_{int}$ with the sign inverted, the power amplification gain for a signal of frequency ω is given by

$$G_{FP}(\omega) = \frac{(1 - R_f)(1 - R_b)G_s}{1 + R_f R_b G_s^2 - 2(R_f R_b)^{1/2} G_s \cos(2\beta L)} \tag{8.7}$$

$$G_s = G_s(\omega) = \exp\{[\Gamma g(\omega) - \alpha_{int}]L\}$$

where R_f and R_b are the reflectivities of the input and output facets, L is the resonator length, and G_s is the one-way gain. Since the propagation constant β in the above expression depends upon ω, $G_{FP}(\omega)$ exhibits an oscillatory spectrum consisting of peaks at the FP modes with separation $\Delta\omega = 2\pi v_g/2L$ and a width narrower than the bandwidth of the gain $g(\omega)$, similar to the transmissivity shown in Fig. 5.20. The maxima are given by

$$G_{FPmax} = \frac{(1 - R_f)(1 - R_b)G_s}{[1 - 2(R_f R_b)^{1/2} G_s]^2} \tag{8.8}$$

and from the above expression and Eq. (5.137) the 3 dB bandwidth of each peak for the high-gain case is given by

$$\delta\omega = \frac{(\Delta\omega/\pi)[(1 - R_f)(1 - R_b)]^{1/2}}{(R_f R_b)^{1/4} G_{FP\,max}^{1/2}} \tag{8.9}$$

The equation shows that increasing the injection current to increase the gain is associated with narrowing of the bandwidth $\delta\omega$ in inverse proportion to $G_{FPmax}^{1/2}$.

The effect of the saturation on the amplification gain can be analyzed by using Eq. (8.4). Assuming uniform carrier and photon densities, the output power P_f and photon density S are correlated by Eq. (6.35), the P_f dependence of the amplification gain G can be calculated using Eqs (8.3), (8.4), and (8.7) or (8.8), and using $P_{\mathrm{f}} = GP_{\mathrm{in}}$ the result can readily be converted into the dependence of G on the input power P_{in} and also into the input–output characteristics.

As the above expressions show, resonant amplifiers have the advantage that the amplification with positive feedback gives a high gain even with a low injection level. A typical gain is 15–25 dB. The output power, however, saturates at a small level, from −10 to −5 dBm (dBm is a power unit defined so that 1 mW is 0 dBm). A high efficiency is difficult to achieve, since the output power is delivered also to the input port. The drawbacks also include a narrow gain bandwidth (around 10^{10} Hz), a gain peak wavelength that varies with the operation conditions, difficulty in wavelength matching required for stable operation, and large noises.

The above discussion was with respect to amplification at subthreshold injection level. When the injection level exceeds the threshold, the laser oscillates. The oscillation, however, is pulled into oscillation at the same frequency as the input wave and with the phase synchronized with it, unless the input optical power is too small. This operation is referred to as injection locking [2]. Increasing the input power gives rise to a reduction in the gain due to saturation. Then, the oscillation stops, and the operation transfers to resonant amplification as described above. In this way, the laser serves as an amplifier even with injection above the threshold. The injection-locking amplifiers offer a gain and output power higher than those obtained with subthreshold injection. The application, however, is limited, because the output power does not reach zero for zero input, and the input–output characteristic exhibits remarkable nonlinearity.

8.3 TRAVELING-WAVE LASER AMPLIFIERS

The drawbacks of resonant amplifiers can be eliminated by depositing an antireflection coating on the input and output facets. Then the optical wave propagates in the laser as a unidirectional traveling wave, as shown in Fig. 8.3, and hence a laser amplifier of this type is called a traveling-wave type amplifier [3].

8.3.1 Amplifier Gain and Amplifier Bandwidth

The gain spectrum given by Eq. (8.4), when the gain saturation is omitted, can be written as

$$g(\omega) = \frac{g_0}{1 + [(\omega - \omega_0)/\Delta\omega]^2} \tag{8.10}$$

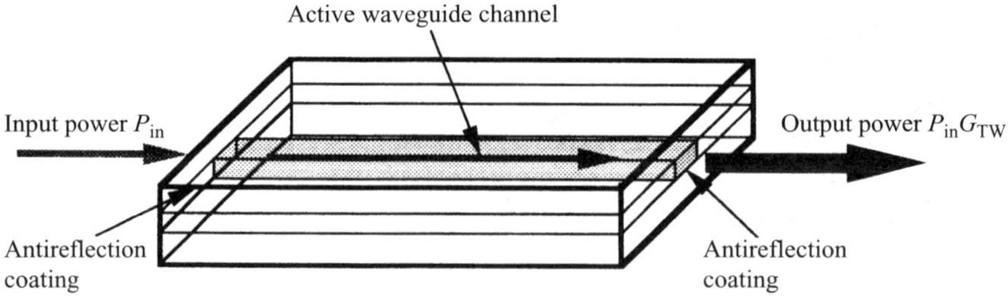

Figure 8.3 Traveling-wave semiconductor laser amplifier.

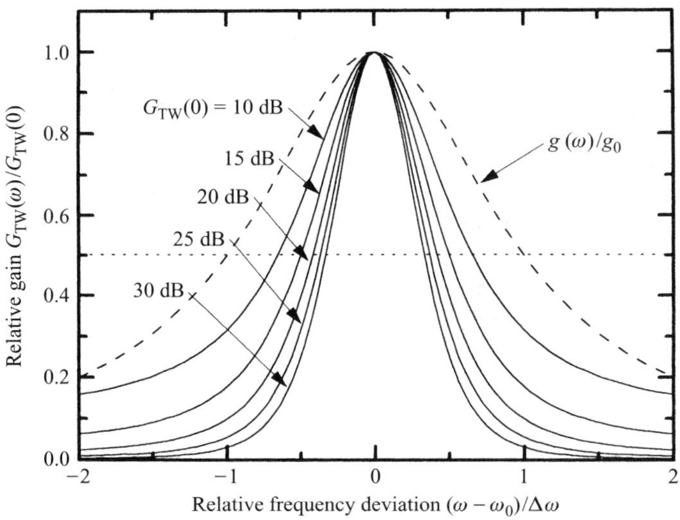

Figure 8.4 Frequency dependence of the gain of traveling-wave amplifiers.

and the effective mode gain is given by $\Gamma g - \alpha_{\mathrm{int}}$, The power gain of an amplifier of length L is given by

$$G_{\mathrm{TW}}(\omega) = G_{\mathrm{s}}(\omega) = \exp\{[\Gamma g(\omega) - \alpha_{\mathrm{int}}]L\} \tag{8.11}$$

and the maximum gain is $G_{\mathrm{TW}}(\omega_0) = \exp[(\Gamma g_0 - \alpha_{\mathrm{int}})L]$. A typical gain is 20–30 dB. Assuming that $\alpha_{\mathrm{int}} = 0$, the amplifier exhibits a frequency response given by Eq. (8.11) and shown in Fig. 8.4, and the 3 dB bandwidth is given by

$$2\,\Delta\omega_{\mathrm{TW}} = 2\,\Delta\omega\left(\frac{\ln 2}{\Gamma g_0 L - \ln 2}\right)^{1/2} \tag{8.12}$$

The frequency bandwidth of an amplifier given by the above expression is different from the 3 dB bandwidth of the gain spectrum $2\,\Delta\omega$ given by Eq. (8.10). A bandwidth of around 10^{12} Hz, much wider than that for a FP laser amplifier, is obtained, although the bandwidth is narrowed when the amplification gain is high. In practical devices, the facet reflectivity cannot be completely reduced to zero; the residual reflection degrades the smoothness of the amplifier gain spectrum, as described by Eq. (8.7) for FP amplifiers. The reflectivity can be reduced to around 10^{-4} by a fine antireflection coating [4], by slanting the waveguide channel with respect to the facet [5], or forming a window structure where the waveguide channel is terminated before the facet of the crystal is reached [6]. Then a reasonably smooth frequency response appropriate for practical applications can be achieved.

8.3.2 Saturated Output Power

We next consider the effect of the gain saturation. The internal loss is omitted for simplicity. The change in the power of an optical wave at the gain peak frequency ω_0 propagating in an amplifier can be written, by using g given by Eq. (8.4), as

$$\frac{\mathrm{d}}{\mathrm{d}z}P(z) = \Gamma g P(z) = \frac{\Gamma g_0 P(z)}{1 + P(z)/P_{\mathrm{sat}}} \tag{8.13}$$

The above equation can easily be integrated to yield an expression for the amplifier gain $G_{\mathrm{TW}} = G = P(L)/P(0)$:

$$G = \frac{P(L)}{P(0)} = G_0 \exp\left(\frac{-(G-1)P_{\mathrm{out}}}{GP_{\mathrm{sat}}}\right) \tag{8.14}$$

$$P_{\mathrm{out}} = P(L), \qquad G_0 = \exp(\Gamma g_0 L)$$

where P_{out} is the output power and G_0 the amplifier gain with the saturation omitted. The relation between the output power and the amplifier gain calculated by the above equation is shown in Fig. 8.5. From Eq. (8.14), the output power at which $G = G_0/2$ is given by

$$P^{\mathrm{s}}_{\mathrm{out}} = \frac{G_0 \ln 2}{(G_0 - 2)} P_{\mathrm{sat}} \approx (\ln 2) P_{\mathrm{sat}} \tag{8.15}$$

The above $P^{\mathrm{s}}_{\mathrm{out}}$ represents the output power level at which the gain saturation starts to appear and is referred to as the saturation output power. A typical value of $P^{\mathrm{s}}_{\mathrm{out}}$ for double-heterostructure lasers is 0–10 dBm.

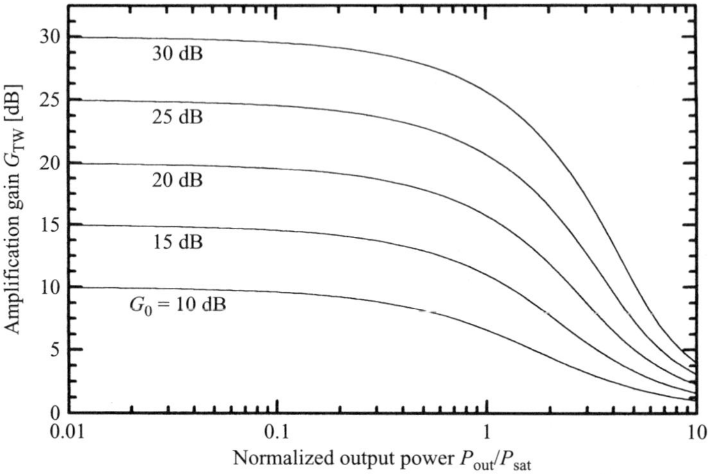

Figure 8.5 Relation between the gain and the output power of a traveling-wave laser amplifier.

For quantum well lasers, P^{s}_{out} can be higher since, as Eq. (8.6) indicates, P_{sat} is inversely proportional to Γ; a P^{s}_{out} value as high as 20 dBm has been achieved. In applications to optical communications, the saturation effect gives rise to problems such as distortion in signal waveform, intersymbol interference, and crosstalk among multiplexed signals.

8.3.3 Noise Characteristics

An important requirement for applications as a preamplifier, where a weak signal is amplified, is a low noise level. The noise characteristics of semiconductor laser amplifiers [7] can also be analyzed by using rate equations similar to those for laser oscillators. Although the rate equation for photon density for FP-type oscillators is given by Eq. (6.21), it must be modified for application to traveling-wave amplifiers. The photon lifetime term on the right-hand side of Eq. (6.21) is $S/\tau_{ph} = v_g(\alpha_{int} + \alpha_{mir})S$, and the second term represents the outgoing photon flow through the mirrors. In traveling-wave amplifiers, there are no mirrors present, and traveling-wave power is amplified. Accordingly, the second term should be replaced by $v_g(dS/dz)$ to represent the net number of photons that flow from a unit volume of the active region per unit time in the form of a traveling-wave power flow. On the other hand, the final term on the right-hand side of Eq. (6.21) representing the contribution of spontaneous emission can be rewritten as $C_s N/\tau_s = n_{sp} v_g \Gamma g/dWL$, where n_{sp} is the population inversion factor given by Eq. (A5.8), dW and L are the cross section and the length of the active region. This term represents the contribution to a mode under oscillation. It should be noted that, in amplifiers, the spontaneous emission that is mixed with the signal wave and gives rise to noise is not limited to a single mode but includes several modes within the signal bandwidth. Although the spontaneous emission in an amplifier consists of many modes, components belonging to lateral modes different from the signal mode can be removed by inserting a spatial filter or coupling to an optical fiber. The components of the same lateral mode as the signal still consist of many longitudinal modes with a frequency separation of v_g/L. Although some of these can be removed by inserting a wavelength filter, the bandwidth of the filter should not be narrower than the signal bandwidth, since the signal must be transmitted. Let $\mathrm{d}\nu(=\mathrm{d}\omega/2\pi)$ be the frequency bandwidth with which the output signal is detected; then this bandwidth corresponds to a mode number $(L/v_g)\mathrm{d}\nu$. Therefore, in the rate equation for amplifiers, it is appropriate to replace the spontaneous emission term by $(n_{sp} v_g \Gamma g/dWL)(L/v_g)\ \mathrm{d}\nu = (n_{sp}\Gamma g/dW)\,\mathrm{d}\nu$. As a result of this modification, we obtain a rate equation for the local photon density $S(z)$ for traveling-wave amplifiers:

$$\frac{\mathrm{d}}{\mathrm{d}t}S = v_g\left[(\Gamma g - \alpha_{\mathrm{int}})S - \frac{\mathrm{d}S}{\mathrm{d}z}\right] + \frac{n_{sp}\Gamma g}{dW}\,d\nu \tag{8.16}$$

For the steady state, using Eq. (8.3), the above equation can be converted into the equation for optical power P:

$$\frac{\mathrm{d}}{\mathrm{d}z}P = (\Gamma g - \alpha_{\mathrm{int}})P + (n_{sp}\Gamma g\hbar\omega)\,d\nu \tag{8.17}$$

Let $P(0) = P_{\mathrm{in}}$ be the input power. Solving the above equation with the saturation omitted, we obtain an expression for the output power:

$$P_{\mathrm{out}} = P(L) = GP_{\mathrm{in}} + (G-1)n_{\mathrm{spef}}\hbar\omega\,d\nu \tag{8.18}$$

$$n_{\mathrm{spef}} = n_{sp}\frac{\Gamma g}{\Gamma g - \alpha_{\mathrm{int}}} \tag{8.19}$$

The amplifier gain G is given by Eq. (8.11). The first term on the right-hand side of the above expression represents the amplified spontaneous emission (ASE) and indicates that the ASE is mixed with the signal. In order to examine the ASE inclusion, let $A_s \cos(\omega_s t)$ be the amplitude of the amplified wave for a coherent input wave, and $a_{\mathrm{ASE}}(\omega)$ be the time-truncated Fourier transform of the incoherent ASE amplitude. Then the amplifier output power is given by

$$P(t) = \left\langle \left(A_s \cos(\omega_s t) + \frac{1}{2\pi} \int a_{ASE}(\omega) \exp(-i\omega t)\, d\omega \right)^2 \right\rangle$$

$$= \frac{A_s^2}{2} + 2S_{ASE}(\omega)\, d\nu$$

$$+ 2A_s \int \mathrm{Re}\{a_{ASE}(\omega) \exp[-i(\omega - \omega_s)t]\} \frac{d\omega}{2\pi} \quad (8.20)$$

where $\langle\,\rangle$ indicates the average over a time interval much longer than the period corresponding to the optical frequency, and $S_{ASE}(\omega)$ is the ASE noise power spectrum based on the definition in Eq. (6.74). From the fact that the first term on the right-hand side of the above expression representing the time average of $P(t)$ equals P_{out} of Eq. (8.18), we have

$$\frac{A_s^2}{2} = GP_{in}, \qquad 2S_{ASE}(\omega)\, d\nu = (G-1)n_{spef}\hbar\omega\, d\nu \quad (8.21)$$

As the last line of Eq. (8.20) shows, inclusion of the ASE gives rise to beat noise due to the interference between the signal and the ASE.

The input light is considered to be a photon flow with a photon number per unit time of $P_{in}/\hbar\omega$. Assuming the noise of the standard quantum limit (SQL), the noise power for P_{in} is given by $(\hbar\omega)^2(P_{in}/\hbar\omega)\, d\nu$. Therefore, the signal-to-noise ratio (SNR) for the input light is given by

$$\mathrm{SNR}_{in} = \frac{P_{in}^2}{(\hbar\omega)^2(P_{in}/\hbar\omega)\, d\nu} = \frac{P_{in}}{\hbar\omega\, d\nu} \quad (8.22)$$

We next calculate the noise power for the output light $P(t)$. Neglecting the second term in Eq. (8.20) under the assumption that $A_s^2/2 \gg S_{ASE}(\omega)\, d\nu$, the major time fluctuation of $P(t)$ is the beat noise represented by the third term, and the noise power is given by $2A_s^2 S_{ASE}(\omega)\, d\nu$. The first term in Eq. (8.20) equals GP_{in} and the associated SQL noise power is given by $G\hbar\omega P_{in}\, d\nu$. Therefore, the total noise power of the output light is given by $[G\hbar\omega P_{in} + 2A_s^2 S_{ASE}(\omega)]\, d\nu$, and using Eq. (8.21) the SNR for the output light is given by

$$\mathrm{SNR}_{out} = \frac{GP_{in}}{[1 + 2(G-1)n_{spef}]\hbar\omega\, d\nu} \quad (8.23)$$

A convenient numerical figure for representing the noise characteristics of an amplifier is the noise figure (NF), defined as $\mathrm{NF} = \mathrm{SNR}_{in}/\mathrm{SNR}_{out}$, which describes the degradation in SNR caused by the amplification. From Eqs (8.22) and (8.23), the NF for the TWA is given by

$$\mathrm{NF} = \frac{1 + 2n_{\mathrm{spef}}(G-1)}{G} \approx 2n_{\mathrm{spef}} \tag{8.24}$$

Quantum theory calculation also gives a result consistent with the above result. As the above result shows, in optical laser amplification, 3 dB degradation of the SNR is inevitable even in ideal cases when $\alpha_{\mathrm{int}} = 0$ and $n_{\mathrm{sp}} = 1$. Typical values for practical traveling-wave amplifiers are 6–8 dB.

8.4 TAPERED LASER AMPLIFIERS

8.4.1 Amplifier Structure

As presented in Section 8.1, since in the region above the saturation power level most of the injected carriers is consumed by stimulated emission recombination, the saturation output power is limited by the injection power. Therefore, to attain a high output power, it is necessary to enhance the saturation power level and to construct a structure that allows injection of a large current. However, it is evident that the heat problem and the catastrophic optical damage (COD) problem limit the achievement of a high output power; for lasers using an ordinary channel waveguide, the maximum output power is 100 mW or so. A structure that allows further enhancement of the output power is a tapered amplifier [8], where the width of the active region of a traveling-wave amplifier is enlarged gradually along the direction of light propagation, as shown in Fig. 8.6. The amplifier serves as a planar waveguide, and the light, coupled in the input port as a guided wave with a narrow lateral width, propagates as a two-dimensional diverging Gaussian-like beam in the waveguide. The propagation is associated with amplification. In this structure, however, the saturation effect is significantly reduced, since the lateral broadening of the guided

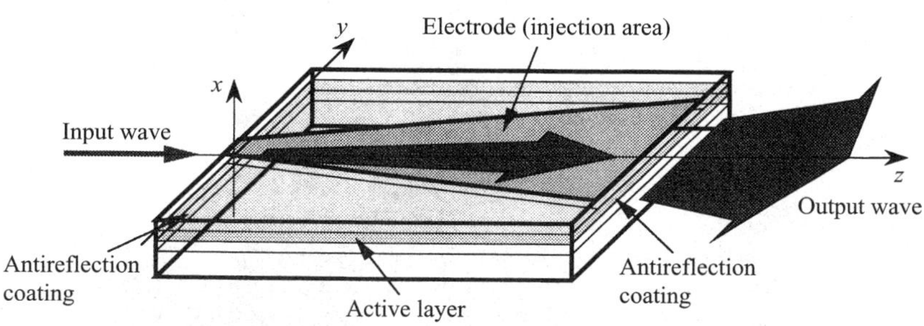

Figure 8.6 Tapered semiconductor laser amplifier.

beam gives rise to suppression of the increase in the optical power density and improvement of the uniformity in the amplifier area. The large area of the injection region and the large width of the output port allow a high current injection and reduce the heat and COD problems.

8.4.2 Beam Propagation Method Analysis

An optical wave in a tapered amplifier having a large-area active region propagates in the planar waveguide with evolution of the lateral distribution. The optical amplitude, the optical power flow, and the carrier density are functions of both the longitudinal coordinate z and the lateral coordinate y. The gain saturation is not uniform, and the carrier-induced refractive index is not uniform. Moreover, the temperature in the plane of the active region forms a nonuniform distribution, which gives a thermal lens effect on the optical wavefront through the temperature dependence of the refractive index. Thus, the nonuniform index changes resulting from the nonuniform distribution of the temperature and the carrier density affect significantly laser performances. The analysis of a tapered amplifier requires treatment of these two-dimensional distributions, and therefore cannot be made using the simple rate equations presented thus far. The analysis of tapered amplifier performances can be made by numerical simulation based on the beam propagation method (BPM) [9].

Let E be the optical field amplitude in the amplifier, P the power flow per unit width in the y direction, T the temperature, and J the injection current density. The complex propagation constant for the optical wave is given by

$$\tilde{\beta}(y,z) = \beta_0 + \delta\tilde{\beta}(y,z), \qquad \beta_0 = k_0 N_{e0} \tag{8.25}$$

$$\delta\tilde{\beta}(y,z) = \frac{-\Gamma(\alpha_c + \mathrm{i})g(y,z)}{2} + \frac{\mathrm{i}\alpha_{int}}{2} + k_0\alpha_T T(y,z) \tag{8.26}$$

$$g(y,z) = \frac{B_g(J - J_0)}{1 + P(y,z)/P_s} \tag{8.27}$$

$$P(y,z) = |E(y,z)|^2, \qquad P_s = \frac{\hbar\omega}{q\Gamma B_g} \tag{8.28}$$

where α_T denotes the temperature coefficient of the refractive index. The dependence of the gain on the injection current density and the saturation are expressed by Eq. (8.27), and the carrier-induced and heat-induced refractive index changes are represented by the first and the fourth terms on the right hand side of Eq. (8.26). Let $Q(y, z)$ be the heat power per unit area given by the power provided by the current injection and the optical absorption, from which the power converted into the stimulated emission power is substracted. Let $G(y-y')$ be the one-dimensional temperature distribution describing the heat response, to a unit line heat source, of the semiconductor multilayer structure constructing the amplifier. Then the temperature distribution $T(y, z)$ can be calculated, by taking a convolution of $Q(y, z)$ and $G(y-y')$, as

$$T(y,z) = \int G(y-y')Q(y',z)\,\mathrm{d}y' \tag{8.29}$$

$$Q(y,z) = VJ + \alpha_{\mathrm{int}}P(y,z) - \Gamma g(y,z)P(y,z) \tag{8.30}$$
$$V = \frac{\hbar\omega}{q} + V_{\mathrm{d}} + \rho J$$

where V is the voltage across the electrodes, V_{d} the diffusion voltage of the junction, and ρ the resistivity of the semiconductor substrate.

The BPM is a split-step algorithm. The positions on the z axis in the propagation direction are represented by equidistant sampling points z_m with a small separation Δz. From the y- distribution $E(y, z_m)$ of E at $z = z_m$, the distribution $E(y, z_{m+1})$, at $z = z_{m+1} = z_m + \Delta z$, is calculated by the two mathematical steps described below. In the first step, $\delta\tilde{\beta}$ in Eq. (8.25) is omitted, and the distribution after the simple propagation with diffraction is calculated by a relation deduced on the basis of plane-wave expansion of optical fields:

$$E'(y, z_{m+1}) = \mathcal{F}^{-1}\{\mathcal{F}[E(y, z_m)]\ \exp[\mathrm{i}(\beta_0^2 - k_y^2)^{1/2}\,\Delta z]\} \tag{8.31}$$

In the above equation, k_y denotes the spatial frequency in the y direction; $\mathcal{F}[]$ and $\mathcal{F}^{-1}\{\}$ denote the Fourier transform and the inverse Fourier transform, and they can be calculated numerically by employing the fast Fourier transform algorithm. In the second step, the above-obtained $E'(y, z_{m+1})$ is used to calculate $E(y, z_{m+1})$, with the effect of $\delta\tilde{\beta}(y, z)$ included:

$$E(y, z_{m+1}) = E'(y, z_{m+1})\ \exp[\mathrm{i}\delta\tilde{\beta}(y, z_m)\,\Delta z] \tag{8.32}$$

The required $\tilde{\beta}(y, z_m)$ can be calculated using Eqs (8.26)–(8.30). For an input field distribution given as an initial distribution, distributions of E, P, T, and other variables at each step in the amplifier and at the output port can be obtained by iterating the above-described procedure. The output power can be calculated by integrating the power density on the output facet $P(y, L)$.

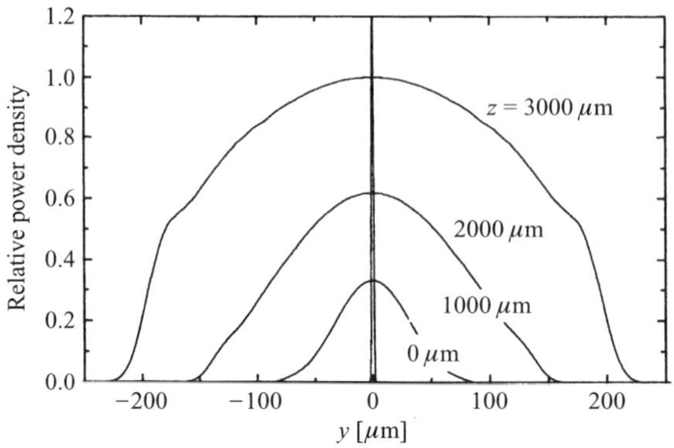

Figure 8.7 Optical intensity profile in a tapered amplifier.

8.4.3 Amplifier Characteristics

A SCH-SQW structure with a small confinement factor Γ offers a high saturation level, and therefore a high output power can be obtained in a tapered amplifier using such a structure. Figure 8.7 shows the result of BPM calculation of the optical power distribution $P(y, z_m)$ at a few positions z_m. The figure shows that, as a result of amplification, the height and the width of the distribution increase with propagation from the input port towards the output port. Since both sides of the distribution are amplified more in comparison with the central part, where the intensity is higher and the saturation is more significant, the distribution evolves from a nearly Gaussian distribution near the input port to somewhat flattened distributions near the output port. The refractive index is increased more on and near the optical axis, and therefore the wavefront is deformed in such a way that the rays are curved so that they are directed towards the central axis. As a result, for a high injection current and a large propagation length, the quality of the output wavefront deteriorates. Thus the output performances of a tapered laser amplifier depend upon many factors, and careful optimizations of the device parameters such as the amplifier length and the taper angle are required to achieve a good performance. In appropriately designed amplifiers, output powers as large as 1 W have been demonstrated in continuous operation [10].

8.5 MASTER OSCILLATOR POWER AMPLIFIER

A combination of a laser oscillator and an amplifier to boost the output power is called a master oscillator power amplifier (MOPA). A semiconductor laser

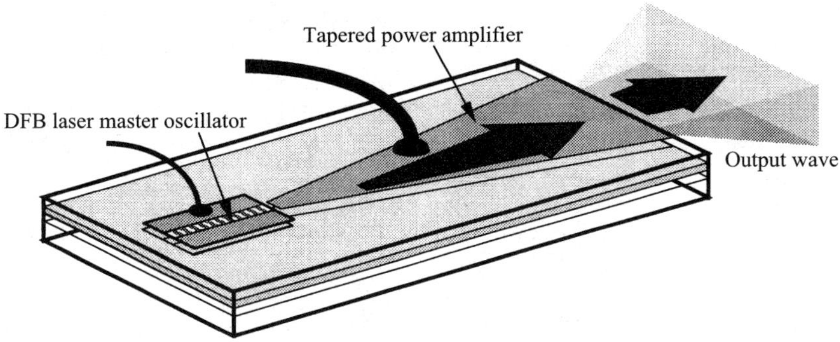

Figure 8.8 Monolithic MOPA.

oscillator and a semiconductor laser amplifier can be integrated together on a same substrate to implement a monolithic MOPA. For this application, a distributed feedback (DFB) laser or a distributed Bragg reflector laser is used as an oscillator, since they do not require a facet mirror and therefore are suitable for monolithic integration. As for the amplifier, a traveling-wave amplifier of channel waveguide type is suitable for applications to optical communications, because of the low noise and ease of coupling to an optical fiber. Such a MOPA laser provides an output power higher than that of a single oscillator. Another advantage is that high-speed modulation of the high output power can be achieved by superimposing a small electric signal on the oscillator driving current [11]. A MOPA using a tapered power amplifier provides an even higher output power. They have applications as a pump source for optical fiber amplifiers and nonlinear optical devices [12]. A monolithic MOPA consisting of a DFB laser oscillator and tapered power amplifier is shown in Fig. 8.8.

REFERENCES

1. Y. Yamamoto *IEEE J. Quantum Electron.*, **QE-16**, 1047 (1980).
2. S. Kobayashi and T. Kimura *IEEE J. Quantum Electron.*, **QE-17**, 681 (1981).
3. T. Saitoh and T. Mukai *IEEE J. Quantum Electron.*, **QE-23**, 1010 (1987).
4. T. Saitoh, T. Mukai, and O. Mikami *J. Lightwave Technol.*, **LT-3**, 288 (1985).
5. C. E. Zah, J. S. Osinski, C. Caneau, S. G. Menocal, L. A. Reith, J. Salzman, F. K. Shokoohi, and T. P. Lee *Electron Lett.*, **23**, 990 (1987).
6. N. K. Dutta, M. S. Lin, A.B. Piccirilli, R.L.Brown, and U. K. Chakrabarti *J. Appl. Phys.*, **67**, 3943 (1990).
7. T. Mukai and Y. Yamamoto *IEEE J. Quantum Electron.*, **QE-18**, 564 (1982).

8. G. Bendeli, K. Komori, S. Arai, and Y. Suematsu *IEEE Photon. Technol. Lett.*, **3**, 42 (1991).
9. R. J. Rang, A. Hardy, R. Parke, D. Mehuys, S. O'Brien, J. Major, and D. Welch *IEEE J. Quantum Electron.*, **QE-29**, 2044 (1993).
10. J. N. Walpole, E. S. Kintzer, S. R. Chinn, C. A. Wang, and L. J. Missaggia *Appl. Phys. Lett.*, **61**, 740 (1992).
11. U. Koren, B. I. Miller, G. Raybon, M. Oron, M. G. Young, T. L. Koch, J. L. DeMiguel, M. Chien, B. Tell, K. Brown-Goebeler, and C. A. Burrus *Appl. Phys. Lett.*, **67**, 1375 (1990).
12. S. O'Brien, D. Welch, R. Parke, D. Mehuys, K. Dzurko, R. J. Rang, R. Waarts, and D. Scifres *IEEE J. Quantum Electron.*, **29**, 2052 (1993).

Appendix 1

Outline of Density Matrix Analysis

A1.1 DEFINITION OF DENSITY MATRIX AND EXPECTATION VALUES

The density matrix offers an effective technique for dealing statistically with a system consisting of many electrons using the quantum theory for an electron. A mixed state consisting of a statistical distribution of various quantum states can be specified by a set of probabilities p_j with which the electron is found in a quantum state $|\psi_j\rangle$. The density operator ρ is defined by

$$\rho = \sum_j |\psi_j\rangle p_j \langle\psi_j| \tag{A1.1}$$

The probability satisfies $0 \leq p_j \leq 1$ and $\sum_j p_j = 1$. The operator ρ is a Hermite operator, and the matrix description of ρ is called the density matrix. Using a system of eigenstate $\{|n\rangle\}$, the elements of the density matrix are given by

$$\rho_{nn'} = \langle n|\rho|n'\rangle = \sum_j \langle n|\psi_j\rangle p_j \langle\psi_j|n'\rangle \tag{A1.2}$$

The diagonal elements of the density matrix

$$\rho_{nn} = \sum_j p_j |\langle n|\psi_j\rangle|^2 \tag{A1.3}$$

give the probability with which the system belongs to the eigenstate $|n\rangle$. The off-diagonal elements represents the correlation of states $|n\rangle$ and $|n'\rangle$. The expectation value $\langle A\rangle$ for a physical quantity represented by an operator A, being the weighted average of the expectation values for states $|\psi_j\rangle$, can be written as

$$\begin{aligned}\langle A\rangle &= \sum_j p_j \langle\psi_j|A|\psi_j\rangle \\ &= \sum_{jnn'} p_j \langle\psi_j|n\rangle\langle n|A|n'\rangle\langle n'|\psi_j\rangle\end{aligned}$$

$$
\begin{aligned}
&= \sum_{nn'} \rho_{n'n} A_{nn'} \\
&= \mathrm{Tr}\{\rho A\}
\end{aligned} \tag{A1.4}
$$

Since $\langle A \rangle$ can be expressed by A and ρ only, it is possible to calculate the value of the macroscopic observable $\langle A \rangle$ without knowing $|\psi_j\rangle$ and p_j, provided that ρ is obtained.

A1.2 EQUATION OF MOTION FOR THE DENSITY OPERATOR

The time variation of a state $|\psi_j\rangle$ can be written by using the system Hamiltonian H as

$$
|\psi_j(t)\rangle = U(t)|\psi_j(0)\rangle, \qquad U(t) = \exp\left(\frac{-\mathrm{i}Ht}{\hbar}\right) \tag{A1.5}
$$

and, if the time dependence of p_j is omitted, the time variation of ρ can be written as

$$
\rho(t) = \sum_j U(t)|\psi_j(0)\rangle p_j \langle \psi_j(0)|U(t)^\dagger = U(t)\rho(0)U(t)^\dagger \tag{A1.6}
$$

Then, calculation of the time derivative of ρ results in

$$
\begin{aligned}
\frac{\mathrm{d}}{\mathrm{d}t}\rho(t) &= \frac{H(t)\rho(t) - \rho(t)H(t)}{\mathrm{i}\hbar} \\
&= \frac{1}{\mathrm{i}\hbar}[H(t), \rho(t)]
\end{aligned} \tag{A1.7}
$$

Thus, the equation of motion for ρ is described by using the commutation relation between H and ρ. When the initial state $\rho(0)$ is given by a matrix representation based on an appropriate eigenstate system, solving the above equation to calculate $\rho(t)$, followed by calculation of $\langle A \rangle$ by Eq. (A1.4), clarifies the behavior of the whole system concerning the observation of the quantity A. The above description is made in the Schrödinger picture using a time-dependent operator $\rho(t)$. However, for cases where the Hamiltonian H can be written as a sum of a Hamiltonian H_0 with the interaction omitted and an interaction Hamiltonian H_i, i.e.,

$$
H(t) = H_0 + H_\mathrm{i}(t) \tag{A1.8}
$$

converting $\rho(t)$ into a density operator in the interaction picture:

$$
\rho_\mathrm{I}(t) = U_0(t)^\dagger \rho(t) U_0(t), \qquad U_0(t) = \exp\left(\frac{-\mathrm{i}H_0 t}{\hbar}\right) \tag{A1.9}
$$

transforms the equation of motion into that in the interaction picture:

$$d/dt\rho_I(t) = \frac{1}{i\hbar}[H_I(t), \rho_I(t)] \tag{A1.10a}$$

$$H_I(t) = U_0(t)^\dagger H_I(t) U_0(t) \tag{A1.10b}$$

where $H_I(t)$ is the interaction Hamiltonian in the interaction picture. Let $E_n = \hbar\omega_n$ be the energy eigenvalues of $|n\rangle$; then the density matrix elements $\rho_{Inn'}$ and $\rho_{nn'}$ are correlated by

$$\rho_{nn'}(t) = \exp(-i\omega_{nn'}t)\,\rho_{Inn'}(t), \qquad \omega_{nn'} = \omega_n - \omega_{n'} \tag{A1.11}$$

In the interaction picture, the expectation value of A is given by

$$\langle A\rangle = \mathrm{Tr}\{\rho_I(t)A_I(t)\}, \qquad A_I(t) = U_0(t)^\dagger A U_0(t) \tag{A1.12}$$

Appendix 2

Density of States for Electrons

A2.1 THREE-DIMENSIONAL STATE

The relation between the wave vector $\boldsymbol{k}$ and the energy E of electrons in the vicinity of a band edge of a bulk semiconductor, based on the effective-mass approximation can be written using a parabolic function as

$$E(\boldsymbol{k}) = \frac{\hbar^2}{2m^*}k^2, \qquad k = |\boldsymbol{k}| \tag{A2.1}$$

where E is measured with reference to the band edge. Consider a cube of side length L in the semiconductor. Application of the periodic boundary condition to the electron wave functions limits the wave vector $\boldsymbol{k}$ to the form

$$\boldsymbol{k} = \left(\frac{2\pi m_x}{L}, \frac{2\pi m_y}{L}, \frac{2\pi m_z}{L}\right), \qquad m_x, m_y, m_z = 0, \pm 1, \pm 2, \ldots \tag{A2.2}$$

Therefore, a electron state occupies a volume of $(2\pi/L)^3$ in the three-dimensional $\boldsymbol{k}$ space. Since, from Eq. (A2.1), $\mathrm{d}E = (\hbar^2/m^*)k\,\mathrm{d}k$, and the states within $k \le |\boldsymbol{k}| \le k + \mathrm{d}k$ occupy a volume $4\pi k^2\,\mathrm{d}k$ in the $\boldsymbol{k}$ space, the number $\mathrm{d}N$ per unit volume of electron states with energy between E and $E + \mathrm{d}E$ is

$$\begin{aligned}\mathrm{d}N &= \frac{2[4\pi k^2\,\mathrm{d}k/(2\pi/L)^3]}{L^3} \\ &= \frac{k^2\,\mathrm{d}k}{\pi^2} \\ &= \frac{1}{2\pi^2\hbar^3}(2m^*)^{3/2}E^{1/2}\,\mathrm{d}E\end{aligned} \tag{A2.3}$$

where the number is doubled considering the spin. Therefore, the density $\rho_{3\mathrm{D}}(E) = \mathrm{d}N/\mathrm{d}E$ of a three-dimensional state of electrons is given by

$$\rho_{3\mathrm{D}}(E) = \frac{1}{2(\pi^2\hbar^3)}(2m^*)^{3/2}E^{1/2} \tag{A2.4}$$

A2.2 TWO-DIMENSIONAL STATE

Consider first a subband (the nth subband) for electrons in a quantum well. The relation between the two-dimensional wave vector $\boldsymbol{k}_{xy}$ and the energy E, with the nonparabolicity omitted, can be written as

$$E(\boldsymbol{k}_{xy}) = \varepsilon_n + \frac{\hbar^2}{2m^*}k^2, \qquad k = |\boldsymbol{k}_{xy}| \tag{A2.5}$$

where E is measured with reference to the band edge and ε_n denotes the confinement energy cigcnvaluc. Consider a square of side length L in the plane of the quantum well, and apply the periodic boundary condition to the electron wave functions; then the wave vector $\boldsymbol{k}_{xy}$ is limited to those in the form

$$\boldsymbol{k}_{xy} = \left(\frac{2\pi m_x}{L}, \frac{2\pi m_y}{L}\right), \qquad m_x, m_y = 0, \pm 1, \pm 2, \ldots \tag{A2.6}$$

Therefore, the electron state occupies an area of $(2\pi/L)^2$ in the two-dimensional $\boldsymbol{k}_{xy}$ plane. Since the states within $k \le |\boldsymbol{k}_{xy}| \le k + \mathrm{d}k$ occupies a area of $2\pi k\,\mathrm{d}k$ in the $\boldsymbol{k}_{xy}$ plane, the number $\mathrm{d}N$ per unit area of electron states with energy between E ($>\varepsilon_n$) and $E + \mathrm{d}E$ and with the spin considered is

$$\begin{aligned}\mathrm{d}N &= \frac{2[2\pi k\,\mathrm{d}k/(2\pi/L)^2]}{L^2}\\ &= \frac{k\,\mathrm{d}k}{\pi}\\ &= \frac{m^*}{\pi\hbar^2}\mathrm{d}E \qquad (E > \varepsilon_n)\end{aligned} \tag{A2.7}$$

Thus we have $\mathrm{d}N/\mathrm{d}E = (m^*/\pi\hbar^2)$ $(E > \varepsilon_n)$ for the density of states of electrons in a subband. Since the total density $\rho_{2\mathrm{D}}(E)$ of two-dimensional state of electrons is given by the sum of the density of states for each subband, we have

$$\rho_{2\mathrm{D}}(E) = \frac{m^*}{\pi\hbar^2}\sum_n u(E - \varepsilon_n) \tag{A2.8}$$

where $u(\xi)$ is a unit-step function.

A2.3 ONE-DIMENSIONAL STATE

Consider a subband for electrons in a quantum wire. The relation between the wave vector magnitude k in the direction along the wire and the energy E

is given by

$$E(k) = \varepsilon_{nn'} + \frac{\hbar^2}{2m^*}k^2 \tag{A2.9}$$

where $\varepsilon_{nn'}$ denotes the confinement energy eigenvalue. Considering a quantum wire section of length L, with the periodic boundary condition applied to the electron wave functions, k must be

$$k = \frac{2\pi m}{L}, \qquad m = 0, \pm 1, \pm 2, \ldots \tag{A2.10}$$

From Eq. (A2.9) we have $\mathrm{d}E = (\hbar^2/m^*)k\,\mathrm{d}k$. Considering the electron states of positive and negative k for a value of E, and considering the spin, the number $\mathrm{d}N$ per unit length of electron states with energy between E ($>\varepsilon_{nn'}$) and $E + \mathrm{d}E$, and with the spin considered is

$$\begin{aligned}\mathrm{d}N &= \frac{2[2\,\mathrm{d}k/(2\pi/L)]}{L} \\ &= \frac{2\,\mathrm{d}k}{\pi} \\ &= \frac{1}{\pi\hbar}(2m^*)^{1/2}(E - \varepsilon_{nn'})^{-1/2}\,\mathrm{d}E \qquad (E > \varepsilon_{nn'})\end{aligned} \tag{A2.11}$$

Since the total density $\rho_{1\mathrm{D}}(E)$ of one-dimensional state of electrons, is given by the sum of the density of states for each subband, we have

$$\rho_{1\mathrm{D}}(E) = \frac{1}{\pi\hbar}(2m^*)^{1/2}\sum_{nn'} u(E - \varepsilon_{nn'})(E - \varepsilon_{nn'})^{-1/2} \tag{A2.12}$$

Appendix 3
The Kramers–Kronig Relation

The electric field $\boldsymbol{E}(\boldsymbol{r}, t)$ and dielectric polarization $\boldsymbol{P}(\boldsymbol{r}, t)$ in a medium can be expressed by using Fourier transformations as

$$\boldsymbol{E}(\boldsymbol{r}, t) = \int \boldsymbol{E}(\boldsymbol{r}, \omega) \exp(-\mathrm{i}\omega t)\, \mathrm{d}\omega \tag{A3.1}$$

$$\boldsymbol{P}(\boldsymbol{r}, t) = \int \boldsymbol{P}(\boldsymbol{r}, \omega) \exp(-\mathrm{i}\omega t)\, \mathrm{d}\omega \tag{A3.2}$$

where $\boldsymbol{E}(\boldsymbol{r}, \omega)$ and $\boldsymbol{P}(\boldsymbol{r}, \omega)$ are the Fourier spectra. Although the polarization includes contributions associated with the electric dipole, electric quadpole, and magnetic dipole, we neglect the latter two, since usually the electric dipole dominates the polarization. Then the polarization at a position $\boldsymbol{r}$ depends upon the electric field at the point $\boldsymbol{r}$ only (locality). However, the polarization at time t depends not only on the electric field at t but also on the electric field at any other time. Assuming that the polarization is related linearly to the electric field, the polarization can generally be written as

$$\boldsymbol{P}(\boldsymbol{r}, t) = \int R(\boldsymbol{r}, t; t')\boldsymbol{E}(\boldsymbol{r}, t')\, \mathrm{d}t' \tag{A3.3}$$

where the integration should be over $-\infty < t' < +\infty$. Because the characteristics of a material do not depend upon the choice of the origin for the time axis (time invariance), the function $R(\boldsymbol{r}, t; t')$ can be rewritten as

$$R(\boldsymbol{r}, t; t') = R(\boldsymbol{r}, \tau), \qquad \tau = t - t' \tag{A3.4}$$

The function $R(\boldsymbol{r}, \tau)$ is referred to as the response function. Since the polarization at time t is determined by the electric field at t and the time before t (causality), we have

$$R(\boldsymbol{r}, \tau) = 0, \qquad \tau < 0 \tag{A3.5}$$

In the following, $\boldsymbol{r}$ is omitted. Substituting Eqs (A3.1) and (A3.4) into Eq. (A3.3) and comparing the resultant equation with Eq. (A3.2), we see

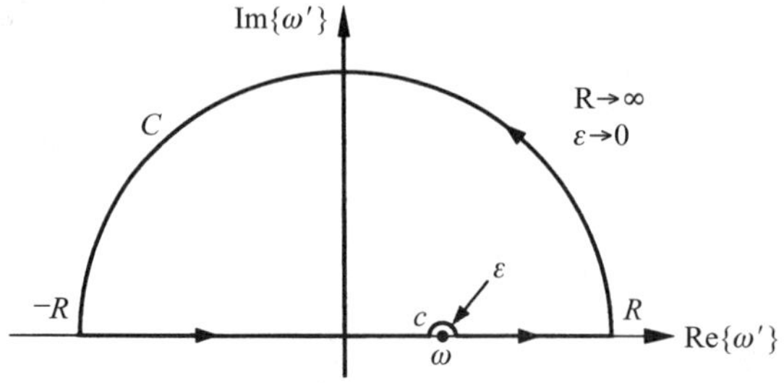

Figure A3.1 Locus for integration.

that the polarization can be written as

$$\boldsymbol{P}(\omega) = \varepsilon_0 \chi(\omega) \boldsymbol{E}(\omega) \tag{A3.6}$$

$$\chi(\omega) = \frac{1}{\varepsilon_0} \int R(\tau) \exp(\mathrm{i}\omega\tau)\, \mathrm{d}\tau \tag{A3.7}$$

Equation (A3.6) gives the definition of the dielectric susceptibility $\chi(\omega)$, and Eq. (A3.7) shows that $\chi(\omega)$ is given by the Fourier transform of the response function $R(\tau)$. The function $\chi(\omega)$ can be considered a complex function of a complex variable ω. Since $R(\tau)$ satisfies Eq. (A3.5), $\chi(\omega)$ is analytical in the upper half of the complex ω plane. Noting this, we integrate a function $\chi(\omega')/(\omega'-\omega)$, of a complex variable ω' with a real constant ω, on a closed loop consisting of the real axis, an upper semicircle of infinitely large radius C, as shown in Fig. A3.1. The pole at $\omega' = \omega$ on the real axis can be excluded from the loop by modifying it with the upper semicircle of infinitely small radius c. Then, since there is no singular point within the loop, the closed-loop integral vanishes to zero. The integration on C vanishes to zero, as the magnitude of $\chi(\omega')$ decreases exponentially with increasing radius of C. On the other hand, the integration on c results in $-\mathrm{i}\pi\chi(\omega)$. We therefore obtain

$$\chi(\omega) = \frac{1}{\mathrm{i}\pi} \mathrm{P}\left(\int_{-\infty}^{\infty} \frac{\chi(\omega')}{\omega' - \omega} \mathrm{d}\omega' \right) \tag{A3.8}$$

Separating $\chi(\omega)$ into real and imaginary parts, we obtain

$$\mathrm{Re}\{\chi(\omega)\} = +\frac{1}{\pi} \mathrm{P}\left(\int \frac{\mathrm{Im}\{\chi(\omega')\}}{\omega' - \omega} \mathrm{d}\omega' \right) \tag{A3.9a}$$

$$\mathrm{Im}\{\chi(\omega)\} = -\frac{1}{\pi} \mathrm{P}\left(\int \frac{\mathrm{Re}\{\chi(\omega')\}}{\omega' - \omega} \mathrm{d}\omega' \right) \tag{A3.9b}$$

In the above expressions, P indicates the principal value obtained by integration over $-\infty < \omega' < +\infty$ excluding the singular point $\omega - 0 < \omega' < \omega + 0$. Equations (A3.9) is referred to as the Kramers–Kronig relations.

It shows that the real and imaginary parts of the complex susceptibility are not independent of each other but are correlated by integral transformations. Noting that $\boldsymbol{E}(t)$ and $\boldsymbol{P}(t)$ are real, and therefore $\mathrm{Re}\{\chi\}$ and $\mathrm{Im}\{\chi\}$ are even and odd functions, respectively, of ω, Eq. (A3.9) can be rewritten as integrations over positive frequencies. Consider a case where there is a change in the absorption (or amplification) spectrum $\alpha(\omega)$ of a material by $\Delta\alpha(\omega)$, and let $\Delta\chi(\omega)$ and $\Delta n(\omega)$ be the changes in the susceptibility and the refractive index, respectively. Then, $\Delta n = (1/2n)\,\mathrm{Re}\{\Delta\chi\}$ and $\Delta\alpha = (\omega/nc)\,\mathrm{Im}\{\Delta\chi\}$ and, from Eq. (A3.9), we obtain

$$\Delta n(\omega) = \frac{c}{\pi}\mathrm{P}\left(\int_0^{+\infty} \frac{\Delta\alpha(\omega')}{\omega'^2 - \omega^2}\,\mathrm{d}\omega'\right) \qquad \text{(A3.10)}$$

Equations (A3.9) and (A3.10) are also referred to as the Kramers–Kronig transformation.

Appendix 4

Experimental Determination of Gain and Internal Loss

The gain (spectrum) and the internal loss of a active layer and the waveguide structure in a Fabry–Perot (FP) type of semiconductor laser can be experimentally determined. The reflectivity R of the cleaved facet mirror and the confinement factor Γ for the guided mode in the active layer can be evaluated through calculation using the parameters of the laser structure and the refractive indexes of each layer. The threshold current I_{th} and the external differential quantum efficiency η_D of a given FP laser can easily be measured. From Eqs (5.6) and (6.6), the inverse of the external differential quantum efficiency $2\eta_D$ for the total output at both edges of a FP laser using cleaved facet mirrors can be written as

$$\frac{1}{2\eta_D} = \frac{1}{\eta_{stm}}\left[1 + \frac{\alpha_{int} L}{\ln(1/R)}\right] \tag{A4.1}$$

By preparing several lasers with different resonator lengths L, measuring their η_D values, and plotting $1/2\eta_D$ against L, a linear graph is obtained. The internal quantum efficiency η_{stm} can be determined by extapolating the linear graph to obtain $1/2\eta_{stm}$ as a value of $1/2\eta_D$ at $L=0$, and the internal loss α_{int} can be determined from the slope of the line.

From the oscillation condition in Eq. (6.1), the threshold mode gain is given by

$$\Gamma g = \alpha_{int} + \alpha_{mir}, \qquad \alpha_{mir} = \frac{1}{L}\ln\left(\frac{1}{R}\right) \tag{A4.2}$$

Measuring the threshold currents I_{th} for several lasers with different resonator lengths L, and correlating the results with the value calculated by using the above expression, the dependence of the maximum mode gain Γg on the injection current density can be determined.

Through the facet of a FP laser with subthreshold current injection, spontaneous emission and amplified spontaneous emission (ASE) are observed. The spectrum of this light emission is such that the broad peak-like spontaneous emission spectrum is modulated with the periodic resonance characteristic of the FP resonator. The resonance characteristic is given by the integration, with respect to the light source elements, of P_f given by Eq. (5.140) with the attenuation factor α replaced by the effective attenuation factor $\alpha_{int} - \Gamma g$ of the subthreshold laser. Accordingly, the ratio R_{ASE} of the maxima to the minima in the observed ASE spectrum is given by

$$R_{ASE} = \frac{1 + R\exp[(\Gamma g - \alpha_{int})L]}{1 - R\exp[(\Gamma g - \alpha_{int})L]} \tag{A4.3}$$

Therefore, by measuring R_{ASE} at various injection currents in the subthreshold range, the mode gain spectrum and the injection current dependence can be determined by

$$\Gamma g(\omega) = \frac{1}{L}\ln\left[\frac{R_{ASE} - 1}{R_{ASE} + 1}\right] + \alpha_{int} + \alpha_{mir} \tag{A4.4}$$

Appendix 5
Spontaneous Emission Term and Factors

Consider the spontaneous emission term describing the contribution of the spontaneous emission to the laser oscillation, which appears as the final term on the right-hand side of the rate equation for the photon density (Eq. (6.21)). Since this term represents the component of the spontaneous emission belonging to the same mode as the laser oscillation (angular frequency ω_m), it is closely related to the gain for the oscillation mode. Consider a laser of index-guiding type, and let $\boldsymbol{E} = \boldsymbol{E}(\boldsymbol{r}) = \boldsymbol{E}(x, y)E(z)$ be the complex electric field of the oscillation mode. The field is normalized in the similar manner as Eq. (2.6), so that the optical energy in the resonator corresponds to the energy of a photon:

$$\int_{\mathrm{c}} \frac{\varepsilon_0 n_{\mathrm{r}} n_{\mathrm{g}}}{2} |\boldsymbol{E}(\boldsymbol{r})|^2 \, \mathrm{d}V = \hbar\omega_{\mathrm{m}} \tag{A5.1}$$

Then, from Eqs (2.50a) and (2.57), the transition probability relevant to the photons of this mode can be written as

$$\begin{aligned} \frac{w_{\mathrm{abs}}}{n} = \frac{w_{\mathrm{stm}}}{n} &\\ = w_{\mathrm{spt}} &\\ = \frac{\pi}{2\hbar} \left| \boldsymbol{E}(\boldsymbol{r}) \cdot \langle \psi_1 | e\boldsymbol{r} | \psi_2 \rangle \right|^2 \delta(E_1 + \hbar\omega_{\mathrm{m}} - E_2) & \end{aligned} \tag{A5.2}$$

where n is the number of the photons in the resonator. Using the direct-transition model, the net number of the stimulated emission transition per unit volume of the active region per unit time can be calculated by integrating w_{stm} multiplied by $(1/2\pi^3)(f_2 - f_1)\,\mathrm{d}\boldsymbol{k}$. The average value of $|\boldsymbol{E}(\boldsymbol{r})|^2$ in the active region is given by

$$\langle |\boldsymbol{E}|^2 \rangle_{\mathrm{a}} = \frac{\Gamma(2\hbar\omega_{\mathrm{m}}/\varepsilon_0 n_{\mathrm{r}} n_{\mathrm{g}})}{V_{\mathrm{a}}} \tag{A5.3}$$

$$\Gamma = \frac{\int_{\mathrm{a}} |\boldsymbol{E}(\boldsymbol{r})|^2 \,\mathrm{d}V}{\int_{\mathrm{c}} |\boldsymbol{E}(\boldsymbol{r})|^2 \mathrm{d}V} \tag{A5.4}$$

where V_{a} is the volume of the active region, Γ is the confinement factor, and use has been made of Eq. (A5.1). Replacing $\boldsymbol{E}$ in Eq. (A5.2) by the average given by Eq. (A5.3), and calculating the relative time variation in the mode photon number in the resonator of volume V_{a}, we obtain an expression for the mode gain:

$$\Gamma G(\omega_{\mathrm{m}}) = \Gamma \frac{\pi e^2}{n_{\mathrm{r}} n_{\mathrm{g}} \varepsilon_0 m^2 \omega_{\mathrm{m}}} |M|^2 (f_2 - f_1) \rho_{\mathrm{r}}(\hbar\omega_{\mathrm{m}}) \tag{A5.5}$$

In the derivation of the above expression, use has been made of Eqs (3.14)–(3.17). This result is consistent with that obtained by rewriting the material gain g given by Eq. (3.16) in $G = v_{\mathrm{g}} g$, and then in the mode gain ΓG.

Let $R_{\mathrm{sp}}(\omega_m)$ be the number of photons spontaneously emitted per unit time in the active region of a volume V_{a}. Then R_{sp} can be calculated by integrating w_{stm} given by Eq. (A5.2) multiplied by $V_{\mathrm{a}}(1/2\pi^3)\, f_2(1 - f_1)\,\mathrm{d}\boldsymbol{k}$ in a similar manner as above, to yield

$$R_{\mathrm{sp}}(\omega_{\mathrm{m}}) = \Gamma \frac{\pi e^2}{n_{\mathrm{r}} n_{\mathrm{g}} \varepsilon_0 m^2 \omega_{\mathrm{m}}} |M|^2 f_2 (1 - f_1) \rho_{\mathrm{r}}(\hbar\omega_{\mathrm{m}}) \tag{A5.6}$$

Accordingly, from Eqs (A5.5) and (A5.6), the spontaneous emission term and the mode gain are correlated by

$$R_{\mathrm{sp}}(\omega_{\mathrm{m}}) = n_{\mathrm{sp}} \Gamma G(\omega_{\mathrm{m}}) \tag{A5.7}$$

$$n_{\mathrm{sp}} = \frac{f_2(1 - f_1)}{(f_2 - f_1)} = \left[1 - \exp\left(\frac{\hbar\omega_{\mathrm{m}} - \Delta F}{k_{\mathrm{B}} T}\right)\right]^{-1} \tag{A5.8}$$

where $\Delta F = F_{\mathrm{c}} - F_{\mathrm{v}}$ is the difference between the quasi-Fermi levels. The parameter n_{sp} in Eq. (A5.8) is referred to as the population inversion factor.

The optical waves in an ordinary laser structure include many radiation modes with the propagation vector not parallel to the waveguide axis. The majority of the spontaneous emissions belong to such radiation modes, and therefore the guided mode component is negligibly small. Therefore, the expression given by Eq. (3.20) for a homogeneous semiconductor can also be used to describe approximately the spontaneous emission spectrum in a laser structure. By integrating it, the total spontaneous emission R_{sp} can be calculated. Approximating the spontaneous emission spectrum by a Lorentzian distribution with a half-width $\Delta\omega$ at half-maximum and using an approximation that spontaneous emission peak frequency $\approx$ oscillation frequency, we obtain an expression for R_{sp}:

$$\begin{aligned} R_{sp} &= \frac{n_r e^2 \omega}{\pi m^2 c^3 \varepsilon_0} |M|^2 f_2(1 - f_1)\rho_r(\hbar\omega_m) \\ &\quad \times \int \frac{(\Delta\omega/2)^2}{(\omega - \omega_m)^2 + (\Delta\omega/2)^2} \, d\omega \\ &= \frac{n_r e^2 \omega \, \Delta\omega}{2m^2 c^3 \varepsilon_0} |M|^2 f_2(1 - f_1)\rho_r(\hbar\omega_m) \end{aligned} \tag{A5.9}$$

The spontaneous emission term $C_s N/\tau_s$ at the end of the right-hand side of the rate equation given by Eq. (6.21) was introduced, by representing the total spontaneous emission per unit volume in the active region approximately as $R_{sp} = N/\tau_s$, and by representing the component belonging to the oscillation mode per unit volume in the active region $R_{sp}(\omega_m)/V_a$ as $C_s N/\tau_s = C_s R_{sp}$. Therefore, we see from Eqs (A5.6) and (A5.9) that the spontaneous emission factor C_s is given by

$$C_s = \Gamma \frac{2\pi c^3}{n_r^2 n_g V_a \omega^2 \, \Delta\omega} = \frac{\Gamma \lambda^4}{4\pi^2 n_r^2 n_g V_a \, \Delta\lambda} \tag{A5.10}$$

It should be noted that the above result does not apply for lasers of gain-guiding type, where the guided mode cannot be described independently to the carrier injection. Since the guided mode in gain-guiding lasers has curved wavefronts, the coupling of the spontaneous emission to the guided mode is stronger, and the value of C_s is several times the value given by Eq. (A5.10).

Index

Absorption factor, 42
Absorption spectrum, 103
Active layer, 3
Active mode locking, 228
Airy function, 133
Alpha parameter, 225
Amplified spontaneous emission, 222, 277, 298
Amplifier, 269
  bandwidth, 272, 273
  gain, 272, 273
Amplitude operator, 22
Anisotropy of effective mass, 90
Annihilation operator, 22
Antiguiding, 145
  factor, 174, 195, 212
ASE, 222, 277, 298
Astigmatism, 161
Attenuation factor, 116
Auger process, 63
Auger recombination, 199
Autocorrelation function, 202

Band-filling effect, 45, 266
Bandgap energy, 10
Bandgap shrinkage, 57, 101
Band mixing, 89, 94
Band structure, 37
Band tail, 48, 101
Barrier, 82
Beam propagation method, 280
Bistable oscillation, 186
Bragg condition, 236
Bragg diffraction, 236
Brillouin diagram, 240
Broad-area laser, 7, 113
Bulk semiconductor, 81
Buried heterostructure waveguide, 145

Carrier density, 61
Carrier-induced refractive index change, 68, 280
Carrier overflow, 63
Catastrophic optical damage, 178, 279
Channel waveguide, 113, 139
Chaotic behavior, 220
Characteristic equation, 126
Characteristic temperature, 99
Charge flow density vector, 84
Chirped pulse, 227
Chirping, 195, 258
Cladding layer, 3
Cleaving, 5, 147
COD, 178, 279
Coherence decay, 220
Coherent state, 23, 209
Completeness, 118
Complex electric susceptibility, 66
Complex frequency, 155
Complex propagation constant, 115, 238

Complex relative dielectric
constant, 115
Complex transfer function, 193
Composite resonator, 219, 233
Composite resonator mode, 219
Conduction band, 38
Confinement factor, 138
Continuous oscillation at
room-temperature, 3
Coupled-cleaved-cavity laser, 233
Coupled mode equation, 237
Coupling coefficient, 239, 240
Creation operator, 22
Critical angle, 112
Cross-saturation coefficient, 78
Crosstalk, 276
Crystal regrowth, 264
Cutoff, 127

DBR laser, 8, 233, 258
$\Delta n = 0$ selection rule, 94
Density matrix, 65, 169, 285
analysis, 65
equation, 67
Density of states, 39, 90, 289
Density operator, 67, 285
Detuning, 255
DFB laser, 8, 233, 243
DFB mode 245
DH structure, 3, 111
Differential gain, 189
Diffusion effect, 170
Direct modulation, 187
Direct transition, 9,
Direct-transition bandgap, 37
Direct-transition model, 41, 92
Disordering, 104
Dispersion curve, 127
Dispersion relation, 126
Distributed Bragg reflector, 258
Distributed Bragg reflector laser, 8,
233, 258
Distributed feedback laser, 8, 233,
243
Double heterostructure, 3, 111
Dynamic single-mode laser, 199, 233

Effective differential gain, 192
Effective index, 117
Effective index method, 141
Eigenequation, 84
Eigenvalue, 85
Einstein relation, 34
Electric dipole moment, 31
Electron polarization, 66
Energy eigenstate, 23
Envelope function, 84
Epitaxal growth, 11
Equation for fluctuation, 203
Equation of motion, 21, 286
Evanescent wave, 125
Excess noise, 210
Exciton, 100
External differential quantum
efficiency, 168, 297
External feedback, 215
External grating resonator laser,
233

Fabry–Perot resonator, 152
Fabry–Perot-type laser, 5, 163
Facet mirror, 147, 153
Facet reflectivity, 151
Far-field pattern, 159
Feedback, 5
Feedback induced noise, 215
Fermi–Dirac distribution, 35
Fermi level, 40
Fermi's golden rule, 29
Filamentation, 113
Finesse, 154
FP-type amplifier, 271
Frequency chirping, 195, 258
Frequency modulation, 195
Frequency noise, 211
Frequency response, 192
Fresnel formula, 148

Gain, 68, 54, 72, 92
Gain-coupled DFB laser, 249
Gain factor, 42, 54, 66, 95, 116, 269
Gain guiding, 7, 143, 186
Gain saturation, 77, 189, 271, 275, 279
Gain saturation coefficient, 176
Gain spectrum, 54, 75, 97, 269
Gain switching, 226
GHLBT model, 50
GHL-SME model, 46, 54
Goos-Haenchen shift, 125
Graded-index waveguide, 128
Grating structure, 235
GRIN-SCH-SQW, 139
Group index, 20, 154
Guided mode, 116
Guiding layer, 120

Hamiltonian, 24
Hamiltonian of electromagnetic field, 20
Heavy hole, 38, 88
High-speed modulation, 194
Hole, 38, 88
Hole burning, 76, 253
Hybrid mode, 141
Hysteresis, 187, 220

Index-coupled DFB laser, 244
Index-guiding, 7, 145, 184
Injection current, 61, 105, 166
Injection laser, 3
Injection locking, 273
Intensity noise, 207
Interaction Hamiltonian, 25
Interaction picture, 26, 287
Intermode competition, 186
Intermode orthogonality, 117
Internal loss, 137
Internal quantum efficiency, 61, 167
Intersymbol interference, 276
I–P characteristic, 167

Joyce–Dixon expression, 55

Kane function, 49
kp perturbation theory, 51
Kramers–Kronig relation 73, 174, 293

Langevin noise source, 203
Large-signal fast modulation, 198
Large-signal transient response, 197
Laser, 35
Laser diode, 3
Laser oscillation, 6
Lateral mode, 7, 156, 181
Lattice matching, 9
Lattice mismatch, 107
Light hole, 38, 88
Line-shape function, 74, 101
Linewidth enhancement factor, 212, 224, 257
Longitudinal mode, 7, 156, 165
Longitudinal mode characteristic, 182
Longitudinal mode separation, 156
Low-dimensional electron, 92
LPE, 12

Marcatili method, 139
Material gain, 138, 165
Matrix element, 30, 50
Maximum gain, 59, 101
MBE, 12
Mode density, 18
Mode expansion, 17
Mode gain, 138, 165
Mode hopping, 8, 212
Mode index, 117
Mode locking, 227
Mode partition noise, 214
Modulation characteristic, 187, 257
Monolithic integrated optical circuit, 104
Monolithic integration, 235, 265

Monolithic MOPA, 283
MOPA, 282
MOVPE, 12
MQW, 82
Multimode noise, 212
Multimode waveguide, 127
Multiple longitudinal mode
oscillation, 184
Multiple quantum well, 82

Near-field pattern, 159
Negative temperature, 35
NF, 278
Noise characteristic, 200, 276
Noise figure, 278
Noise power spectrum, 278
Nominal current density, 61
Nonlinear gain, 77
Nonlinear polarization, 76
Nonparabolic band, 90
Nonradiative recombination, 63
Normalized effective index, 127
Normalized frequency, 126
Normal mode, 239

Occupation probability difference, 45
OEIC, 15
Ohmic electrode, 163
One-dimensional state, 290
Optical phase, 173
Optical transition, 30
Optical waveguide, 111
Optolectronic integrated circuit, 15
Orthonormal relation, 118
Oscillation condition, 164, 243, 261
Oscillator strength, 100
Output coupling coefficient, 178
Output power, 167, 177, 252, 278

Passive mode locking, 228
Periodic boundary condition, 19,
90, 289
Perturbation, 134
Phase modulation, 195
Phase-shift index coupling DFB
laser, 246
Photon, 17, 22
Photon density, 172
Photonic integrated circuit, 15, 235
Photon lifetime, 158, 172, 250, 256,
264, 276
Planar waveguide, 112
Poissonian state, 24
Polarization 65, 115
Polarization characteristic, 255
Polarization dependence, 100
Population inversion, 35
Population inversion factor, 300
Potential well, 82
Power flow, 118
Power spectrum, 201
Poynting vector, 118
Preamplifier, 276
Principle of laser, 2
Propagation constant, 117

Q switching, 226
Quantization of optical wave, 17, 20
Quantum box, 92
Quantum confinement effect, 81
Quantum noise, 24, 200
Quantum number conservation
rule, 94
Quantum well, 8, 81
laser, 9, 95
Quantum wire, 92
Quasi-Fermi level, 41, 54, 96
Quasi-single-mode oscillation, 184
Quasi-thermal equilibrium
approximation, 40
Quasi-thermal equilibrium state, 41

Radiation decay factor, 242
Radiation mode, 117
Rate equation, 168, 250

Recombination rate, 199
Reduced density of states, 44, 93
Reflectivity, 149
Refractive index change, 72, 116
Relative intensity noise, 210
Relaxation effect, 68, 100
Relaxation oscillation, 190
Relaxation oscillation frequency, 191
Resonance condition, 165
Resonant laser amplifier, 271
Rib waveguide, 145
Ridge waveguide, 145
RIN, 210
Rotational wave approximation, 68

Saturable absorber, 229
Saturable absorption, 226
Saturated output, 275
Saturation effect, 75
Scalar potential, 17
SCH, 96, 139
Schawlow-Townes linewidth, 221
Schrödinger equation, 25
Screening, 49, 57
Self-pulsation, 220, 225
Self-saturation coefficient, 78
Semiconductor laser amplifier, 269
Separate-confinement heterostructure, 96, 139
Shot noise, 202
Shot noise limit, 210
Side mode, 197
Side mode suppression, 184, 214, 256
Side mode suppression ratio, 198, 256
Signal to noise ratio, 278
Single-mode model, 176, 199
Single-mode waveguide 127
Single quantum well, 81
Snell's law, 112
SNR, 278
Spatial hole burning, 182
Spectrum width, 197, 222
Spin-orbit interaction, 38
Spin selection rule, 51, 94
Split-off band, 38
Spontaneous emission, 2, 33, 60, 92, 104, 178, 299
Spontaneous emission coupling coefficient, 180, 301
Spontaneous emission recombination, 270
Spontaneous emission spectrum, 60, 298
SQL, 210, 278
SQW, 81
Standard quantum limit, 210, 278
State transition equation, 27
Steady-state oscillation, 176
Step-index waveguide, 120
Step response, 190
Stern matrix element, 53
Stimulated emission, 2, 33
Stimulated emission recombination, 170, 270
Stop band, 240
Strained quantum well, 101, 107
Stripe laser, 113
Subband, 88
Sublinear dependence, 103
Superlattice, 83
Surface emitting laser, 9

Tapered amplifier, 279
TE mode, 119
Three-section-type wavelength-tunable DBR laser, 265
Threshold current, 6, 165, 177
Threshold current density, 7, 177
Threshold gain, 245, 263, 297
Time-averaged spectrum, 196
TM mode, 120
Total internal reflection, 112
Total noise power, 209
Transition matrix element, 93
Transition probability, 29
Transparency carrier density, 59, 103
Transparency current density, 270

Transparency nominal current density, 62
Traveling-wave amplifier, 273
Turning point, 129
Turn-on delay, 199
2DEG, 87
Two-dimensional electron gas, 87
Two-dimensional state, 290

Ultra-short pulse generation, 225
Uncertainty relation, 23, 28

Valence band, 38
Vector potential, 17, 25

Waveguide, 111
Waveguide Fabry–Perot resonator, 152
Wavelength selectivity, 261
Wavelength-tunable DBR laser, 265
Wavelength width, 196
Wave-vector diagram, 235
Wiener-Khintchene theorem, 202
Window structure, 275
WKB method, 129